How to Do *Everything* with Your

How to Do *Everything* with Your

Todd W. Carter
Michael Bellomo

McGraw-Hill/Osborne

New York Chicago San Francisco Lisbon
London Madrid Mexico City Milan New Delhi
San Juan Seoul Singapore Sydney Toronto

The McGraw·Hill Companies

McGraw-Hill/Osborne
2100 Powell Street, 10th Floor
Emeryville, California 94608
U.S.A.

To arrange bulk purchase discounts for sales promotions, premiums, or fund-raisers, please contact **McGraw-Hill**/Osborne at the above address. For information on translations or book distributors outside the U.S.A., please see the International Contact Information page immediately following the index of this book.

How to Do Everything with Your TiVo®

1234567890 FGR FGR 01987654

ISBN 0-07-223140-8

Publisher	Brandon A. Nordin
Vice President & Associate Publisher	Scott Rogers
Senior Acquisitions Editor	Jane Brownlow
Project Editor	Lisa Wolters-Broder
Acquisitions Coordinator	Agatha Kim
Copy Editor	Andy Carroll
Proofreader	Lisa Wolters-Broder
Indexer	Valerie Perry
Composition	International Typesetting & Composition, Melinda Lytle
Illustrators	International Typesetting & Composition, Melinda Lytle
Cover Series Design	Dodie Shoemaker
Cover Illustratior	Tom Willis

This book was composed with Corel VENTURA™ Publisher.

Dedication

Once again to Zelda.
—Todd W. Carter

My thanks go out to my friends and fellow TiVo aficionados, networking wizard James Marchetti, and my fellow UCI MBA alum and connoisseur of the online auction, Joel Elad.
—Michael Bellomo

About the Authors

Todd W. Carter is a full-time author and freelance writer, writing mostly about business and technology. He's the author of *Microsoft OneNote 2003 Visual QuickStart Guide* and *Teach Yourself Visually Wireless Networking*. His newest book is *Wireless All-in-One Desk Reference for Dummies*.

Todd's articles have appeared in *The Wall Street Journal, Detroit Free Press,* MSNBC.com, *Home Office Computing* magazine, *Pages* magazine, *The Grand Rapids Press,* Office.com, and other outlets.

He's covered the Internet since 1993, including policy actions by the Clinton Administration and Congress. Not wanting to miss out on the dot-com craze, he served as Editorial Director of two high-tech news Web sites from 2000-2002. He once flipped a dot-com stock after holding it for two hours, making $21,000.

Todd graduated with a bachelor's degree in journalism from Michigan State University in 1988, and is currently working on a master's degree in literary studies. He lives in Allendale, Mich., which is halfway between Grand Rapids and Lake Michigan.

His Web site is www.toddcarter.com.

Michael Bellomo holds an M.B.A. from UC Irvine, a J.D. from the University of California, San Francisco, and a Black Belt certification in Six Sigma project management.

In the dot-com boom days in Silicon Valley, he was an IT Manager and Consultant in the field of financial e-commerce. During his M.B.A. tenure, he worked with The Knowledge Labs, a technology-centered think tank in Irvine, Calif., which studied how growing companies could attract or develop top-notch business leaders, and how elements such as TiVo, eBay, and the Blackberry would change the consumer market and the pop-culture mindset.

Upon Michael's graduation from business school, he joined ARES Corporation, a project and risk management firm that works with the Department of Defense, NASA, and the Department of Energy. Since then, he has worked on projects relating to the International Space Station and been featured as the narrator for a multimedia presentation sent to Congress on the development of NASA's Orbital Space Plane.

Michael has written 12 books on business and technology, including *Linux Blueprints* and *Windows 2000 Administration for Dummies*. His books have been published in Italian, Portuguese, French, Dutch, German, Russian, and Chinese. His latest business-related book, *Itanium Rising,* was ghostwritten with Jim Carlson, former Vice President of Marketing at Hewlett Packard.

He is also the co-author of McGraw-Hill's forthcoming release *eBay Your Business: Maximize Sales and Get Results.*

About the Contributing Authors

Josh Lehman is a renowned home theater consultant and is considered one of the top authorities on HDTV in the U.S. today. His home theater designs are showcased in some of the largest homes in the U.S. and more than seven countries around the world. Josh is also one of a handful of Imaging Science and Home Acoustics Alliance technicians working today. He is the publisher of DigitalAudioVideo.com, as well as DocDVD.com, providing 24-hour-a-day, 365-days-a-year consumer electronics technical support.

Jason Kaplan, author of this book's chapter on upgrading your TiVo's hard drive, is a writer and former sound editor. He received his first computer, a TRS-80, at the impressionable age of 13. He currently lives in Los Angeles with his HDR312 modified to hold 90 hours. He has upgraded more than a thousand TiVos.

Evan Young currently works for a DVR company in the San Francisco Bay Area. He previously worked as a product manager at SkyStream Networks, where he developed hardware and software for broadcast video and digital media services. He also previously worked on advanced consumer technologies for Geocast Network Systems and Philips Electronics. Evan has an M.B.A. from Stanford University and a B.A. from Harvard University. While at Stanford, he founded "The Future of Content," an annual academic and professional conference on media, entertainment, and technology. He has also worked as a management and information technology consultant at PricewaterhouseCoopers and American Management Systems. He is a composer and conductor, and has previously edited books on digital music composition and recording. He has been an avid TiVo user since 2000.

Amy Hoy has been writing about Mac topics since 1998. She ran the now-extinct Web site The Daily Mac, and guest-wrote on other sites. She has also been a technical editor for seven Mac-related books from McGraw-Hill/Osborne.

Contents at a Glance

PART I **Hook It Up and Get Started**

1 Set Up Your TiVo 3
2 Watch TV and Record Programs with Your TiVo 29
3 Train Your TiVo 73

PART II **Try This at Home**

4 Hook TiVo Up to Your Home Network 93
5 Listen to Music on Your TiVo 119
6 Remotely Schedule Programs 137
7 Transferring Programs Between TiVos 165
8 Setting Up a Pioneer Combination TiVo and DVD Recorder 183
9 Playing and Recording Programs on a Pioneer TiVo-DVD Recorder 205

PART III **Don't Say I Told You So: Hacking TiVo Systems**

10 Easy Hacks 227
11 Upgrading Your TiVo's Hard Drive 241

PART IV **What's Next?**

12 Finally... High Definition TiVo 297
13 The Future of DVRs 325

Glossary 343
Index 349

Contents

Acknowledgments xvii
Introduction xix

PART I **Hook It Up and Get Started**

CHAPTER 1 **Set Up Your TiVo** **3**
Unpacking Your TiVo 4
What Comes with Your TiVo 5
Hooking Up Your TiVo 5
Cables, Connection Types, and Your TV 6
Home Entertainment Center Wiring Basics 6
Hooking Up Your TiVo to the Essentials 8
Using a Higher Quality Connection for a Better Picture 15
Testing & Using Your Configuration 16
Using Your TiVo 17
Activating Your TiVo 17
Configuring Your TiVo with Guided Setup 19

CHAPTER 2 **Watch TV and Record Programs with Your TiVo** **29**
Introduction to Using TiVo 30
How TiVo Works 30
TiVo's Remote 32
TiVo's Menus 33
TiVo Central Icons 35
Watching and Manipulating Live TV 36
Watching Live TV 36
Manipulating Live TV 41
Finding and Recording Shows 43
Recording Basics 43
Recording Icons 48
Recording While Watching TV 49
Setting Up Recordings from TiVo Central 50

Playing and Managing Recordings 65
  Using Now Playing 65
  Now Playing Icons 66
  Managing Scheduled Recordings and Season Passes 69

CHAPTER 3 **Train Your TiVo** **73**
How TiVo Learns 74
  TiVo Services 75
TiVo's Suggestions 80
Using the Thumbs-Up and Thumbs-Down Options 81
  Suggestions on Thumbing Through TiVo 82
  Changing Your Ratings 83
  Resetting Your Ratings 83
Understanding the Sharing of Your Personal Information and Viewing Habits 84
  TiVo's Privacy Policies 86
Determining Whether to Opt Out of TiVo's Information-Collection System 88
  TiVo's Positive Record on Privacy 88
  No One Can Predict the Future 89
How to Opt Out of TiVo's Information-Collection System 89

PART II **Try This at Home**

CHAPTER 4 **Hook TiVo Up to Your Home Network** **93**
Reasons to Hook Up TiVo to Your Home Network 94
What Makes Up a Home Network? 95
Choosing to Go Wired or Wireless 96
Setting Up a Wired Connection 97
  Creating a Network Connection on a Wired DHCP Network 98
  Creating a Network Connection on a Wired Static IP Network 100
Setting Up a Wireless Connection 103
  Configuring Network Settings for Wireless Home Networks 104
  Creating a Network Connection on a Wireless DHCP Client ID Network 107
  Creating a Network Connection on a Wireless Static IP Network 108
Using TiVo's Desktop Software 109
  Downloading TiVo Desktop 111

CHAPTER 5 **Listen to Music on Your TiVo** **119**
Organizing Your Music Files 120
Publishing Your Music 121
Listening to Your Music 123
Music Play Options 127
Music Browsing Options 129
Using MoodLogic 130

CHAPTER 6 **Remotely Schedule Programs** **137**
Remote Scheduling Over the Internet 138
Limitations of Remote Scheduling 139
Phone Line versus Broadband 140
Searching TV Listings on TiVo Central Online 141
Doing a Basic Search 143
Doing an Advanced Search 144
Browsing TV Listings by Channel 146
Scheduling TV Shows 148
Scheduling the Recording 149
Confirming Recording Options 151
Receiving E-mail Confirmations 153
Getting a Season Pass Remotely 154
Using AOL's Customized Version of Remote Scheduling 157

CHAPTER 7 **Transferring Programs Between TiVos** **165**
TiVo's Multi-Room Viewing Feature 166
Setting Up Two TiVos for Multi-Room Viewing 166
Transferring Programs Between Two TiVos 168
Setting Up Programs for Transfer 168
Estimating Transfer Times for Programs 172
Playing a Transferred Program 175
Connecting TiVos Via a Wireless Connection 176
Stacking Your TiVos to Increase Storage Capacity 178

CHAPTER 8 **Setting Up a Pioneer Combination TiVo and DVD Recorder** . . . **183**
Is the TiVo-DVD Recorder the Right Device for You? 184
Making Sense of the DVD Maze 185
Setting Up Your Pioneer TiVo-DVD Recorder 187
Cables and Accessories 188
A Word on Audio/Video Cables 188
Connecting Your TiVo-DVD Recorder to the Phone Line . . . 191
Connecting Your TiVo-DVD Recorder to a Cable or Satellite Box 198

Example Setups . . . 200
Connecting to a Satellite Box and an RF Program Source . . . 201
Connecting to a Cable or Satellite Box, an A/V Receiver, and a Game Console . . . 202

CHAPTER 9 **Playing and Recording Programs on a Pioneer TiVo-DVD Recorder** . . . **205**
Reducing Noise and Filling Jagged Edges . . . 207
Go For the Burn: CD or DVD Compatibility . . . 207
Recording Speed . . . 208
Disc Compatibility for Recording . . . 208
Disc Compatibility for Playback . . . 208
MP3 Format Issues . . . 210
Troubleshooting: Physical Problems that can Occur to your Pioneer TiVo-DVD Recorder . . . 211
A Word on Copy Protection Issues . . . 211
How to Play a DVD or CD . . . 212
Inserting and Playing a Disc . . . 212
DVD Information and Playback Settings . . . 213
MP3/Audio Information and Playback Settings . . . 215

PART III **Don't Say I Told You So: Hacking TiVo Systems**

CHAPTER 10 **Easy Hacks** . . . **227**
Identifying Your TiVo . . . 228
Unlocking the TiVo Backdoor Mode . . . 229
TiVo OS Version 3.0 and Earlier (Series 1 TiVos Only) . . . 229
TiVo Remote Control Codes . . . 230
Select-Play-Select (SPS) Codes . . . 230
Triple-Thumb Codes . . . 232
Clear-Enter-Clear Codes . . . 233
Other Software Hacks . . . 235
AutoTest Mode . . . 235
Sort Now Playing in OS 3.0 . . . 236
Enable Advanced WishLists . . . 236
Adjust Fast Forward and Rewind Speeds . . . 237
Remote Shortcuts . . . 237
Using an External Modem . . . 238
Materials . . . 238
Configuring the Modem . . . 239
Connecting the Modem and Configuring the TiVo . . . 240

CHAPTER 11 **Upgrading Your TiVo's Hard Drive** **241**

Reasons for Upgrading 243
- Reason 1—More Space 243
- Reason 2—Dead Drive 243

Can My TiVo Even Be Upgraded? 248

Planning for a New Hard Drive 248
- How Much Hard Drive Space Do I Need? 248
- Add Another Drive or Replace the Old One? 250
- Preformatted Kits vs. Doing It Yourself 251

Precautions and Warnings 256
- Safety Measures 256
- A Note on Warranties 257
- Drive Handling 257

Surgery 258
- Architecture 1 260
- Architecture 2 262
- Architecture 3 265
- Architecture 4 268
- Note for TiVo TCD23x and TCD24x Owners 270
- Architecture 5 271
- Note for TiVo TCD540140 and TCD540040 Owners 271
- Architecture 6 273

Installation Details 275
- Parts Summary 275
- MFS Tools 280
- Hard Drive Formatting 282
- Unlocking TiVo Hard Drives 291

Useful Web Resources 292
- Information Sites 292
- Kit and Part Dealers 293

PART IV **What's Next?**

CHAPTER 12 **Finally... High Definition TiVo** **297**

The Basics of High Definition TV 299

Why It Took So Long to Market HD DVRs 302

Purchasing a High Definition TiVo 305

Setting Up a High Definition TiVo 306
- Before You Start 306
- Satellite Dishes and Antennas 309
- Connecting the Equipment 310

Finding High Definition and Dolby Digital 5.1 Shows 320

Getting the Most Out of Your Hard Drive 323

CHAPTER 13 **The Future of DVRs** **325**
The Numbers Behind the TiVo Phenomenon 326
Some Speculation on TiVo's Strategies 327
All's Well (Mostly) in TiVo-Land 327
Patent Issues 328
Threats to the TiVo 329
TiVo's Place on the Home Network 333
Upcoming Changes in DVD Technology 333
Unsafe at Any (Okay, Just 4X) Speed 334
Double-Layer DVDs 335
Optical Discs 335
High-Definition Television and TiVo 335
The Good and Bad of the Coming HDTV Tide 336
The Up-and-Comers: Emerging HDTV Devices 336
The Advertising Wars 337
The Marketing Empire Strikes Back! Video-to-Video 339
What About Privacy Concerns? 339
What's in Store for the Longer Term 340

Glossary **343**

Index **349**

Acknowledgments

Thanks to Jane Brownlow, who brought me on this project. Also, kudos to Michael Bellomo for contributing a good chunk of the book's chapters. Together, we worked to create the most comprehensive and up-to-date book on TiVo. As always, my appreciation goes to Neil Salkind, my Kevlarian agent. He's a writer's Season Pass.

—Todd Carter

I'd like to acknowledge the help and resources provided by Jane Brownlow, Agatha Kim, Lisa Wolters-Broder, and especially Todd Carter's assistance with the tricky screen shots in making this book a reality.

—Michael Bellomo

Introduction

Welcome to *How to Do Everything with Your TiVo*. It's been a while since a consumer electronics product has changed the way people do something fundamental. In this case, that something is watching TV. TiVo has entered our living rooms like a hurricane, reigning supreme among the new generation of digital video recorders, or DVRs.

If you haven't already done so, it's time to throw out your VCR. Those messy tapes and complicated timers are history with a TiVo in the house. You can schedule a recording at a time of your choosing or you can tell TiVo to record every episode of a show. The power is shifting, from the hands of the giant broadcasters and studios to your TiVo remote control.

You're in charge now.

TiVo, to the company's dismay, has become a verb, too. We no longer record something with our TiVo, we *TiVo* it. Even if something happens in our lives outside of the world of television, we may still exclaim that it was a moment to be *TiVo'ed*.

We hope this book helps you enjoy your TiVo even more. We've divided the book into four sections:

In Part I, Hook It Up and Get Started, we help you set up your TiVo in Chapter 1, show you how to watch TV and record programs with your TiVo in Chapter 2, and train your TiVo with your likes and dislikes in Chapter 3.

In Part II, Try This at Home, we start by helping you hook up your TiVo to your home network in Chapter 4. That's essential for playing music from your PC through your TiVo, which we discuss in Chapter 5. In Chapter 6, we tell you about an exciting feature of TiVo that lets you remotely schedule programs over the Internet. In Chapter 7, we write about operating multiple TiVos before moving on to Chapter 8, where we discuss Pioneer's combo TiVo and DVD recorder. In Chapter 9, we cover the playing and recording of programs on the Pioneer.

In Part III, Don't Say I Told You So: Hacking TiVo Systems, hardcore readers will appreciate our coverage of hacks. In Chapter 10, we cover the easy hacks before moving on to the more difficult hack in Chapter 11 of upgrading of your TiVo's hard drive.

In, Part IV, What's Next?, we talk about the next generation of TiVos: ones capable of receiving and recording high-definition TV, or HDTV, in Chapter 12. We also discuss the future of DVRs in Chapter 13, which includes much more than just TiVos.

The middle-of-the-book purple Spotlight section, Display Digital Photos on Your TiVo, explores another of TiVo's features. Viewing photos on your TiVo is one of the Home Media Features, a package that is now included free with your monthly or lifetime subscription fee. Until mid-2004, displaying photos and listening to music were among features that cost extra.

To make the book more useful for you, we've included some helpful design elements. They are:

- **How To sidebars** How to sidebars provide step-by-step instructions on completing a task.
- **Did You Know sidebars** Did You Know sidebars contain extra information that may not fit into the main part of a chapter. They contain background information and other trivia.
- **Voices from the Community** Voices from the Community sidebars are short interviews with people just like you who use their TiVos in different ways.
- **Notes** Notes point out things that are important. Be sure to read any notes you come across.
- **Tips** Tips are items that provide nuggets of advice on things that will help you make the most of your TiVo's functions and features.
- **Cautions** Cautions are just that: Areas of concern. Heed the cautions!

We've included Web links from the book at co-author Todd Carter's Web site, http://toddcarter.com/tivo. If you have any thoughts about the book, we would love to hear them. Please write us at tivo@toddcarter.com.

Part I

Hook It Up and Get Started

Chapter 1

Set Up Your TiVo

How to…

- Identify the Items that Came with Your New TiVo
- Hook your TiVo up to your TV and Cable, Satellite, or Antenna
- Hook Up Other Recording Devices (Such as DVD Burners or VCRs) for Archiving TiVo Shows
- Hook Up your TiVo so You Can Watch One Channel While Recording Another
- Get the Best Quality Picture Out of Your TiVo
- Activate your TiVo Service

NOTE *Already have your TiVo up and running? You can skip this chapter. If you don't have your TiVo set up, or would like more information about connecting your TiVo to a complicated home entertainment center, this chapter is for you.*

Unpacking Your TiVo

TiVos come with a lot of goodies, most of which you'll need to get your TiVo hooked up. If you're not an audio-video expert, or you bought your TiVo secondhand, it's a good idea to run through this list to make sure you can identify everything that came with your TiVo, and to make sure you have all the required parts.

A newly opened TiVo box from the factory

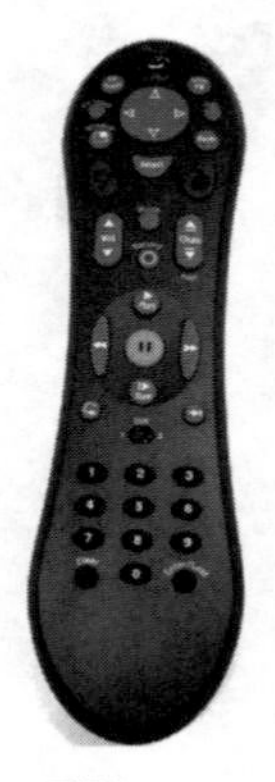

The TiVo remote

What Comes with Your TiVo

Obviously, the most critical piece of hardware here is the TiVo itself. Your TiVo should also come with the following:

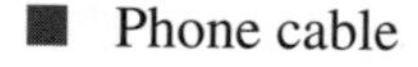

- TiVo remote (also critical—you can't use your TiVo without it)
- Power cable
- Phone cable
- Phone splitter
- Composite (RCA) video cable
- Coaxial (RF) cable
- Serial control cable
- Infrared (IR) controller
- S-Video Cable (Most Series 2 DVRs and first generation Sony models come with the standard S-Video cable)

If you bought your TiVo secondhand, it's possible that you may be missing one or more of these pieces. The power cable, TiVo remote, infrared controller (if you need it), and serial control cable (if you need it) can be replaced by visiting TiVo's online store at http://store.tivo.com/. Any of the other cables can be replaced easily at your local AV or electronics store. Not sure whether you need the IR controller or serial control cable? That's all explained later in this chapter.

Hooking Up Your TiVo

You have a lot of choices when it comes to how you hook up your TiVo, even if your home entertainment center consists of just a TV and plain old analog cable. Once you add DVD players, VCRs and DVD burners, cable or satellite boxes, AV receivers, and game systems, things get really complex. But don't worry if we've just described your private AV nightmare, because helping you figure it all out is the purpose of this chapter.

CAUTION *For safety's sake, it's smart to leave everything turned off while hooking up your TiVo. If you want to really cover all the bases, you can unplug everything from the power as well. Also, do not move your TiVo while it's on! The TiVo has a hard drive inside that is almost always spinning, so jostling it while it's on could permanently damage it at worst, and could easily cause data loss or a crash at the less catastrophic end of the spectrum.*

Cables, Connection Types, and Your TV

In order to decide how to configure your home entertainment system with a TiVo, you need to consider what kind of connections your TV offers and which connection type strikes the best balance between convenience and quality:

- **Coaxial (RF)** These are the familiar cables with the annoying screw-on connectors, used to route cable or satellite TV through your walls and to your TV. These have the lowest quality signal.
- **Composite (RCA)** This is the standard connection used for home entertainment, with three plugs on one cable—red (right audio), white (left audio), and yellow (video). These cables offer a medium quality signal.
- **S-Video** This is the next step up from composite cables. S-Video has a 4 pins inside a circular-shaped connector. These cables provide a high quality signal.

You'll be spending a lot of time looking at the signal output from your TiVo, so it's a good idea to pick the highest quality connection type available. Of course, not all TVs come equipped with all of the these connection types, so you'll need to either dig out your TV manual or get behind your TV and eyeball the connections to find out what options you have if you don't know off the top of your head.

Home Entertainment Center Wiring Basics

There can be no doubt that wiring up a home entertainment center is tough. You have to carefully plan your connections in advance, and sometimes the rules about what works and what doesn't can seem pretty complex. The goal of this chapter is not only to tell you what to do in a small handful of specific situations, but to make you an AV master so you can figure out the solutions to new situations as they arise.

In order to have your wiring work out right the first time, and do it in the most efficient way possible, you need to plan before you start. The shortest path between two points is a straight line, and when it comes to analog signals, taking the shortest path is important because the signal degrades, or attenuates, over distance. And you don't want that!

Take into account everything that you need to wire up. You have a TV, obviously, but what inputs does it have, how many overall, and how many of each type? How many TV sources, TiVos, VCRs, DVD burners, DVD players, AV receivers, and game systems do you have? What kinds of inputs do they offer, and which ones benefit

Did you know?

Video Quality

Digital cable or satellite TV, TiVos, DVD players, and game systems benefit the most from higher quality connections because they either put a lot of text on the screen or simply have more visual data in the video signals that could get lost or degrade. VCRs, analog cable television, and broadcast TV are low quality to begin with.

the most from the higher quality? Once you've figured out these details, you may want to make a game plan, a quick and dirty sketch showing what to plug into where.

You may find that you have, or will, run out of inputs on your TV, even if you shuffle some devices to a lower quality connection. In that case, you have a few work-arounds to choose from. One option is to "chain" several items via coaxial cable, wiring item A's output into item B's input, and then item B's output to the TV's input or even to item C, and so on. Such chaining will eventually cut down on picture quality, so it's best to keep the chain to a maximum of two items plus the TV. Another option is to purchase an input switch, which will let you hook up several devices to it, and in turn take up only one input on your TV. But then you will have to manually "switch" to the device you want, as well as tuning to that input on the TV when you wish to change from one device to another.

Another option is to use an AV receiver, such as the one in Figure 1-1, which is something you may already own. AV receivers are kind of like big input switches,

FIGURE 1-1 An AV receiver can help solve your cabling problems.

but with audio processing hardware to drive your speaker system. You can read more about using a receiver to solve your cabling problems in the "Hooking Up to a Home Theater" section later in this chapter.

Hooking Up Your TiVo to the Essentials

No matter how simple or complex your home entertainment center, there are a few things you have to do to get your TiVo up and running. Namely, you need to hook your TiVo up to both power and a phone line.

Hooking the TiVo up to power is really self-explanatory, but there are still a few things you should know. Your TiVo needs power all the time to function as advertised—it can't very well record shows while it's turned off, after all! For this reason, the TiVo does not have a power switch, and you must unplug it to turn it off. It's a good idea to use a power outlet that you can reach easily so that you can unplug the TiVo before you attempt to move it, in case you want to rewire or rearrange your entertainment center. Never move your TiVo while it's on because you could damage it.

The other critical connection for your TiVo's operation is phone access. The TiVo needs access to a phone line so it can dial in and download your programming lineup, which is really the foundation for all of the really cool features TiVos employ. Luckily, you don't have to give up a phone jack because your TiVo came with a phone splitter that allows the TiVo to share a phone jack with a modem or phone. Please keep in mind that this does *not* mean that you can make a call or use a modem while your TiVo is dialed in.

If you're not using the phone jack for any other purpose, you can simply take the phone cord that came with your TiVo (or another if you prefer), and plug one end into the TiVo's "Phone Line" jack and the other end into the phone jack. If you already use the phone jack for some other device, you can use the splitter: simply unplug the phone cord that is currently plugged into the wall (if there is one), plug in the phone splitter, and then plug the old phone cord into one of the splitter jacks. Then plug the TiVo into the free jack on the splitter. Give the phone cords a little tug to make sure everything is firmly seated, and you're done.

Hooking Up to Just a TV with No Cable Box

We'll start by looking at the most basic configuration, and work up from there. Even if you have a gaggle of extra gadgets, you should consider reading about these less-complicated configurations to gain a solid understanding of how to wire up entertainment centers.

Voices from the Community

The TiVo Revolution Begins

Scott Eaton works in the booming tech sector that is Cupertino, Calif., and he is also an early adopter of the TiVo technology. "I don't know exactly how long I've been using TiVo," he says. "I bought one not long after they came out. I had seen one at a friend's house; he was 'in the know' and had some prerelease model and after he demo'd it to me, I knew I was getting one."

Eaton is not alone in being introduced to TiVo in this way. In fact, he says most early users were introduced to TiVo through word of mouth by friends and coworkers. "For a long time, every TiVo owner became, in effect, a TiVo salesperson," he says. "I've done more TiVo demos than I can remember. The concept is so simple, yet when newbies see you do your own instant replay, they flip. Having used one for years now, I can't do without it."

Eaton has owned the same model TiVo the entire time he's been a TiVo user and hasn't seen a reason to replace it with a newer model—a fact he attributes to the unit's useful life span and expandability. "It still works beautifully. It doesn't have folders in the UI like the Series2 units do, but that's the only real difference for me." He has, however, upgraded his TiVo (and you can, too, with the help of Chapter 12 in this book): "I think it was originally a 30 hour unit, but I've changed hard drives a few times, so right now it's somewhere over 100 hours—way more than I personally use.

"I think Pause is the single biggest feature of these devices," Eaton says, describing his favorite aspect of life with TiVo. "It has what is probably the best remote control of any consumer device ever. I also watch less TV. A lot less—and I think that's a good thing. I like how I can have my own *Futurama* marathon on a Saturday, plus I can now watch shows that are broadcast at odd hours. I can't even tell you off the top of my head when some shows I regularly watch are broadcast.

"The TV is off right now. Sunday night at 9 P.M. and it's off. I think it's recording *The Simpsons* or something. I'm not sure, so I think my TiVo experience is best described by what I'm doing instead of revolving around the TV."

The simplest way to hook up a TiVo is to plug the coaxial cable from your wall jack into the port marked "RF in" on your TiVo, and then run a second coaxial cable from the "RF out" port on your TiVo to your TV's cable or antenna port. Switch your TV to channel 3 or 4, and voila!

Alternatively, instead of using coaxial cable—or in addition to coaxial—you can use composite or S-Video to hook up your TiVo. See the "Using a Higher Quality Connection for a Better Picture" section later on in this chapter. Be careful when lining up the 4-pin connector though as the pins do easily bend and occasionally break. Lining up the different size pins and holes without pushing in the cable at an angle will prevent damage. Never yank an S-Video, or any other cable, out of your display.

NOTE *Using a coaxial connection to your TV? You can choose which TV channel your TiVo broadcasts on. There's a switch on the back of your TiVo, next to the "RF out" jack, where you can choose between channel 3 and channel 4.*

Hooking Up to a Cable Box or Satellite

It's no problem getting your TiVo to play nice with your cable box or satellite receiver, but it can certainly look intimidating. If you have digital cable, the first step is to figure out which cable box you own. If you have a DirecTV satellite receiver or a Motorola or General Instruments cable box, such as the one shown in Figure 1-2, the TiVo can control it using the serial control cable. If you have a regular cable box, a non-DirecTV satellite receiver, or another model of digital cable box, you'll need to break out the IR control cable and jury-rig it so that the TiVo can change channels that way.

TIP *Not sure which cable box you own? Check on the back, the sides, and the bottom for the maker's name and model number; if you still can't find it and don't know where the manual is, give your cable company a call—they'll know for sure. You may even be able to request a newer cable box for no charge if yours is a few years old.*

FIGURE 1-2 Some satellite receivers and cable boxes, such as the Motorola DCT2500, can be controlled by the TiVo.

Once you've determined whether you'll be using the serial control cable or IR control, it's time to get wiring. Your cable box should already be connected to the source of the TV signal, but if it's not, run a coaxial cable from the wall jack to the cable or satellite box's "RF in" or "TV in" port.

To connect the cable or satellite box to the TiVo, you need to run either a coaxial or composite cable from an output on the cable box to an input on your TiVo. (If your cable box offers S-Video out, you could use that for the best quality.) If you're using coaxial cable, simply screw one end of the cable into the output jack on the cable or satellite box and the other into the "RF in" jack on the TiVo. If you're using composite cable, plug all three of the composite plugs into the cable box's outputs and the other ends into the set of jacks under "Input" on the TiVo.

The next step is the trickier one: hooking up either the serial control cable or the IR control.

Connecting the Serial Cable If you have a satellite receiver, or a Motorola or General Instruments DCT2000-series cable box, grab the serial cable out of your pile of TiVo stuff, screw the multipin end into the back of your cable or satellite box, and insert the little pin end into the "Serial" port on your TiVo under "Channel Change."

Connecting the IR Control The IR control cable is what is known as an "IR blaster," which basically emits IR signals just like a TV remote, letting you control something that uses a remote—without a remote. You may be tempted to skip this step because it sounds like no fun at all, but your TiVo won't work properly if it can't change channels. Figure 1-3 shows an older cable box that will require the IR blaster to function with your TiVo.

To install the cable, first locate the IR sensor on your cable or satellite box. Grab a flashlight and look for a little bulb behind a translucent plastic window (the IR window) on the front of the unit.

Once you've found the IR sensor, take the two IR emitters and position them on top and on the bottom of your satellite or cable box, centered on the little IR

FIGURE 1-3 TiVo can change channels on an older cable box with the IR control cable.

sensor you located. They should stick out about 1.5 inches beyond the edge of the box so they have enough room to transmit the signal. Plug the other end of the IR control cable into the "IR" port under "Channel Change" on the back of your TiVo. That's all you have to do for now; you'll complete the configuration and testing when you're running the TiVo guided setup.

After you've gone through the TiVo Guided Setup and confirmed that your IR emitter is working, use the double-sided sticky tape that came with the IR control cable to secure the IR emitters in place. If you have consistency problems with this setup later on, try covering the IR emitters and IR window with a magazine or some other opaque material; if that solves the problem, use a piece of thick fabric or paper to fashion a permanent covering. This will block out any extraneous light that might be interfering with the IR sensor in your cable or satellite box, and increase the performance of the IR emitters.

Hooking Up to Multiple TV Sources

It's true—you can hook your TiVo up to both satellite and cable, or satellite and antenna. Just hook up your cable or antenna via the "RF in" port, and connect your satellite via composite cables as described previously. If both your satellite receiver and cable box require IR control, however, your TiVo won't be able to change channels on one of them, so the TiVo recording capability won't work on that TV source.

Hooking Up to a VCR or DVD Burner to Archive Shows

Your TiVo doesn't have unlimited hard drive space, so what to do when you want to save a show indefinitely? The good folks at TiVo thought of this and designed the TiVo with the ability to archive shows with your VCR or DVD burner. If you just want to watch tapes or DVDs on your DVD burner and not archive shows, you can skip this section and wire up your entertainment center as described in the next section, "Hooking Up to a Home Theater."

The goal is to put the VCR or DVD burner in between your TiVo and TV—you can simply daisy-chain everything via coaxial cable. Run a coaxial cable from the wall (or from your digital cable box or satellite receiver) to your TiVo's "RF in" jack, and then run a second coaxial cable from your TiVo's "RF out" jack to your VCR or DVD burner's coaxial input. Then connect the last coaxial cable to your VCR or DVD burner's coaxial output and to your TV's "Cable" or "Antenna" in. You may need to buy an extra coaxial cable or two to do this. Figure 1-4 shows an example DVD burner that you might use to archive shows.

It's likely that you will see a noticeable degradation in video quality from chaining coaxial devices like this, so it's a good idea to run a second connection via composite or S-Video straight from your TiVo to your TV. Alternatively, if your VCR or DVD

FIGURE 1-4 A DVD burner can be used to archive shows that would fill up your TiVo hard drive.

burner offers both composite inputs and composite outputs, you can chain with composite cables instead; as long as the VCR or DVD burner is in between the TiVo and the TV, it should work. For instructions on how to hook up your TiVo via S-Video or composite cable, see the "Using a Higher Quality Connection for a Better Picture" section later on in this chapter.

Finally, make sure to set your VCR to broadcast on channel 3 if you have your TiVo set to channel 4, or vice versa. Then, when you want to watch something using your VCR, all you have to do is switch to the cable or antenna input on your TV and go to that channel.

Hooking Up to a Home Theater

DVD player, AV receiver, video game consoles, oh my—does this sound like your home entertainment center? If so, this is the section for you. It will help you figure out how to put everything together with your TiVo.

The important thing to remember is that the TiVo needs direct access to the TV source, regardless of whether that TV source is plain old cable straight from the wall or something fancier from a cable or satellite box. You need to run a cable from the TV source directly to your TiVo's "RF in" or composite input, and then you can run another cable (composite or better!) from one of the TiVo's outputs to one of the AV receiver's inputs.

NOTE *If you want to incorporate a VCR or DVD burner into this setup* and *have it be able to archive shows from the TiVo, you will need to connect it between the TiVo and AV receiver, as described in the previous section.*

You'll also need to run a cable (composite or better!) from one of the AV receiver's outputs to one of the TV's inputs, so you can get video from the TiVo on the TV screen.

How to ... Hook Up Your TiVo to Watch One Channel While Recording Another

If you have a DirecTiVo, you can skip this section because you already have the capability to watch one channel while recording another without doing anything special at all. If you have satellite connection and a regular TiVo, sorry, you're out of luck—this method only works with broadcast TV and cable.

Here's what you need to do to set it up:

1. Go out and buy yourself a one-to-two-jack coaxial splitter, an additional coaxial cable (you might need two more, depending on your setup), and an additional composite cable if you have a digital cable box. These items can all be had for just a few dollars apiece at your local electronics or home entertainment store.
2. Unplug the the coaxial cable that comes from the TV source's wall jack, and connect the single-jack end of the splitter to the wall jack instead. Then hook up a coaxial cable to each of the two split jacks.
3. If you use regular (analog) cable or broadcast TV, connect a coaxial cable to one of the split jacks and to the "RF in" connector on the back of the TiVo. Connect another coaxial cable to the other split jack and to your TV's "Antenna" or "Cable" jack, and run a composite cable from the TiVo's composite output to your TV's composite input (if you haven't done so already).
4. If you have a cable box, connect a coaxial cable to one of the split jacks and to the cable box, and run a composite cable from the composite output on the cable box to the composite input on the TiVo. Hook up a composite cable from the TiVo's composite output to the composite input on your TV (if you haven't done so already). Connect another coaxial cable to the other split jack and to the "Antenna" or "Cable" jack on the TV.

Keep in mind that if you use this method with digital cable, you will only get regular TV channels via the direct hookup to your TV because you bypass the digital tuner. The extra digital channels are also not available for TiVo recording because you also bypass the TiVo.

If your cable signal is weak or snowy after installing the cable splitter, but it wasn't before, you may need to purchase a powered splitter, also known as an inline signal booster or inline amplifier (be sure to get one for coaxial cable). You can expect to pay about $15 for one at most electronic stores, although your cable company may offer you one for free if you give them a call.

Finally, to use the TiVo you'll need to switch the receiver to the TiVo input and then switch the TV to the receiver input. Don't worry, though. It's not as bad as it sounds.

If you use a receiver and you still run out of inputs to plug your devices into, consider getting an input switch as described earlier in the chapter. We have three game systems hooked up to one composite input switch (described in the "Home Entertainment Center Wiring Basics" section earlier in the chapter) so that they don't entirely fill up the receiver. For example, you could have just one receiver input dedicated to video games and use the input switch to switch between games; the switch is then hooked up to the receiver in turn.

Using a Higher Quality Connection for a Better Picture

If you're connecting your TiVo to your TV through a VCR or DVD burner, and you want to avoid signal degradation, or simply want a crisper picture, you should consider hooking your TiVo up to your TV via composite or S-Video cable instead of coaxial cable. Of course, if your TV only has coaxial inputs, there's nothing you can do but purchase a newer TV.

Using S-Video Cables

If your TV has S-Video inputs, you're in luck—for the highest quality picture possible with your TiVo, hook it up to your TV using S-Video. S-Video connections are easy: once you have purchased an S-Video cable, plug it into the S-Video connector under "Output" on the back of your TiVo, and plug the other end into the S-Video input on your TV.

S-Video cables carry only video signals, so you will also need to run composite audio from the TiVo to the TV, or you'll have a gorgeous picture but no sound. Use only the red (right channel) and white (left channel) composite plugs, and be sure to plug them into the composite jacks that are specifically grouped with the S-Video

Did you know?

S-Video Cables Are Not All Equal

Don't skimp on the cable! S-Video cables can cost upwards of $10 to $15 a foot, but if you're a stickler for quality it's worth it. Try to pick a brand-name cable with gold pins and a solid cable sheath. One company particularly known for great cables is Monster; you can purchase Monster cables online at http://www.monstercable.com/ or locally at any store with a focus on home entertainment (including Best Buy).

port on both the TiVo and TV. If you plug in the yellow composite plug, your TV may end up using the composite video signal instead of the S-Video signal, which would defeat the purpose of using S-Video; if you don't make sure to use the composite jacks that are grouped with the S-Video jacks, you may not get audio and video together.

Using Composite Cables

Composite cable is the way to go if your TV doesn't have S-Video in, or if you simply don't want to shell out $30 to $50 for a cable. Simply run the composite cable from one of the TiVo's composite outputs to one of your TV's composite inputs (unlike when using S-Video, you'll need to use all three composite plugs).

Testing & Using Your Configuration

Once you're all wired up, turn on your TV and get ready to test out your wiring. If you've wired your TiVo to your TV using coaxial cable, try channel 3 or 4 (whichever you picked) and see if you have a TiVo welcome screen.

If you're using a composite or S-Video connection, use your TV's remote control to switch from the cable or antenna input to the composite or S-Video input. The button you use to switch inputs should be labeled something like "Input" or "Source" or "TV/Video"—when in doubt, consult your TV's manual. Depending on how many inputs your TV has, you may need to press the button several times before you get to the one your TiVo is using.

If you cycle through all the inputs and still don't have a picture, check that all of the connections are secure, and that the TiVo and any other components in between the TiVo and TV (such as a receiver) are powered on, and try again.

Using Your TiVo

Now that you're all wired up, you're ready to begin using the TiVo itself!

Activating Your TiVo

To get your TiVo up and running, you need to activate your TiVo service by registering your TiVo on TiVo's Web site:

1. Go to http://www.tivo.com/ and click on the Set Up TiVo link.
2. Click on the Activate or Upgrade TiVo Service link, and you will see the screen pictured in Figure 1-5.

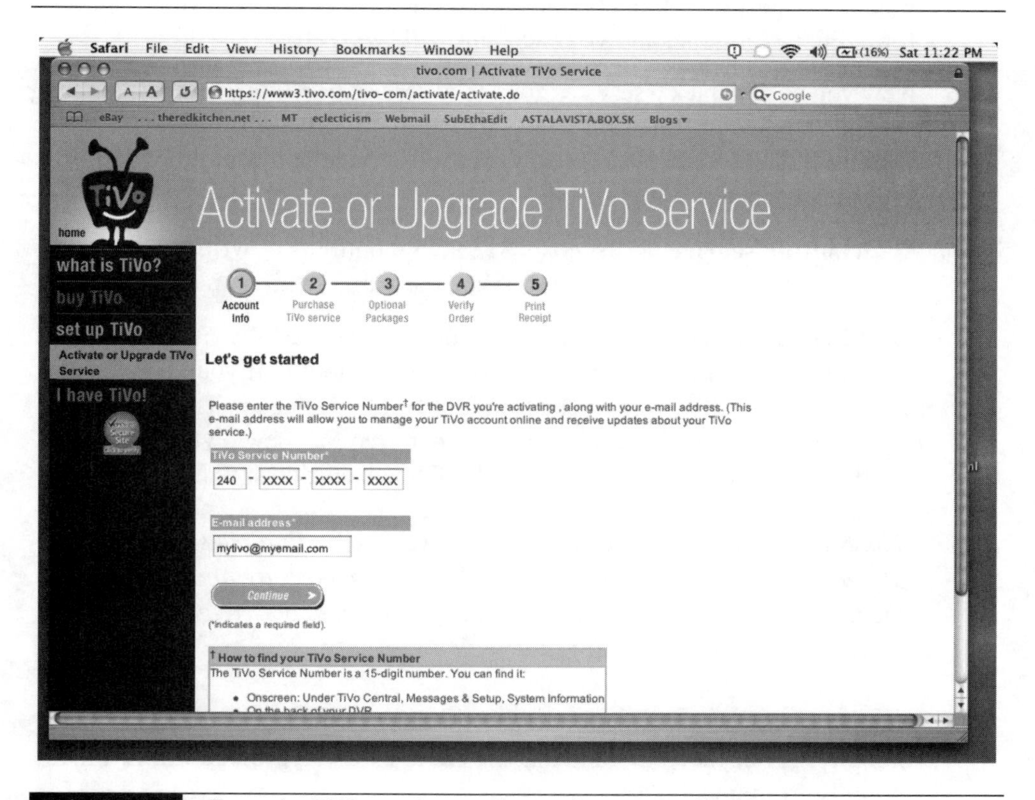

FIGURE 1-5 Enter the TiVo service number and your e-mail address.

3. Enter your TiVo service number and your e-mail address. The TiVo service number is a 15-digit number that you can find on the back of your TiVo unit or on the box at the bottom of the shipping label. If you don't see it in either of those places, turning on the TiVo and beginning the setup process onscreen will reveal your service number. The e-mail address is needed so you can log in to the TiVo site later on. Click Continue when you're finished.

NOTE *If you have a DirecTiVo, you can't register your TiVo online. You need to call 1-800-DIRECTV instead.*

4. On the next page, you'll be asked to provide a lot of personal information—your name, address, and so on. If you're concerned about your privacy, you might want to review TiVo's privacy policy by clicking the Privacy Policy link at the bottom of the left sidebar. When you're finished filling out the form, click Continue, and your account will be created.
5. Now you need to pick a service. Your TiVo service provides your programming lineup so you can schedule shows to record—you can't use the TiVo without it. Choose the level of subscription service you want, monthly or lifetime, as shown in Figure 1-6.
   - **Monthly service** This costs $12.95 a month (as of writing), but you can cancel or switch to a new TiVo at any time with no penalty.
   - **Lifetime service** This is a one-time $299 fee (as of writing), which is good for the life of the TiVo that you are registering. If your TiVo croaks or you replace it, you'll need to subscribe all over again. The good news is that the lifetime service can be transferred to a new owner if you sell the TiVo, and it does add value to the TiVo on auction sites like eBay.com.

TIP *The break-even point of monthly service versus the lifetime subscription is about 23 months, so keep that in mind! If you're not sure about dropping the nearly $300 on the lifetime service, go ahead and sign up for monthly service—you can always upgrade with no penalty.*

6. Enter your billing information to pay for your service. If you choose the monthly service, keep in mind that the credit card you provide will be billed monthly from now on.

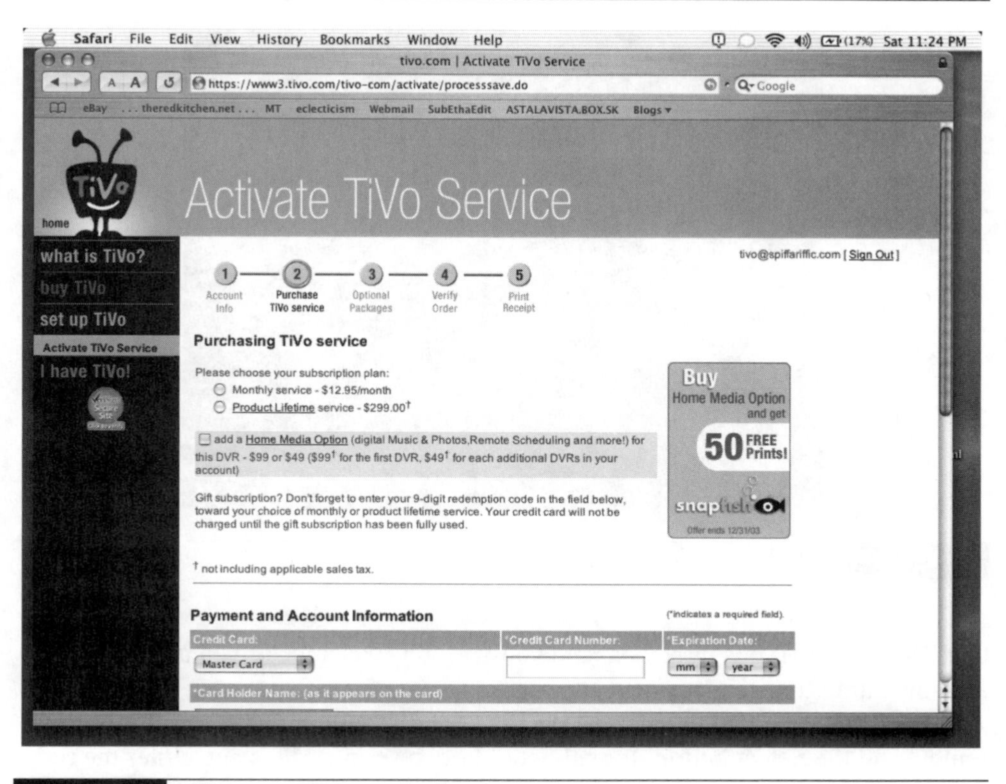

FIGURE 1-6 Choose a TiVo service level.

7. At the bottom of the page, you get to name your TiVo. Pick a short name that makes sense to you, and click Continue.

8. On the next page, verify your billing and contact information, and click Continue.

That's it! You're done. Make a note of the e-mail address and password you gave TiVo so that you can log in to your account later.

Configuring Your TiVo with Guided Setup

Now it's time for your TiVo's maiden voyage. If your TiVo is not already plugged in to power and phone as described earlier in the chapter, now's the time to remedy

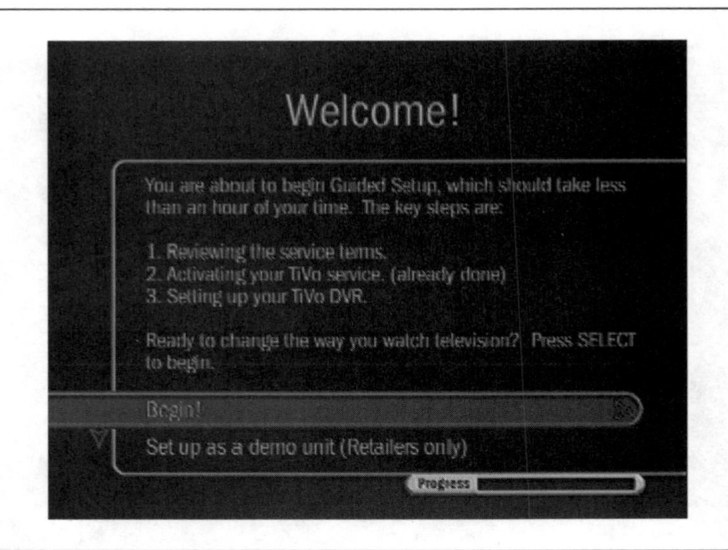

FIGURE 1-7 TiVo's Welcome screen

that. Once your TiVo is turned on, and you've got its welcome screen up on the TV, as shown in Figure 1-7, you're ready to go.

To navigate the menus on the TiVo, use the directional pad at the top of the remote and the Select button. It really couldn't be simpler. Pressing either the right arrow or Select button will choose the selected item and move you forward in the process, unless the TiVo specifically says you must press Select to continue; if you need to go back in the process and change or review something, simply press the left arrow.

NOTE *Setup will take about 20 to 30 minutes of your active time, but with the phone calls to download programming data, the TiVo will not be available for use for up to six hours, depending on how many channels you receive.*

From the TiVo welcome screen, press the right arrow or Select button to move to the next screen, and follow these steps:

1. This screen gives you a chance to review the brief service terms you agree to when you operate your TiVo. Understanding the service terms is important, especially if you intend to "hack" your TiVo, so don't just skip through this screen—as tempting as that may be. Your TiVo should have come with a paper copy of these terms, in case you want to review them later.

2. On the following screen, TiVo explains why you need to have your TiVo hooked up to a phone line: so you can download your programming lineup information. The TiVo will need to make two calls, so you might have to wait until the phone isn't needed for about an hour. Press the right arrow or Select button to continue.
3. Next your TiVo asks you what your TV source is, also known as the "program source": broadcast TV (antenna), cable, satellite, satellite and cable, or satellite and antenna. Pick the one that best matches your configuration.
4. On the next few screens, you'll be asked to enter your ZIP code (Figure 1-8), your time zone, whether or not you observe daylight savings time, and your area code. This information is required for the TiVo to configure itself properly; your TiVo needs to know where and when you are so that it can properly match you up with local phone numbers and correct programming lineup data.

NOTE *Feeling queasy about entering all this identifying information? You can review TiVo's fairly strict privacy policy online at http://www.tivo.com/. Click on the Privacy Policy link at the bottom left.*

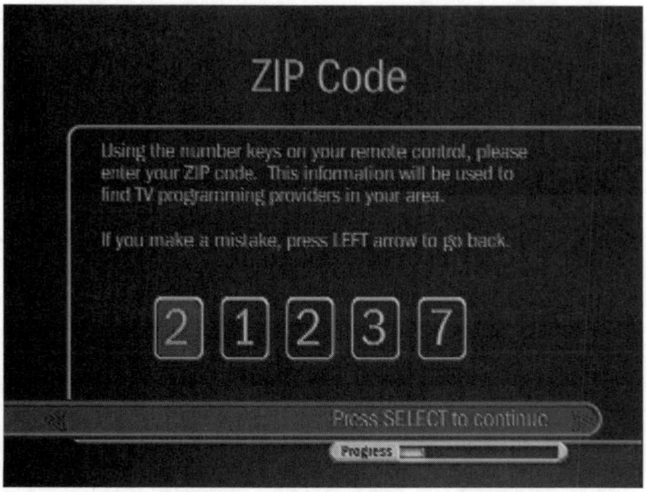

FIGURE 1-8 The ZIP Code screen

5. The next step is to configure the way the TiVo dials, if necessary. By choosing to go to the phone-dialing options (Figures 1-9 and 1-10), you can add a dialing prefix, such as 9 for an outside line, and set the call-waiting prefix to disable call waiting while your TiVo is on the phone. Other settings let you specify whether to use pulse or tone dialing and whether or not the TiVo should attempt to detect the phone line and dial tone before attempting to dial. If you think you're likely to have problems, you can have the TiVo do a test dial and then tweak the settings from this screen.

NOTE *In most areas in the United States, the prefix to disable call waiting is *70.*

6. Your TiVo will now make its first call, a setup call (Figure 1–11). During the setup call, the TiVo dials into a toll-free number and downloads information, such as local TV providers and local TiVo phone numbers for future dialing. As the screen says, this call will take up to 3 minutes, plus up to 15 minutes to "crunch" the data afterwards. While dialing, downloading, disconnecting, and processing, the TiVo will keep you updated with onscreen visual cues—a spinning wheel when it's working and a yellow checkmark when each part of the process is complete.

FIGURE 1-9 Phone Dialing Options main screen

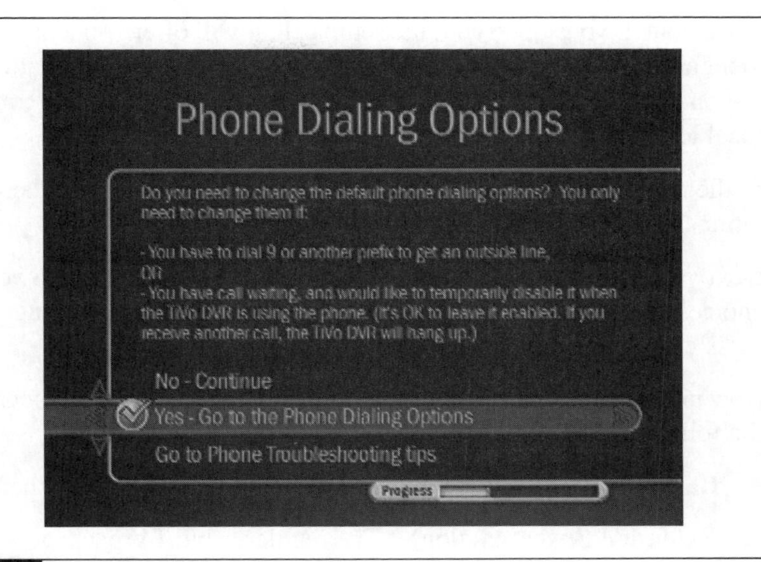

FIGURE 1-10 Go to the Phone Dialing Options.

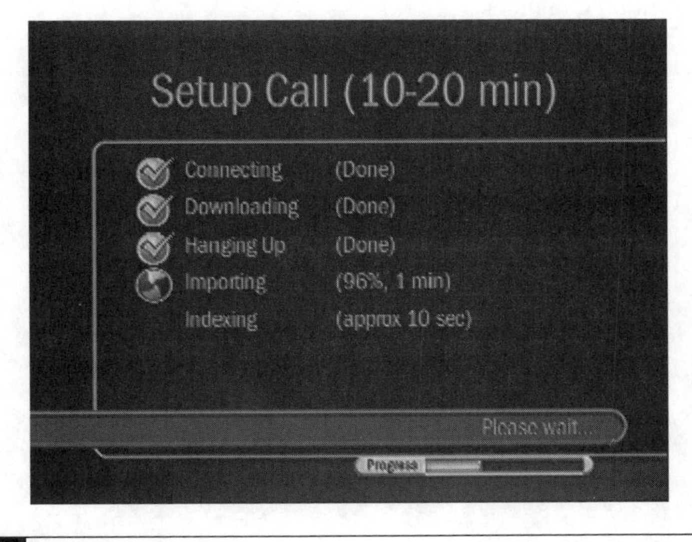

FIGURE 1-11 Setup Call options

7. Once your TiVo is done with the setup call, it will offer you a list of potential local numbers for you to pick from. Be sure to pick the one that is nearest to you geographically, since the list is based on area code and could contain local long distance numbers.
8. In the next screen, specify what format the TiVo should use when dialing numbers.
9. Next you'll need to pick your TV provider. If you have several very similar choices, check your bill to be absolutely sure—picking the wrong one could mess up your TiVo's scheduling.
10. Now pick your TV lineup, as shown in Figure 1-12. Your choices may include the following:
    - **Basic (Analog)** The cable TV package with the fewest channels.
    - **Extended Basic (Analog)** The regular cable TV package with up to about 90 channels.
    - **Digital Basic** The digital cable TV package with the fewest channels.
    - **Digital Extended Basic** The digital cable TV package with more premium channels.

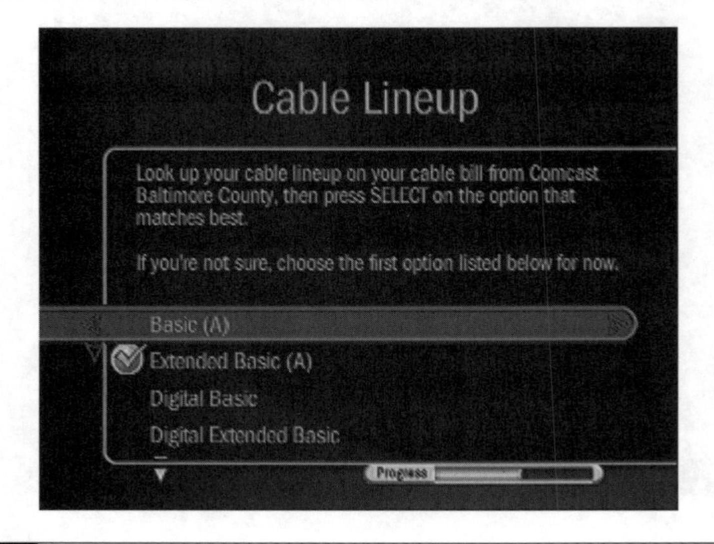

FIGURE 1-12 Cable Lineup

Your choices may also differ slightly depending on your provider. When in doubt, check your bill to see which package you have.

11. Next, pick which channels you actually receive; press Select to add a checkmark to a channel you get that wasn't already marked, and press Select to remove a checkmark next to a channel you don't get that was already marked. Don't worry if you make a mistake, because you can always fix it later.

12. Next, if you have a cable or satellite box, you must specify how it is connected to your TiVo, as shown in Figure 1-13. Use the arrows to move through the list, and be sure to wait long enough for the TiVo to test the video source once you've made your choice. Your cable or satellite receiver must be turned on to continue the setup.

13. You will now be asked to choose whether you have hooked up your cable or satellite receiver via the serial control cable or the IR control cable.

14. You will be asked to pick the brand of cable box you have. You may need to hunt around, looking on the back, bottom, or sides of your cable box to find the brand name.

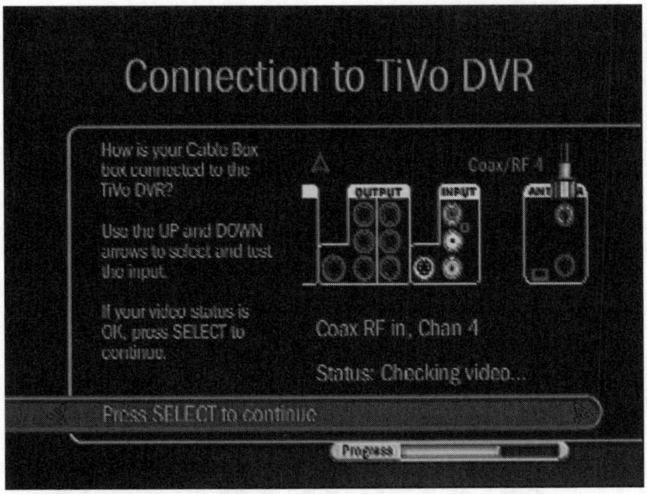

FIGURE 1-13 Connection to TiVo DVR

15. If you have an infrared connection, you will be asked how many digits (two or three) your cable box displays for channels, and whether or not you have to press Enter on your cable box's remote to switch channels, and then you will be given the IR Test Checklist and the IR Test Instructions (Figure 1-14). If you're using serial, you'll go straight to the Serial Test Checklist and the Serial Test Instructions.

16. For an IR connection, you will need to pick from a list of codes until you find a code that fully controls your cable box. For a serial connection, you don't have to pick codes; it will either work or it won't, due to the limited number of units that work with serial control.

17. Next, you'll get a screen explaining the program call the TiVo will now make (see Figure 1-15). This call will take significantly longer than the setup call earlier—up to about 30 minutes—and it may take up to eight hours to crunch your program lineup data! You can't record any shows with your TiVo until this is finished, but once it is you will have no further interruptions.

18. Once the program call is complete and the TiVo begins working on the data, you will be shown a Congratulations screen. You're done! Now you can sit back and relax, watch live TV with your TiVo, and even use the pause, rewind, and instant replay features with live TV. You just can't record programs until the TiVo is done crunching the data.

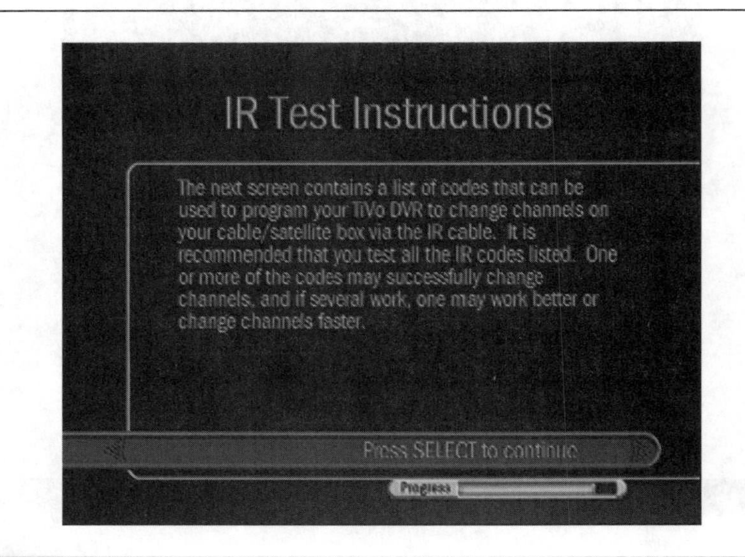

FIGURE 1-14 IR Test Instructions

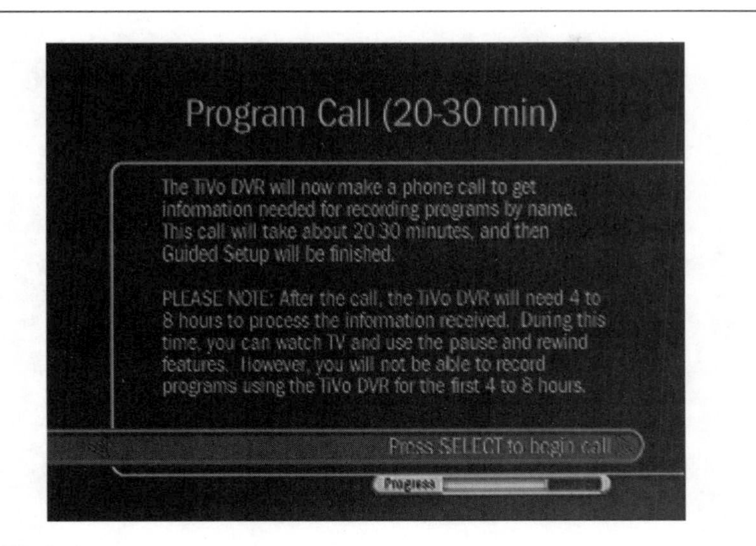

FIGURE 1-15 Program Call information

NOTE

If you'd like to check on the indexing (crunching) progress, press the TiVo button at the top of your TiVo remote, select Messages and Setup, choose Phone Setup, and check out the information under Status. It will show you a percentage completed.

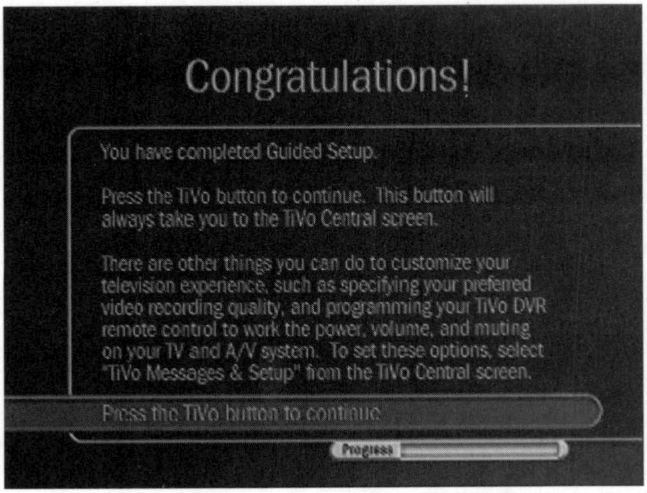

FIGURE 1-16 "Congratulations!" says it all.

Chapter 2

Watch TV and Record Programs with Your TiVo

How to...

- Watch and Manipulate Live TV with TiVo
- Search For and Record Programs with TiVo
- Manage Upcoming and Current Recordings
- Get the Most Out of Your TiVo's Storage Space
- Program Your TiVo Remote to Control Your TV and Stereo

Introduction to Using TiVo

The best way to use your TiVo to the fullest is to learn how it works. This chapter covers not only the practical side of things, but also the background information that will help you truly understand the whats and whys of using TiVo.

NOTE *From here on out, we use the words "TiVo" and "DVR" (digital video recorder) interchangeably. But just like with squares and rectangles, all TiVos are DVRs, but not all DVRs are TiVos.*

How TiVo Works

TiVos work by recording TV signals onto an internal hard drive. When you use TiVo, you interact with the TiVo's operating system, which lets you access the data on that hard drive. The hard drive inside your TiVo has limited space—that's why it's advertised as 40 hours or 80 hours—so you have to strike a balance between the quality of a recording and how much space the recording takes up. Higher quality recordings take more space on the hard drive.

Your TiVo records shows in a variety of ways. The first involves the Live TV buffer, which records up to half an hour of the TV channel you're currently watching live. That means that if you've been watching ESPN for 10 minutes, you'll have 10 minutes that you can rewind through and watch again; if you've been watching it for an hour, you only have the previous 30 minutes. The buffer allows you to use all sorts of nifty features, including rewind, fast forward, instant replay, and pause. Without the buffer, which is pictured in Figure 2-1, none of these would work on live TV.

FIGURE 2-1 TiVo's Live TV buffer

The second way to record content is for you to tell your TiVo what you want to record. You can tell it to record a show in many different ways.

- Tell it to record a single, specific episode or movie.
- Tell it to record all the episodes in a season.
- Tell it to record any show or movie that contains a certain actor or that is in a certain genre.

Finally, based on what it thinks you like, TiVo picks shows it thinks you may enjoy, called TiVo Suggestions. These recordings have a low priority when it comes to hard drive space, so they'll never prevent you from recording something you actually tell the TiVo to record, but they're handy if you're bored and want to check out something new. Keep in mind that TiVo Suggestions are recorded at the default quality setting for the TiVo, so if you find they are too low quality (or they take up too much space), you can change the default setting, and that will affect any Suggestions recorded from then on.

TiVo's Remote

The TiVo remote is crucial—you can't use your TiVo without it. So the first item on the agenda is to familiarize yourself with all the buttons on the remote. There are a lot of buttons, but don't worry: using the remote will become second nature to you very quickly.

- **TiVo button** Pressing this button once takes you to TiVo Central (TiVo's main menu). When you press it twice, it jumps you directly to Now Playing, where you can see all the shows you currently have recorded.
- **TV Power button** This button will turn your TV on and off if you have programmed the remote to work with the TV.
- **Live TV button** Pressing this button takes you to Live TV when you are watching something that is recorded, or when you are looking at the TiVo menus.
- **Directional pad** The big gray button with the four arrows is the directional pad that you use to navigate the menus of the DVR. It's not very sensitive, so don't worry too much about always being spot-on; you'll still get the cursor to go the direction you want.
- **TV Input button** Pressing this button switches inputs on your TV (if you have programmed the remote to work with your TV).
- **Info button** Pressing this button will bring up a little information window on the show you're watching when you press it.
- **Window button** This button doesn't do anything. It's "reserved for future use"—possibly a software upgrade will make use of it in the future, but for version 4.0 it does nothing.
- **Guide button** Pressing this button brings up the window that lets you see all the shows on all the channels that are playing now and in the near future.
- **Select button** Pressing this button is much like clicking the mouse on your computer—it chooses the item you currently have selected.
- **Thumbs-Up and Down buttons** The red thumbs-down button and green thumbs-up buttons let you rate shows, telling TiVo whether you like or dislike the show you're watching.
- **Mute button** This mutes the TiVo output.

- **Vol button** This is the volume control. If you've programmed your TiVo remote to control your TV or stereo, this button will control the volume.
- **Chan button** This is the channel control.
- **Record button** Lets you record whatever it is you're watching simply by pressing it.
- **Play, Fast Forward, Slow, Rewind, and Pause buttons** This constellation of buttons let you control the show you're watching. Clockwise from the top they are Play, Fast Forward, Slow, and Rewind. The big yellow button in the middle is Pause.
- **Jump to End button** This button is to the lower right of the Slow button. If you press it once, you will jump to the end of the current recording; if you press it twice, you will jump to the beginning of the recorded show or buffer.
- **Instant Replay button** This button is to the lower left of the Slow button. Pressing Instant Replay jumps you back eight seconds.
- **DVR switch** This switch has the options 1 and 2, which lets you switch between two DVRs. You can use the one remote for two different TiVos. We'll cover multiple TiVo setups later on in this book.
- **Number buttons** The number buttons are self-explanatory: use them to type in channels, and then press the Enter button at the bottom right of the remote if you need to.
- **Enter button** Press Enter after entering a channel number to change the channel. If you press the Enter button on its own, it functions as a Jump or Last Channel button.
- **Clear button** Pressing this button will clear away on-screen menus when you want to watch a program. It is also a shortcut for deleting shows you don't want to keep any longer.

TiVo's Menus

At the heart of your interaction with your TiVo is TiVo Central, from which you can get to all the menus and options you use to make the DVR do what you want. You'll learn all about these menus in this chapter and others, but here's a quick overview to get you started.

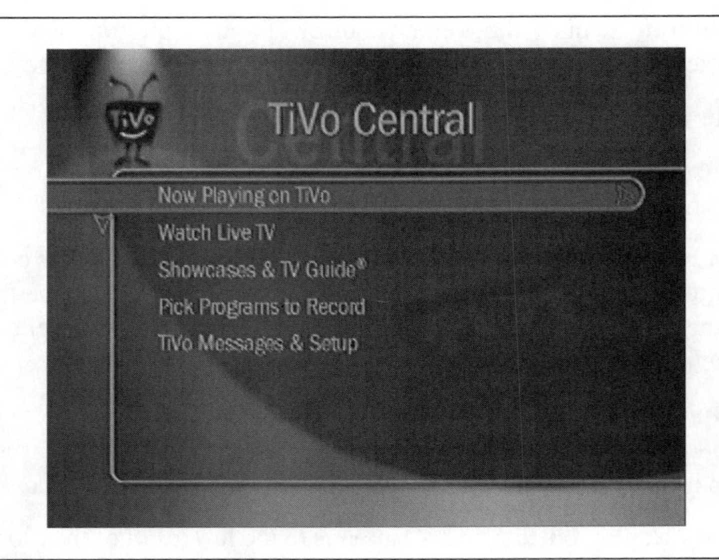

FIGURE 2-2 The TiVo Central screen gives you access to all the TiVo menus.

TiVo Central

Press the TiVo button at the top of your remote to go to TiVo Central (shown in Figure 2-2). This works whether you're watching live TV, a recording, or whatever. (If you're in the middle of configuring something, though, you may need to save your choices first.)

In TiVo Central, you have the following choices:

- **Now Playing on TiVo** This option gives you a list of what's saved on your DVR's hard drive right now.
- **Watch Live TV** This option takes you to live TV. It is the same as pressing the Live TV button on the remote.
- **Music & Photos** This option lets you access your shared music and photos over your home network, if you have the Home Media Features.
- **Showcases and TV Guide** This option takes you to a menu of special features that TiVo wants you to check out (paid placements, usually).

- **Pick Programs to Record** This option is the main way to set up recording for shows.
- **TiVo Messages & Setup** This option takes you to a menu where you'll find TiVo system and company messages to you and the settings for your DVR (including default video quality, channel lineup, etc.).

Now Playing

Now Playing is where you go to watch and manage your recorded shows. To get there quickly, press the TiVo button twice in a row.

Music & Photos

You can browse your music files and photos and play them through your TiVo.

Showcases & TV Guide

One of the ways TiVo makes money is paid promotions: someone pays TiVo, and TiVo downloads a little preview of something to your DVR in the hopes that you'll see it, love it, and then watch the show or event. TiVo distinguishes this content from the content that you choose and from TiVo Suggestions (which TiVo records for you) by keeping it in the Showcases area of TiVo Central.

TiVo Central Icons

Your TiVo uses icons to give you information about the menu options. Here's a list of the icons you may see in TiVo Central, and what they mean:

- **Mail** This icon tells you that you have new TiVo messages to read.
- **Director's Clapboard** This icon appears next to selections with "special video content" (also known as paid promotions).
- **Star** This icon indicates that the associated menu option contains a special preview or showcase.
- **TiVo Character** This icon indicates news or special offers from TiVo.

Did you know?

Save Time by Using the Directional Pad

When navigating through TiVo menus, you can almost always use the right arrow on the directional pad to choose the highlighted option instead of using the Select button. You will probably find that this is less work for your thumb!

Pick Shows to Record

Pick Shows to Record is the place where you'll be scheduling most of your recordings. You can search for programs you want to record by title, by time or channel, by WishList (for instance, by subject matter or by actor), or you can enter a recording time and channel manually. This is also where you set up Season Passes to get all or most of the episodes from any given series.

Messages & Setup

Messages & Setup is probably the TiVo Central section you will use the least. Occasionally your TiVo will need to tell you something technical (for instance, it may be having a problem dialing in), or promotional (TiVo may want you to know about some new feature), and you will be able to read those messages here. The setup section of Messages & Setup allows you to configure all the details of your TiVo, including running Guided Setup again if you've moved or need to change all those settings for some other reason.

Watching and Manipulating Live TV

TiVos don't just let you record and play back shows, they let you watch—and control—live TV, too. The first time the phone rings and you simply press Pause to avoid missing a single second of your favorite prime-time drama, you'll see why this is a killer feature in its own right, and not just an afterthought.

Watching Live TV

There are three ways you can get to live TV when using your TiVo. The first and easiest is to just press the Live TV button on your remote, and you'll jump right to it.

The second option is to choose Live TV from the TiVo Central menu, covered earlier in the chapter. The third is to simply not use the TiVo for about twenty minutes—as long as it isn't paused, it will switch to live TV by itself.

Changing Channels and Volume

Change channels using the Chan rocker button on the right side of the remote, and once you have programmed your remote to work with your TV or stereo, you can use the Vol rocker button to change the volume.

Using the Info Banners

If you're not sure what you're watching, or you'd like more information than you already have, press the Info button once or the right arrow on the directional pad three times. Pressing the Info button will bring up a large window with all the information about the show, as shown here, and so will pressing the right arrow three times.

Pressing the right arrow only once will bring up a little info banner that shows only the time, channel name, and channel number; pressing it twice will turn the little banner into a larger banner across the top of the screen that will also have the show name, run time, and TV rating, like this one.

NOTE *When you press the right arrow on the directional pad, the info banner format that comes up is the last one you used, so if you've used it before, the order may be different than described here.*

You can press Info or the left arrow on the directional pad to dismiss the banner; otherwise it will go away on its own in just a few seconds.

Using the Ipreview Button

While watching live TV (or checking out the showcases or TiVolution Magazine) on your TiVo, a black oblong indicator may be displayed, telling you to "Press Thumbs-Up" and showing an icon of the Thumbs-Up button. This indicator, called the Ipreview icon, indicates that there is a promotion for the show; if you press the Thumbs-Up button, you will be able to get more info on the upcoming show, and also set it to record right from that screen.

The Ipreview icon works even in recorded shows. If you miss it, you can rewind back to where you saw it, and it will work.

Using the TiVo Guides

Does it feel like you've got a hundred channels but nothing's on? Press the Guide button. The TiVo Guide lets you see what's on now and in the near future, in two formats. The default guide format has a listing of channels and the current show on the left (or the upcoming show, if you're close to a changeover); on the right it lists the shows in the next eight time slots. The default Guide style, the TiVo Live Guide, is shown in Figure 2-3.

Alternatively, you can press the Enter button and tweak the Guide options, shown in Figure 2-5, including changing the style from TiVo Live Guide to Grid Guide (Figure 2-4), which makes it look very much like TV Guide. Other options allow you to limit the channels listed to only those you have marked as favorites, or you can have the Guide only list shows that match a certain type of show (for instance, movies) or a particular subject.

Choose the options you want using the directional pad, and then press Select twice to save your choices.

In either style of Guide, you can use the directional pad to move up or down one line at a time, or the Chan rocker button to move up or down one full screen at a time. You can jump to a show that's currently playing by selecting it using the directional pad and pressing Select.

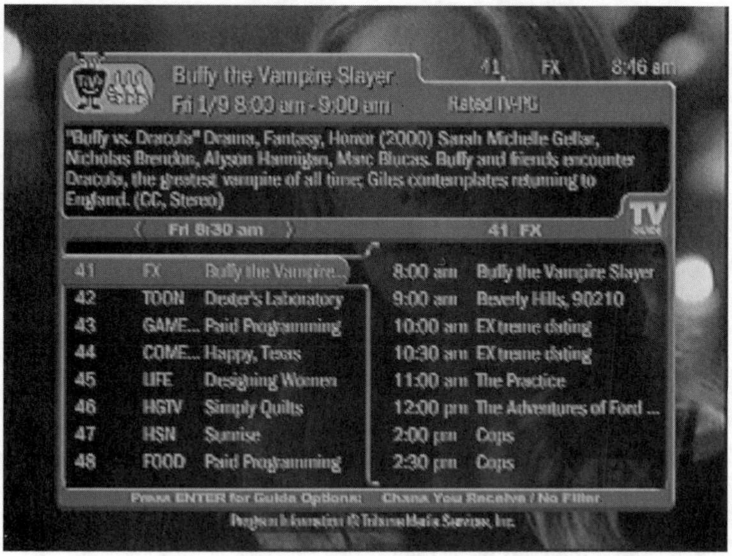

FIGURE 2-3 The TiVo Live Guide

Buffy the Vampire Slayer 41 FX 8:47 am
Fri 1/9 8:00 am - 9:00 am Rated TV-PG
"Buffy vs. Dracula" Drama, Fantasy, Horror (2000) Sarah Michelle Gellar, Nicholas Brendon, Alyson Hannigan, Marc Blucas. Buffy and friends encounter Dracula, the greatest vampire of all time; Giles contemplates returning to England. (CC, Stereo)

Fri 1/9	8:30 am	9:00 am	9:30 am
41 FX	Buffy the Vampir...	Beverly Hills, 90210	
42 TOON	Dexter's Laboratory	Hamtaro	Baby Looney Tunes
43 GAMESH	Paid Programming	Let's Make a Deal	Family Feud
44 COMEDY	Happy, Texas		
45 LIFE	Designing Women	The Golden Girls	The Nanny
46 HGTV	Simply Quilts	The Carol Duvall ...	The Carol Duvall ...
47 HSN	Sunrise	18-Hour Fashion Wearhouse	
48 FOOD	Paid Programming	Paid Programming	Trivia Unwrapped

Press ENTER for Guide Options: Chans You Receive / No Filter
Program information © TribuneMedia Services, Inc.

FIGURE 2-4 The TiVo Live Guide to Grid Guide

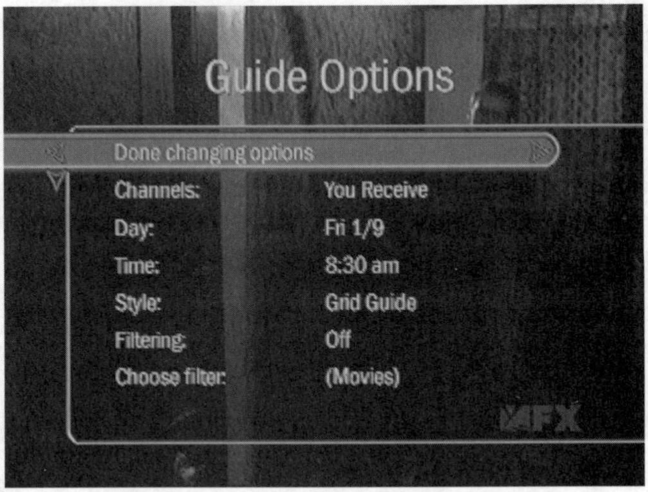

Guide Options

Done changing options

Channels:	You Receive
Day:	Fri 1/9
Time:	8:30 am
Style:	Grid Guide
Filtering:	Off
Choose filter:	(Movies)

FX

FIGURE 2-5 The Guide Options screen

On the TiVo Live Guide style of display, you can also use the fast forward and rewind buttons to change the time shown on the top.

Manipulating Live TV

Thanks to the buffer, your TiVo will not only revolutionize the way you interact with TV through pain-free recording, it will also revolutionize the way you watch live TV.

As mentioned before, the Live TV buffer will hold up to 30 minutes of programming; if you're watching live TV, the buffer will contain the last 30 minutes of the show you've been watching (assuming you've been watching the channel for 30 minutes or more). The status bar (shown in Figure 2-6) illustrates the buffer as well as the timeline for recorded programs, and it will show up at the bottom of your screen when you pause, rewind, or fast forward a program, either live or recorded. The green part of the status bar shows the recorded or buffered portion of the show.

The status bar is marked with little lines showing 15-minute increments, and it also has a thick white line that indicates what you are currently watching. If the

FIGURE 2-6 The Live TV buffer status bar

white line is in the green section, you are watching the buffer or recording and are time-shifted behind live TV.

Pausing Live TV

You can pause live TV at any time—just use the big yellow Pause button in the middle of the remote.

If you plan to leave the show paused for longer than the buffer will hold, you might want to tell the TiVo to record instead. If you're paused when the buffer runs out, the TiVo will revert to live TV, and you will lose the prior contents of the buffer (all those important scenes you were hoping not to miss!).

Rewinding Live TV

Did you just miss something important? Rewind! You have three rewind speeds available to you when using the Rewind button: press once for a slower rewind, twice for a moderately fast rewind, and three times for a very fast rewind. You'll notice that the icon on the status bar indicates how fast you are rewinding by showing one, two, or three green arrows under the white indicator.

Press Play again when you want to resume watching the show.

NOTE *Keep in mind that you can only rewind as far back as the buffer goes, so if you just switched to the channel a minute ago, you only have a minute of saved programming available to you via rewind.*

Using Instant Replay

Your second option for reviewing live TV is to use the Instant Replay button, which jumps you back just eight seconds in the timeline and immediately resumes playing with no more button-pressing from you. The Instant Replay button is the perfect choice when you just want to see that winning catch, romantic kiss, or death-defying car crash once more.

TIP *For even more fun, press Instant Replay and then immediately press Slow for a slow-mo replay! Press Play once more to resume full-speed video.*

Jumping to the End or Beginning

Press the Advance button to jump to the end of the buffer; press it twice to jump to the beginning of the buffer.

Finding and Recording Shows

The real magic of TiVo is the genuinely innovative way it helps you locate and record the shows you want. Once you grow accustomed to having hours of quality programming at your fingertips, programs that you actually want and can watch at any time, you'll never be able to go back to watching TV the old-fashioned way.

Recording Basics

There are many ways to choose a show to record, but with every method you will be using the same Program Information and Recording Options screens. We'll look first at these parts of the TiVo interface, and then at the ways to choose shows.

Viewing Program Information

The Program Information screen (Figure 2-7) shows you the title of the show or movie—or the title of the series and the name of the specific episode—plus date, actor data, and the summary of the plot, as shown here.

FIGURE 2-7 The Program Information screen

To the right, you will find information on the show's time, channel, duration, TV rating, categories, and the video quality it is set to record as (or the default setting, if it's not yet set to record). To the left is a menu with the following options:

- **Record This Showing (Options)** This sets the episode to record, with or without altering the options. (To change recording options, select Record this Showing, and then select Options with the directional pad and press Select.)
- **Get a Season Pass** This option, which is shown only for TV series, allows you to configure a Season Pass for the TV show and record multiple episodes automatically.
- **View Upcoming Showings** This shows you future showings of this program.
- **Don't Do Anything** This option discards any settings you made and takes you back to the previous screen.

Setting Recording Options

The Recording Options screen, shown in Figure 2-8, lets you configure recording settings for a show or series.

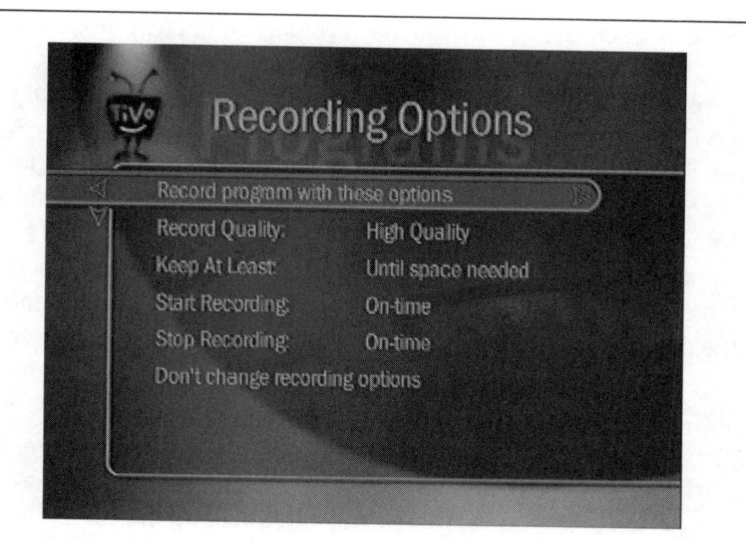

FIGURE 2-8 Recording Options

NOTE *For repeat recordings, such as Season Passes, the Recording Options screen has two extra options: Keep at Most and Show Type. These are discussed in the "Creating a Season Pass" section later in this chapter.*

- **Record Quality** The quality level you pick determines both how much space the show takes up, and how it looks when you play it back.
- **Keep at Least** This specifies the minimum amount of time the TiVo should keep the recording before deleting it to make more room for new recordings.
- **Start Recording** This option lets you start recording earlier or later than the posted start time of the show.
- **Stop Recording** This option lets you stop recording sooner or later than the posted stop time of the show.

Use the directional pad to change the options, and when you're done press Select twice to save your settings and back out of the screen. Next you'll get a confirmation screen, or possibly a screen that tells you of a recording conflict, as shown in Figures 2-9 and 2-10.

In the case of a conflict, you have two choices: you can decide to not record one of the shows (highlight your choice with the directional pad and press Select), or you can look for alternative airing times and record them both. To look for upcoming showings, back out of the menu by pressing the left arrow on the directional pad, highlight View Upcoming Showings, and press Select (see Figure 2-11).

If the show in question has an upcoming rerun, you can easily schedule one of those to be recorded instead of the one that conflicts with your other show; simply highlight the date and time you want, press Select, and configure the recording.

Hard Drive Space vs. Recording Quality

Your TiVo lets you record shows in four quality settings: Basic, Medium, High, and Best. When it advertises "40 hours" or "80 hours" on the box, that assumes the lowest quality setting, Basic, which really isn't usable most of the time. Upgrading to Medium quality will cut your available recording hours, and upgrading to High will reduce them even more. Best quality will give you slightly less than a third of the advertised capacity.

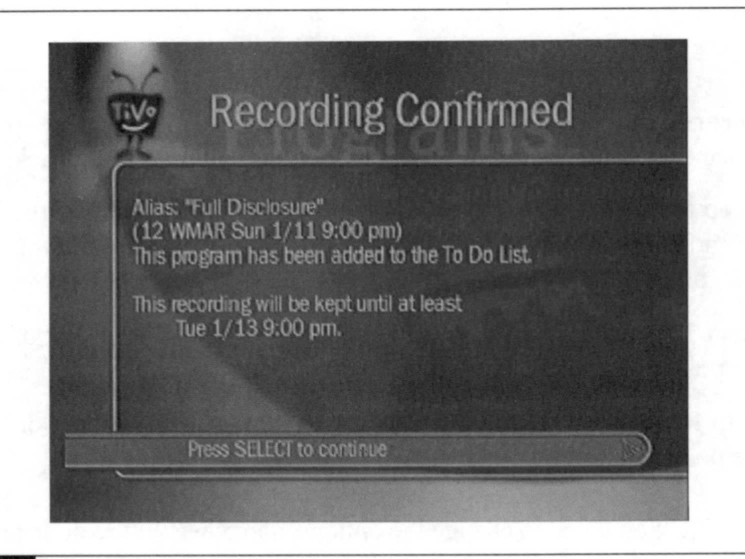

FIGURE 2-9 Recording Confirmed

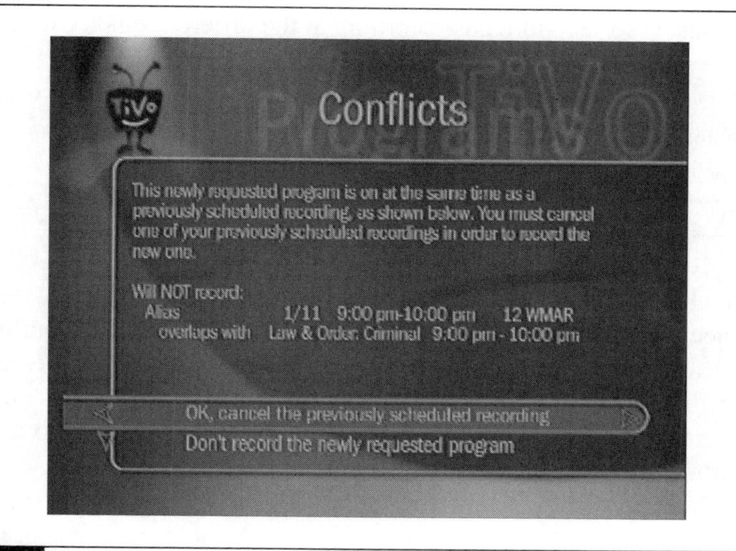

FIGURE 2-10 Conflicts screen

FIGURE 2-11 Upcoming Showings

Here's what TiVo recommends recording at the different quality levels:

Basic Quality	News shows (where you don't care much at all about the looks of the thing)
Medium Quality	Cartoons, daytime talk shows
High Quality	Dramas, films
Best Quality	Sports, action films (where every little detail counts)

You can change the quality settings on a per-show basis, as well as setting the default choice for all shows. To change the default, press your TiVo button to go to TiVo Central, and choose Messages & Setup, then Settings, then Preferences and finally, Video Recording Quality. The current quality level is indicated with a checkmark in a yellow circle, and how many hours the TiVo will hold when set to that level of quality is shown to the right.

You should experiment with different quality settings for a wide range of shows to determine your preferences. We use Medium quality most of the time, and occasionally High quality, but never Basic or Best if we can avoid it. Keep in mind that you can't change the quality of a recording once it has been recorded—you can only change the quality of future recordings. If you want to change the quality for all future recordings, use the instructions above to change the default setting.

Recording Icons

TiVos use icons to help you see useful information at a glance, such as which shows are scheduled for recording. Here's a quick guide to icons you may see next to program names.

- **Single check mark** This show will be recorded, singly.
- **Double check mark** This show will be recorded as part of a Season Pass or repeat manual recording.
- **Star** This show will be recorded as part of a WishList.
- **Red circle** This show is currently recording.

Recording While Watching TV

TiVo offers you many choices when it comes to setting up shows to record, including two that you can do while watching live TV or looking through the Guide.

Recording from Live TV

When you'd like to record a show you've been watching on live TV, simply press the red Record button (Figure 2-12). If you have any of the show in the buffer, you'll have to record the show in Best quality or set up the recording using another method, such as going to the Pick Programs to Record menu. If you do decide you don't want to record in Best quality, setting up the recording from Pick Programs to Record will cause the buffer to be discarded.

If you leave the TiVo paused long enough for the buffer to fill up, it will completely empty out the buffer and start again fresh with whatever's on live TV at that moment. So if you're planning on being away from the TiVo, or you're not sure how long you'll be leaving it paused, it's a good idea to set the program you are watching to record. Otherwise you may return to find that you lost most of the TV show you were hoping to preserve by pausing.

FIGURE 2-12 In Progress screen

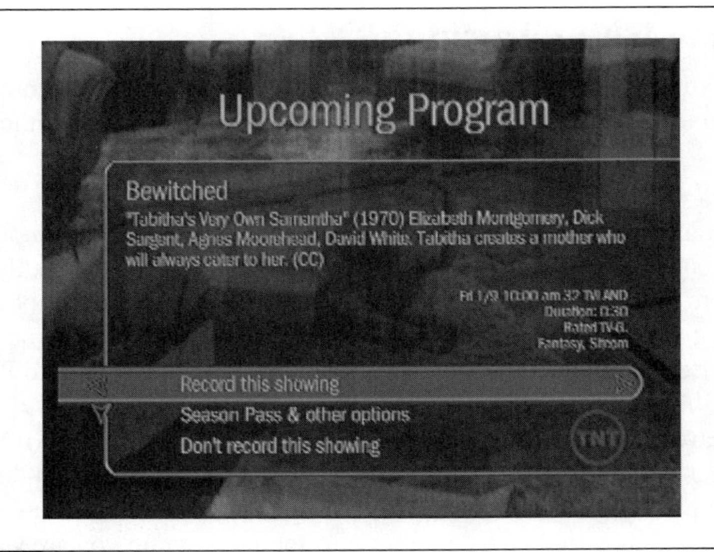

FIGURE 2-13 Upcoming Program screen

Recording from the Guide

You can pick a show to record right from the Guide. Simply highlight it using the directional pad, and press Record (Figure 2-13).

Setting Up Recordings from TiVo Central

The other way to set up recordings is to go to TiVo Central and choose Pick Programs to Record from the menu, as shown in Figure 2-14. From here you have several options for locating and picking shows and for managing your recordings.

- **Search by Title** Look for shows based on their names.
- **Search Using a WishList** Search using defined criteria, such as an actor's name or a genre (such as "Tom Cruise" or "Movies—Action")
- **TiVo Suggestions** View and manage the shows the TiVo plans to record on your behalf.
- **Record by Time or Channel** Browse a Guide-like interface by channel and time.

- **Season Pass Manager** Control and configure your Season Passes to record multiple episodes in a series.
- **To Do List** View and manage all upcoming scheduled recordings.

Searching by Title

You will likely find that Search by Title (Figure 2-15) is the method you'll use most often to select shows to record. Choose it from the Pick Programs to Record menu using the directional pad and pressing Select. You'll be greeted by a screen with a grid of letters on the left and the controls CLR (Clear), SP (Space), and DEL (Delete) at the top. Move around the grid using the directional pad and press Select to choose a letter or to use one of the controls.

The show names on the right are automatically filtered out as you enter each additional letter; you don't have to enter the full name. Once you see the show you want, or get to a point where it is very close, you can use the directional pad to move to the show listings at the right, and then use it to go up and down through the list of shows (or use the Chan rocker button to move up and down by a whole screen at once). Highlight the show you want and press Select, and then configure the recording the way you want.

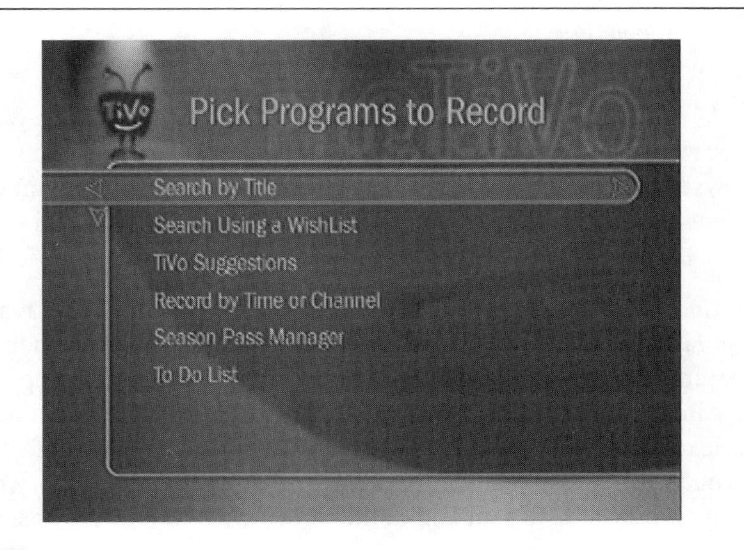

FIGURE 2-14 Pick Programs to Record screen

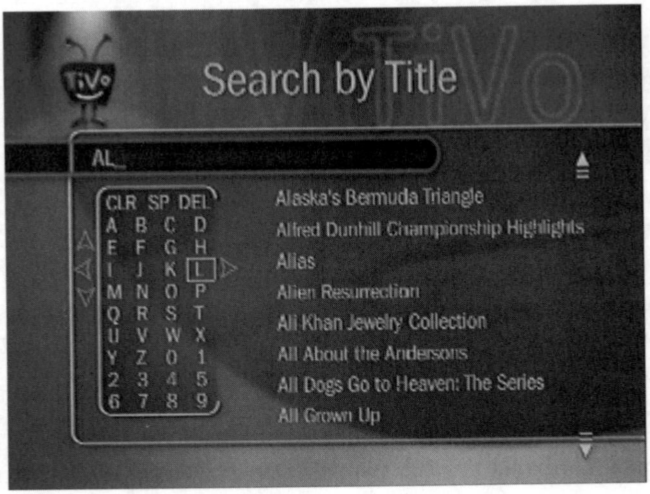

FIGURE 2-15 Search by Title

NOTE *If the list stops changing when you enter more letters that should change the current results, that means the show you're looking for is not in the TiVo's lineup data.*

Recording by Time and Channel

Choose Record by Time or Channel and you'll be greeted by a screen with another menu: Browse by Time, Browse by Channel, and Manually Record Time/Channel. You will generally use the first two options, since setting up a manual recording defeats the whole point of the TiVo.

Browse by Time Browse by Time will show you all the programs available on a particular day and time. Once you select Browse by Time, you can choose whether to see programs from All Channels, just your Favorite Channels, or programs in a specific category (such as comedies or movies). Some categories have many subcategories to help you pin down the kind of show you're looking for.

Once you've made your choice from this menu and the subcategory menu, if applicable, you'll be able to set the timeframe you want to browse, as shown in Figure 2-16.

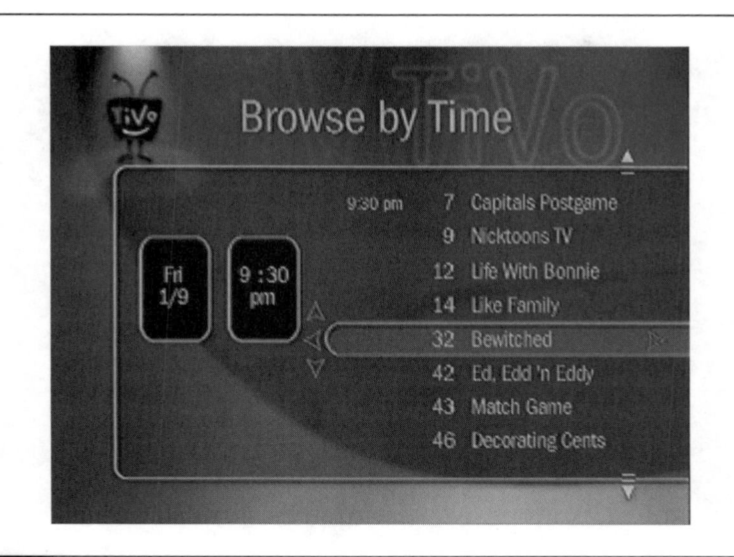

FIGURE 2-16 Browse by Time

Use the up and down buttons on the directional pad to choose the date and time, and then move all the way to the right to bring up the list of shows. Use the directional pad to navigate through the list of shows and to move between the three columns. You can also use the Chan rocker button to jump through a whole screen of TV listings at once.

When you want to view more information about a show or to record it, highlight the show and press Select, or press the right arrow button on the directional pad.

Browse by Channel Browse by Channel (Figure 2-17) lets you see what's coming up on a certain channel. You can choose to browse all channels you receive or only your favorite channels.

Use the up and down arrow buttons to choose a date and channel, and then press the right arrow button until the TV listings show up on the right half of the screen. Use the directional pad or Chan rocker button to scroll through the program listings, and press Select to pick a show.

Creating a Season Pass

Season Passes allow you to set up a recurring recording to get multiple episodes of a series, whether you want to snag them all or just the new episodes. Setting up

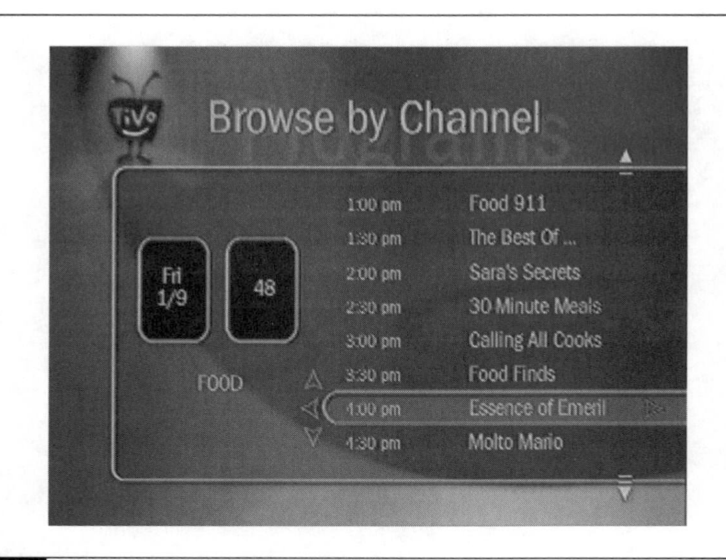

FIGURE 2-17 Browse by Channel

a Season Pass is easy: just select the program from the TiVo Live Guide or using the search methods described in the previous sections.

The Season Pass Recording Options are almost identical to the regular Recording Options (Figure 2-18), save for the addition of two new settings: Keep at Most

Recording Options
Record Season Pass with these options
Record Quality: Medium Quality
Keep At Most: 5 episodes
Show Type: Repeats & first run
Keep Until: Space needed
Start Recording: On-time
Stop Recording: On-time
Don't change recording options

FIGURE 2-18 Recording Options screen

How to ... Record Programs Manually

Generally you should avoid setting up recordings manually, because it limits your access to TiVo's biggest benefit: all those features built around the program guide data. But sometimes it can't be helped. You should set up a manual recording in the following situations:

- You have DirecTV, and program data doesn't show up for sports channels or Pay Per View shows.
- You want to record a show where the program data or description does not change, resulting in a Season Pass that records many duplicates of the same episode (for example, *The Daily Show* has this problem because individual episodes lack differentiated titles or descriptions).
- You've run out of program data and for some reason are unable to get the TiVo to call in and get more program data before the show comes on.

Setting up a manual recording may be no fun, but it's not difficult. From TiVo Central, go to Pick Programs to Record, then Record by Time or Channel, and then, finally, choose Manually Record Time/Channel.

Use the directional pad to change the options and to move from one column of options to another. Set the recording details for your show:

1. Pick the recording type: One Time or Repeating.
2. In the Day(s) column, select a day or days. For a One Time recording, this column will show a list of single days; for a Repeating recording, it will give you a list of patterns to choose from.
3. In the Channel column, pick the channel to record from. You can either scroll through the list of channels, or type in the channel number using the number buttons on the remote and pressing Enter to jump right to it.
4. Press the right arrow button to move to Start Time, and set it. Set the Stop Time.
5. Press the Select button to finalize the recording.

specifies the upper limit of how many episodes you want to keep on your TiVo at once. Show Type offers three choices for narrowing down the scope of your Season Pass:

- **Repeats & First Run** Records both new episodes and reruns of older episodes, but not duplicates of either.
- **First Run Only** Records only brand new episodes.
- **All (with Duplicates)** Record *all* showings, regardless of duplication.

Once you've made your choice, press Select twice to save your settings and back out of the menu. Don't worry if you're not sure about whether the settings you've chosen will always be what you want; you can easily go back and edit your Season Pass whenever you want.

Resolving conflicts with Season Pass recordings is a little bit trickier than with single recordings simply because they repeat. The Conflicts screen (Figure 2-19) will show you what episodes may not be recorded due to the conflict; it will give precedence to the other programs you have set to record, instead of the Season Pass items.

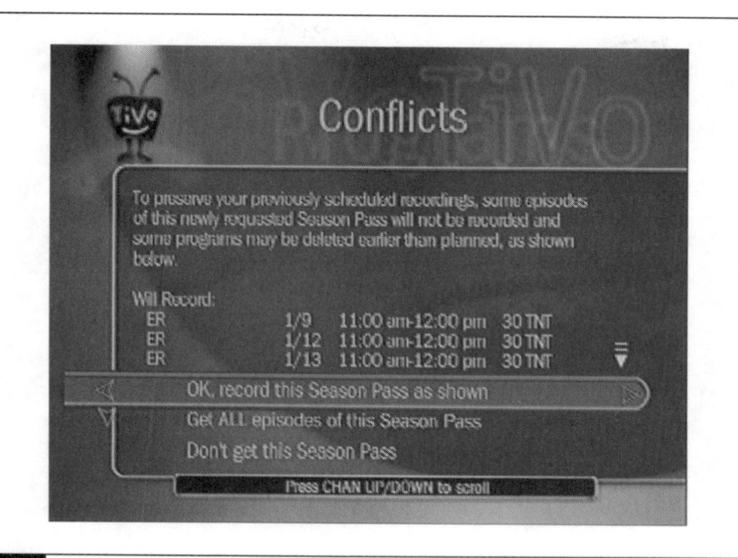

FIGURE 2-19 Conflicts screen

You can either agree with its method (by choosing OK, Record This Season Pass as Shown), tell it to record all episodes of the Season Pass and forget the other scheduled programs (by choosing Get ALL Episodes of This Season Pass), or you can choose to cancel the Season Pass altogether (by choosing Don't Get This Season Pass).

Alternatively, if you have two conflicting Season Passes, you may be able to iron out the situation by using the Season Pass Manager, described in the "Using the Season Pass Manager" section later in this chapter.

Creating and Searching Using WishLists

The WishList is a powerful TiVo feature that lets you record any shows that fit a certain criteria. For example, they may star your favorite actor, or fit a particular genre or show type.

Creating a WishList Follow these steps to set up the WishList, which is shown in Figure 2-20:

1. From TiVo Central, select Pick Programs to Record, and then Search Using a WishList.
2. Choose the Create New WishList option (see Figure 2-21).

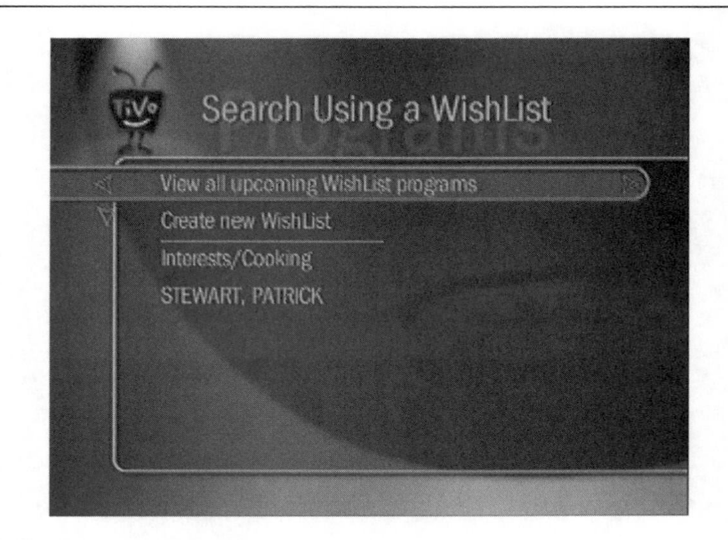

FIGURE 2-20 The WishList

3. Choose from the following types of WishLists:
   - Actor WishList
   - Director WishList
   - Category Only WishList

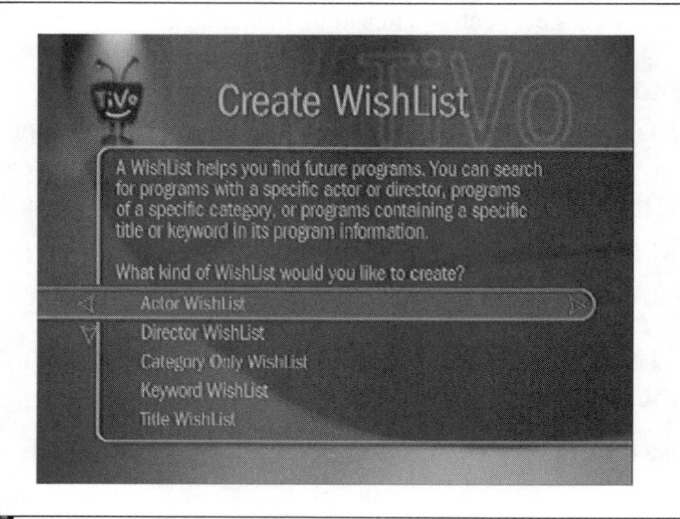

FIGURE 2-21 Create WishList screen

- Keyword WishList
- Title WishList

4. Choose a subcategory to refine your search, if you like. Many of the categories have subcategories. For example, the Movies category has Comedy and Action subcategories, among others, so you're able to further narrow it down once you pick a category.

NOTE *Actor, Director, Keyword, and Title WishLists are all based on the same interface as the Search by Title interface for locating shows (described earlier in this chapter in the "Searching by Title" section): a list of letters on the left, with CLR (Clear), SP (Space), and DEL (Delete) controls at the top. Like the Search by Title interface, you can start spelling out the title or the name of the person by using the directional pad and the Select button to pick letters or commands. The list on the right will be narrowed down as you spell the name or word.*

5. If you are creating an Actor or Director WishList, enter the person's name, with the last name first. Use the comma character found under the letter W to separate the last and first name. Once you see or are alphabetically close to the name you're looking for, navigate all the way to the right using the directional pad, highlight the name, and press Select. If the name does not come up on the right, just finish spelling it out, and you should be able to snag future programs that way—but be warned, if you don't spell it properly, your WishList will never find anything.

TIP *For Keyword and Title WishLists, you can create multiple keywords or title words by using the SP (space) function or pressing Fast Forward on the remote to create a space.*

6. If you are making a Keyword WishList, as shown in Figure 2-22, enter the keywords, separated by spaces. All keywords must be associated with the show for it to be recorded, but it doesn't matter to the TiVo what order they're in or if there are words between them. You can also use the following punctuation to clarify your keywords:

   - Use an asterisk (*) at the end of a partial word so that you get all versions of that word: the term "sit*" would give you results that included sit, sitting, sitter, site, sites, and so on. (Type an asterisk by pressing the Slow button on the remote.)

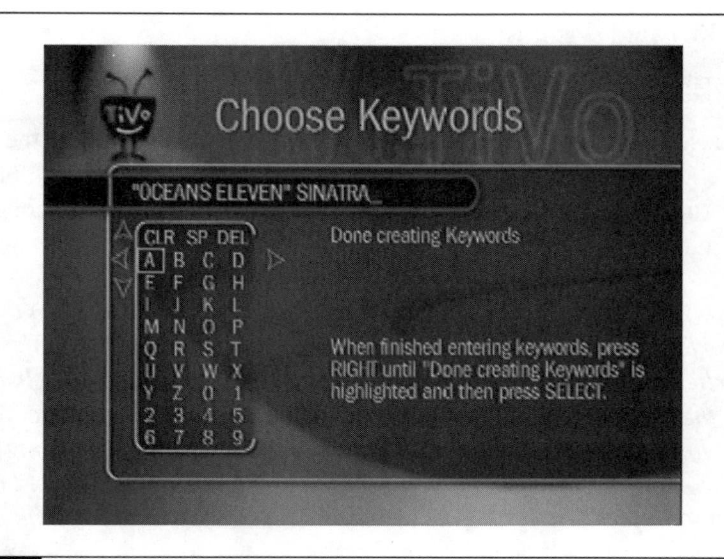

FIGURE 2-22 Choose Keywords screen

- Use quotes around a phrase if you want to search for those words in order, without any words between them. (Enter quote marks by pressing Pause.) So if you only want the original *Ocean's Eleven* with Frank Sinatra, you can use the following keywords: "Oceans Eleven" Sinatra.

7. If you are creating a Title WishList, enter the title starting from the beginning, not including "the," and do not leave out words.

NOTE *When entering a show name that includes a slash, such as* 20/20, *type it as 20 20. Other punctuation, such as apostrophes, are simply ignored, so if you wanted to look for* Rosemary's Baby *just type Rosemarys Baby.*

8. Select a category to limit your search. For instance, you might want to limit your search for keywords "Giants" and "baseball" to the Sports category so you don't get a documentary on "Baseball Giants."

9. You'll now see a screen with options somewhat like a Program Information screen, titled Done Creating WishList. You have the following menu choices:

- **View Upcoming Programs** Use this option to search using your newly created WishList. From the list of programs that come up, pick and choose any that sound good using the directional pad and Select button.
- **Auto-Record WishList Programs** Use this option to have your WishList actively record matching programs for you. This choice works best when you've created a very narrow search (for instance, the category Movies—Romance and actor Kevin Costner, or a genre with multiple keywords). If you use it on a broad WishList search, you may find that it's taking over your TiVo!
- **Edit WishList** Use this option to go back and change the WishList you just configured. You can also come back and edit it later.
- **Delete This WishList** Use this option to delete the WishList that you just created and go back to the main WishList screen.
- **Done Creating WishList** Use this option to save your changes and go back to the main WishList screen (Figure 2-23).

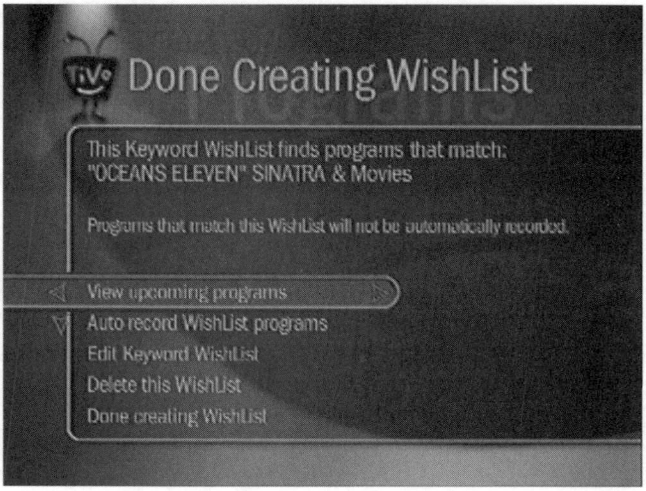

FIGURE 2-23 The Done Creating WishList screen

Searching Using a WishList Once you have WishLists set up, you can use them to search the TiVo's lineup data at any time:

1. Go to TiVo Central, select Pick Programs to Record, and then select Search Using a WishList.
2. Select View All Upcoming WishList Programs to see all your WishList search results, or choose an individual WishList from the menu to see its results alone.
3. On the next screen, select View Upcoming Programs.

Editing and Deleting WishLists You can also edit or delete a WishList at any time:

1. Go to TiVo Central, select Pick Programs to Record, and then select Search Using a WishList.
2. Select the WishList you wish to edit or delete from the menu.
3. Edit or delete the WishList by choosing the pertinent options.

TiVo Suggestions, Showcases, and TiVolution Magazine

TiVo gives you several ways to find out about new programs that you might not have otherwise considered: TiVo Suggestions, TiVo Showcases, and TiVolution Magazine.

TiVo Suggestions Of the three, TiVo Suggestions (shown in Figure 2-24) are the only ones that are automatically recorded by your TiVo if you have free space. You can see all the shows TiVo thinks you may like by going to TiVo Central, selecting Pick Programs to Record, and then TiVo Suggestions.

The Suggestions are listed according to how much TiVo thinks you will like them, with the most likely ones at the top. If you see a Suggestion that sounds interesting, highlight it and press Select to get more information about it or schedule it to record. If you do schedule a TiVo Suggestion to record, it is treated like a regular recording instead of a TiVo Suggestion.

Showcases TiVo describes Showcases (Figure 2-25) as an "interactive content area," filled with information and the opportunity to easily schedule "featured programs," which usually means programs that sponsors of TiVo paid to have featured. Showcases groups its content by channel and then by show title or subject matter, and it includes a wide variety of information.

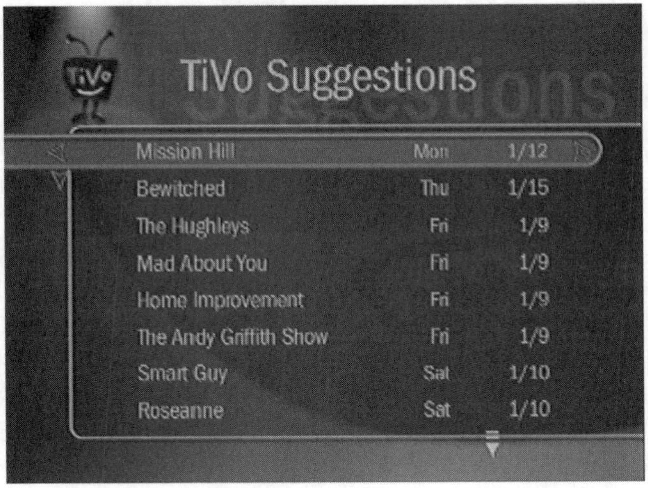

FIGURE 2-24 TiVo Suggestions screen

You can get to Showcases by choosing Showcases and TV Guide from TiVo Central. Use the directional pad to highlight TV Guide, TiVolution Magazine, or any of the TV channels with special features, and press Select. On the next screen you'll be able to pick from a list of themes or some other grouping method, and

FIGURE 2-25 The Showcases screen

Voices from the Community

Norman Ellis has been a TiVo user since August 2003. After receiving his DVR, he found himself almost immediately hooked. "It's like I'm the programming exec for my own network," he says. Although, as he points out, this new power is not without its detractors: "I sadly watch more TV because now I can. I never have to miss a program. But then I have to schedule time to watch the backlog of shows I've taped."

Ellis also found out early on that sharing your TiVo is not the easiest thing to do—especially when you have kids. "I christened my DVR by recording the last MTV Awards. I'm a big Madonna fan and I'd heard there was going to be a fabulous opening segment, so of course I was hyped. Needless to say, I wasn't disappointed, since that was the night Madonna and Britney tongued it down for all the world to see. I saved this recording for months, always planning to transfer it to VHS or even DVD if I could borrow a friend's recordable machine. Then one day I came home and the recording was missing from my saved list!

"I couldn't figure out how this could have happened. I was in a state; I went ballistic as any die-hard Madonna fan would have. Finally my 11-year-old son sheepishly confessed that he had erased it. He couldn't figure out why I was keeping this old show—surely MTV would air it again, and besides, it was taking up space on the hard drive. He thought he was doing me a favor. Little did Mr. 11-Year-Old realize that months later, Janet's 'wardrobe malfunction' would cause MTV to edit future showings of the infamous kiss.

"After that fateful incident I vowed this breach of recording etiquette would not happen again," he says. "The Ellis DVR Rules were ratified and instituted."

then you'll get a list of individual shows that you can choose to learn more about or to schedule for recording.

TiVolution Magazine TiVolution Magazine is a section of Showcases, found next to TV Guide. Instead of this content being grouped first and foremost by TV channel, the content in TiVolution Magazine is eclectic, but otherwise it is identical to the other Showcases.

Playing and Managing Recordings

Once you've recorded all those shows, you have to be able to do something with them, and TiVo lets you do quite a bit more than just play them. Give yourself a few moments to get acquainted with the way TiVo lets you deal with recorded shows, and you'll be managing your hard drive space like a pro.

Using Now Playing

There would be little reason to have a TiVo if you couldn't play back and delete the shows you have recorded! To view the shows you've got at your disposal at any given time, either press the TiVo button on your remote twice, or press it once and choose Now Playing on TiVo (Figure 2-26) from the TiVo Central menu.

Playing and Resuming Shows

To play a show, choose it from the listing by using the directional pad and pressing Select. If you have shows grouped into folders, choose the folder first, and then pick from the list of shows. (Grouping shows into folders is covered later in this section.)

FIGURE 2-26 Now Playing on TiVo

Now Playing Icons

TiVo puts icons next to programs or groups of programs to give you visual clues about the kind of show or the status of the show.

- **Green circle** The program will be saved until you delete it manually.
- **Yellow circle** The program may be deleted in less than one day, if space is needed.
- **Yellow circle with "!"** The program may be deleted at any time, if space is needed.
- **Red circle** The program is currently being recorded.
- **TiVo logo** The program is an automatically recorded TiVo Suggestion.
- **Plain folder** The folder contains one or more episodes from the same series.
- **Red-dot folder** The folder contains a program that is currently being recorded.
- **Blue-star folder** The folder contains at least one program recorded due to a WishList search.
- **Thumbs-Up folder** The folder contains all TiVo Suggestions.

All of the controls on the remote work the same the same way with recorded shows as they do with live TV, except that you have no time limits on how long you can leave a show paused, and you can fast forward and rewind through the entire show. Don't forget that you can fast forward through commercials!

Keep in mind that while you can leave a recorded show paused indefinitely, you shouldn't leave your TV on. If the image on the TV screen doesn't change for a long period of time, it's possible that your screen could suffer from "burn in"—where the prolonged exposure leaves a permanent ghostly after-image on your TV.

If you stop playing a show before it's over, just go to Now Playing, choose it again, and select Resume Playing from the list. You can do this indefinitely until you're done with the program.

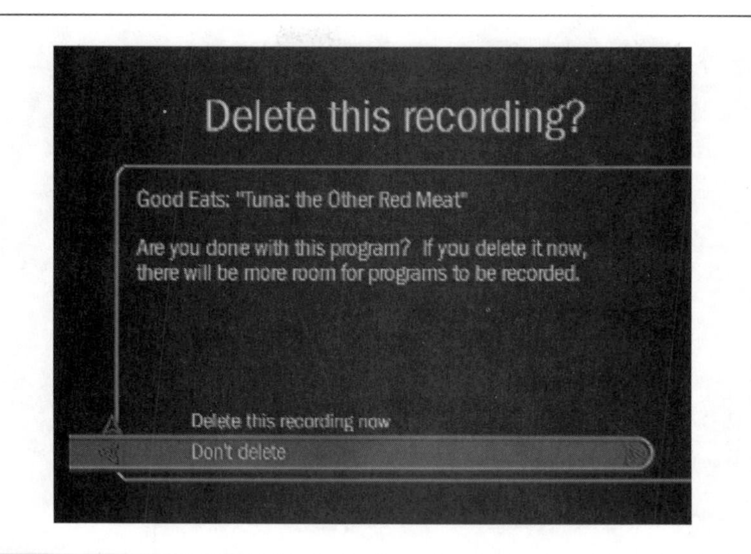

FIGURE 2-27 TiVo asks whether or not you want to delete your recording.

When you've reached the end of a show, the DVR will ask whether you'd like to delete the show or keep it, as shown by Figure 2-27. It's a good idea to delete shows you've watched unless you want to watch them again, in order to make room for fresh content on your TiVo for more recordings.

Managing Recordings

In addition to playing recordings, you can also delete and change other recording settings from the Now Playing screen.

Delete programs by highlighting the show in the list and pressing the Clear button; on the confirmation screen, press Select or the right arrow on the directional pad if you're sure you want to delete it. See Figure 2-28. (You can also delete programs by choosing the show from the list using Select or the right arrow, and then picking Delete Now, but the Clear button provides a shortcut.)

You can also change the length of time the show will be kept before being deleted to make room (if necessary) for newer shows. Select the program from the Now Playing list, press Select, and choose Keep Until on the Program Information screen. You can choose to keep the recording until you delete it manually or until a certain date.

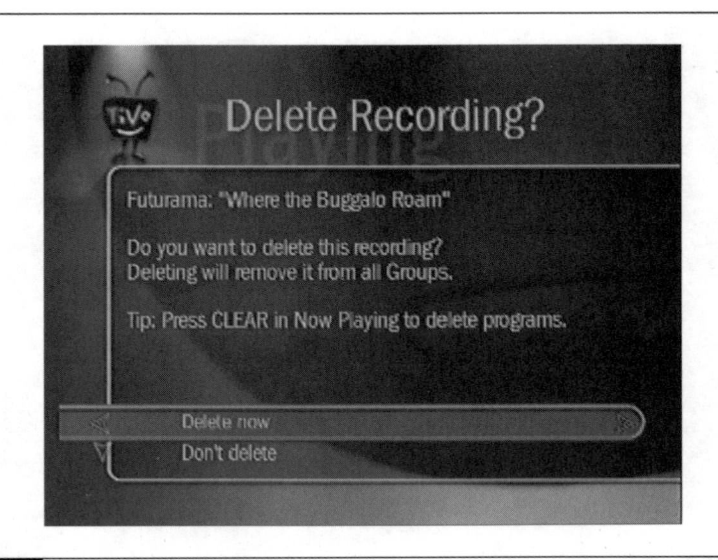

FIGURE 2-28 The Delete Recording screen

Setting Now Playing Display Options

When you're at the Now Playing screen, press Enter to display the Now Playing Options (see Figure 2-29). Using the directional pad, you can change the settings so that the Now Playing listing of shows is sorted by recording date or alphabetically,

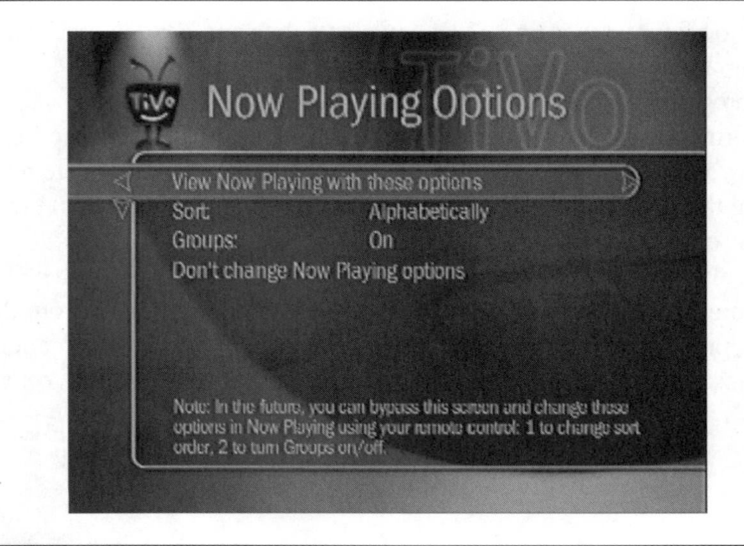

FIGURE 2-29 Now Playing Options

and you can turn Groups on or off. With Groups on, folders will be created for any series that has more than one recording saved; turning Groups on will significantly clean up your Now Playing screen if you have any Season Passes!

Managing Scheduled Recordings and Season Passes

Once you've scheduled a show to record or created a Season Pass, you're not committed to it for perpetuity. You can go back at any time and change the recording settings.

Viewing Upcoming Recordings and Recording History

What your TiVo is about to record doesn't have to remain a surprise. You can view the TiVo's To Do List (Figure 2-30) to look at upcoming recordings and nix the ones you don't want recorded, whether they're WishList items or just programs you changed your mind about.

Follow these steps to check your To Do List:

1. Go to TiVo Central and select Pick Shows to Record.
2. Select To Do List from the bottom of the menu.

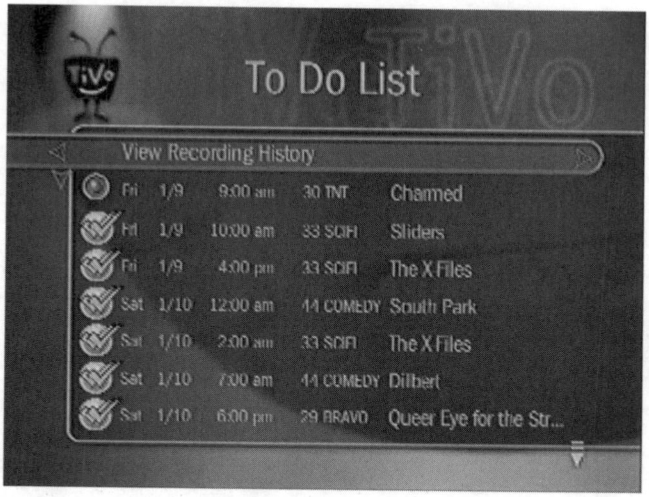

FIGURE 2-30 TiVo's To Do List

NOTE *The icons used to differentiate the shows in the To Do List are the ones described in the "Recording Icons" section earlier in this chapter: a single check mark indicates an individual recording that you set up, a double check indicates a Season Pass or repeat manual recording, and a star indicates a WishList recording.*

3. To view a show's settings and change them or set it not to record, highlight the show in the list and press Select. If the show will not be recorded due to a conflict, you can choose More Recording Options from the menu and search for other showtimes, as described in the "Setting Recording Options" section earlier in this chapter.
4. To see which shows were deleted, were not recorded, or will not be recorded, select View Recording History from the top of the program list.
5. To back out of these screens without changing anything, highlight Don't Do Anything and press Select, or press the left arrow on the directional pad.

Using the Season Pass Manager

Using the Season Pass Manager, shown next, you can edit, delete, or change the priority of your Season Passes. To get to the Season Pass Manager, shown in Figure 2-31, go to TiVo Central, select Pick Programs to Record, and choose Season Pass Manager from that menu.

FIGURE 2-31 The Season Pass Manager

Did you know?

Handling Season Pass Conflicts

Got two conflicting Season Passes? Set Season Pass A to record only first-run episodes of show A, and Season Pass B to record all episodes of show B. Then use the Season Pass Manager to set Season Pass A to a higher priority than B. As long as the lineup data is accurate, Season Pass B will record show B whenever series A is airing a rerun instead of a new show, so you won't miss many episodes of either.

The Season Passes are listed in order of their priority. Priority is an important concept that makes Season Passes work very well—if you have a WishList item for Steven Spielberg movies and a Season Pass for *The Simpsons,* what should the TiVo do to resolve the conflict? It looks to your priorities to decide which program takes precedence. To change a Season Pass's priority, highlight it in the list, press the right arrow on the directional pad, and then use the up and down arrow buttons to raise or lower its priority. Press left on the directional pad or Select to save the new position.

If you want to edit a Season Pass, just highlight it and press Select, and the familiar options screen will be displayed. To delete a Season Pass, highlight it and press the Clear button, and then choose Yes, Cancel This Season Pass on the confirmation screen if that's what you really want to do.

Chapter 3 Train Your TiVo

How to…

- Use TiVo's Suggestions
- Use Thumbs-Up and Thumbs-Down
- Rate Programs Within TiVo
- Change Your Ratings
- Reset Your Ratings
- Understand the Sharing of Your Personal Information and Viewing Habits
- Determine Whether to Opt Out of TiVo's Information-Collection System
- Opt Out of TiVo's Information Collection

Your TiVo is more intelligent than you may have thought at first. You can "train" your TiVo to make the best use of its user-information collection functions. By using the data TiVo collects from you, its Suggestions function can find programs for you to watch that it thinks you'll like. The suggestions that TiVo offers are generated by its built-in suggestion engine, which juggles the data that is fed into it and comes up with related results.

How TiVo Learns

The suggestions engine is a complex piece of software developed by TiVo to alert you to related programming it believes will interest you. Just how it searches and prioritizes new information is, of course, proprietary. There's a lot of speculation about just what goes into the mix as TiVo learns about the kind of programming you like, but once the urban legends are removed, the only things known for sure are these:

- The software is kept under wraps at TiVo HQ.
- The machine employs algorithms (procedures for solving mathematical problems) in its code.
- The algorithms are used to search on several thousand key details (favorite actors, movie and TV genres, and so on) of any user selection to generate as close a tailored fit to a suggestion as is possible.
- And finally, it's an extremely complicated piece of coding!

Given this level of complexity, it's probably best to sum up with the words of TiVo's general manager, Brodie Keast, who explains that the TiVo box is merely "reacting to feedback you give it."

NOTE *If you're wondering what kind of motor oil this suggestions engine uses, don't worry about it. An engine, in computer-speak, is a simply a program that processes data that is fed into it and spits out a result based on the inputs. (The quality of the data you feed into the engine will determine the quality of what comes out. That's why a lot of programmers jokingly tag such processes as pure GIGO: garbage in, garbage out!)*

Interestingly enough, programmers and developers wouldn't call this technology new, at least by Silicon Valley standards. Instead, the latest development really concerns fitting the technology of soft-coded engines to different applications. Today there are a surprising number of gadgets and Web sites that feature so-called *personalization* technologies that also generate suggestions according to your tastes.

For example, the personalization technology in your supermarket or department store scanner can easily profile consumers by the items they purchase, and by the quantity. Prior to TiVo, the Nielsen ratings company used wired boxes to determine what various audience segments would potentially like by determining what they watched or what they listened to. The software files known as cookies, embedded in sites such as Amazon.com and CDNOW.com, track what you purchase and where you surf, and then the sites recommend other books, videos, and music based on a customer's tastes.

Since TiVo's software engine recommends programs based on your preferences, you may be wondering what information TiVo will collect about you. Among other things, TiVo can determine what was watched on a particular TiVo box, when it was watched or recorded, and the number of times a particular moment in a show was rewound and played again. This information is transmitted back to TiVo HQ in Alviso, Calif., over the phone line used to download show schedules to the DVR.

TiVo Services

There are a couple of ways that you can train your TiVo to find and record new programs. Half the fun is stumbling across great shows you hadn't heard of before. And TiVo makes it even easier, by bringing the new selections, often tailored to your taste, right to your screen.

Did you know?

What's the Number-One Re-viewed Moment on TiVo?

In case you're curious, as of this writing, the record for the number of times that TiVo subscribers hit their rewind button is held by the 2004 Super Bowl, where a "wardrobe malfunction" at halftime became the most re-watched moment of a broadcast in three years of measuring audience reactions.

Runner-up for the most-rewound moment: the infamous kiss shared by Britney Spears and Madonna, followed in third place by the only slightly less infamous kiss shared by Christina Aguilera and Madonna, both during the same musical number performed at the 2003 MTV Video Music Awards.

TiVo Showcases

The Showcases section is a sort of one-stop-shopping area with interactive user content. TiVo Showcases enables brands to deliver several minutes of video to the hard drive of subscribers across the country. The content remains on the hard drive for the period of the campaign so that subscribers can opt-in to view the content at any time.

Once you've completed your initial TiVo subscription and Guided Setup (as explained in Chapter 1), you can do either of the following with Showcases:

- Schedule recordings of featured programs.
- View programs that are grouped in "theme" packages. These packages often feature content that is exclusive to DVD collectors or to TiVo subscribers.

The Showcases feature has been a success for both TiVo and TiVo's user community, leading to a winter 2003 deal with Universal Pictures. As of this writing, TiVo and the movie studio have signed a multi-film promotional agreement where the TiVo service will be promoted on DVD releases of major feature films from Universal. Universal Pictures will, in turn, promote more than 20 films and DVD titles using TiVo Showcases. The current plans are to continue with the creation of special promotional content that will include interviews, trailers, and behind-the-scenes footage for many of Universal's major theatrical and DVD releases.

NOTE *Showcases won't reduce the amount of recording space on your TiVo's hard drive. Showcases also won't preempt your scheduled recordings or cause your recordings to expire any earlier.*

To view TiVo Showcases, do the following:

1. Go to TiVo Central by pressing the TiVo button on your remote
2. Select Showcases. The screen will look something like Figure 3-1.
3. Choose the selection you want by using the arrow keys on the remote and pressing Select.
4. When you are done making your selections, select Done Choosing Programs or press the right arrow on your remote.

TiVolution Magazine

TiVolution Magazine is a subsection under the Showcases area. It's rather like having *TV Guide* built right into your TV, listing the best shows coming up on TiVo in the next few days. It's an offbeat guide to the available programming, published in electronic form by TiVo, and while it's not as comprehensive as a *TV Guide,* it's got plenty of TiVo-specific material worth looking at for any real TiVo enthusiast.

FIGURE 3-1 The TiVo Showcases screen

Did you know?

Showcases Grab the Audience's Attention

In 2002, Universal began promoting its new releases in TiVo Showcases, including special promotional content for *8 Mile* and *Red Dragon*. TiVo's in-house research has found that entertaining promotional content consistently succeeds in getting the viewers' attention.

For example, more than two-thirds of TiVo subscribers viewed the promotional content for the 2003 Jim Carrey film *Bruce Almighty*. More important to the advertising and content gurus, subscribers spent an average of three minutes interacting with the clips, interviews, and other content in the Showcase, proving that TiVo's exclusive features generated a higher level of excitement and interest for the film.

To view TiVolution Magazine, do the following:

1. Go to TiVo Central by pressing the TiVo button on your remote.
2. Select Showcases.
3. Choose the TiVolution Magazine selection by using the arrow keys on the remote and pressing Select. The screen will look like Figure 3-2.
4. You can navigate the TiVolution Magazine section by using the arrow keys on the remote and pressing Select to make selections.
5. When you are done making your selections, press the right arrow on your remote.

Ipreview

One very interesting service that TiVo offers is Ipreview, which was first introduced as a bonus TiVo feature in mid-2000 for subscribers to either the Showtime or Home & Garden TV channels. Ipreview enables users to program their TiVo units to record programs specifically advertised in upcoming promotional spots on participating channels.

For example, each time Showtime airs a promotional spot for an upcoming event, an Ipreview thumbs-up icon will appear onscreen, signaling TiVo viewers that they can schedule a recording at that moment by clicking a single button on their TiVo remotes.

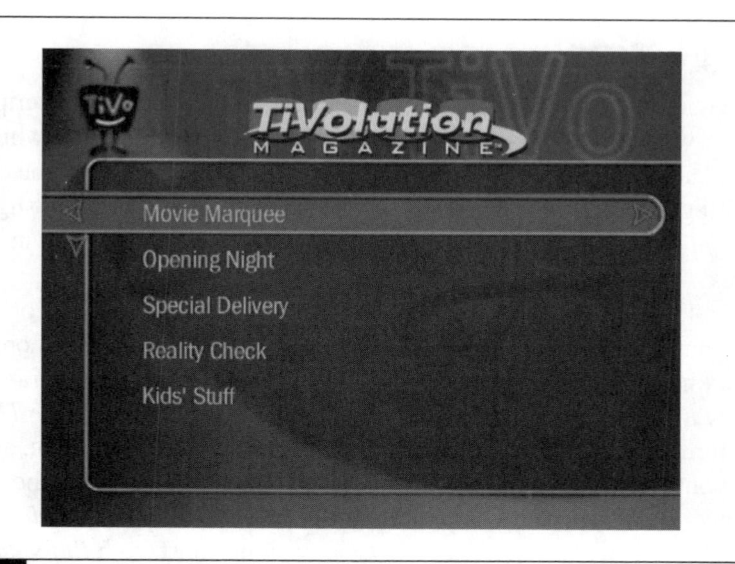

FIGURE 3-2 The TiVolution Magazine screen

You may even see these signals during some commercials. An icon will appear, recommending that you "Press thumbs up to record." At that point, if you're interested, you can set up your TiVo box to schedule the recording of the program being advertised.

When you see an Ipreview icon, follow these steps to set up the recording process:

1. Press the thumbs-up button or the Select button on your remote to bring up the recording screen.
2. Choose Record by using the arrow keys on the remote and pressing Select.
3. Press the Clear button to make the icon go away until either the next preview comes on, or until you rewind past the point where the preview began.

Although Showtime was among the first to make use of this feature, Ipreviews are most commonly seen on NBC and HBO as this is being written. TiVo plans to have it available on more shows as time goes on.

NOTE *The Ipreview feature was known by most early TiVo subscribers as "TiVomatic." If you visit various TiVo online chat rooms, don't be surprised to hear both terms used interchangeably.*

TiVo's Suggestions

If you have DirecTV or a combination DVD-R unit, and you have a subscription to the TiVo Plus service (not to be confused with the Home Media Features, which allows you different and more complex functionality on your TiVo box), then you can access the TiVo Suggestions area. TiVo suggestions are computer-generated suggestions of shows that match, as closely as possible, your television and movie watching preferences.

Best of all, it encourages you to use the built-in TiVo rating system on everything you watch and everything it suggests. This allows the system to collect more data, and the better it knows your preferences, the better it'll be at selecting programs for you.

Let's say that you give a positive rating to the action-adventure movie *The Mummy*. Since this film involves ancient Egypt, action-adventure, science fiction, and a dose of horror, you may find that your TiVo picks up (based on this and other picks of yours) a special on Egyptology, a special broadcast of *The Young Indiana Jones Chronicles,* the sci-fi series *Stargate SG-1,* or even the latest from *Buffy the Vampire Slayer*.

It's also worth noting how Suggestions are recorded. As a rule, they will be recorded at the default quality level that you last selected. Of course, if you've never gotten around to selecting or changing this level, the default is the factory standard.

You can change the recording quality level by performing the following steps:

1. Go to TiVo Central by pressing the TiVo button on your remote.
2. Select Messages & Setup from the TiVo Central screen.
3. Select Settings.
4. Select Preferences.
5. Select Video Recording Quality.

Keep in mind that whenever you alter this setting, any Suggestions that TiVo has already scheduled under the prior quality setting are still recorded at that old quality setting. Only new Suggestions will use the new default.

CAUTION *The Suggestion service won't record any of your suggestions for one half-hour after you press any button on the remote The unit will assume that you are watching live TV.*

Did you know?

Training Your TiVo Inspires a Sitcom Episode

Of course, even a neat system like TiVo's Suggestions can have a few quirks. Like many other technologies that are constantly evolving (or as software engineers like to put it, are in a state of permanent re-invention or improvement), the TiVo Suggestion function can be prone to its share of amusing behavior.

This isn't because the function doesn't work—TiVo's software is by and large bulletproof, which is a testament to the work the designers perform. Instead, the very nature of TiVo's information-collection system is imperfect, because it can only make the narrowest of judgments from the choices that you've entered via its system. Therefore, on rare occasions, TiVo could make an assumption about a subscriber that is way off base, which is a phenomenon that was spoofed in a fall 2002 episode of HBO's *The Mind of the Married Man*.

The creator and star of that show, Michael Binder, had set his home TiVo to record a movie that featured an adult sexual theme. After that, Binder's TiVo assumed he would enjoy suggestions featuring the same kind of programming. He counteracted the onslaught by recording different fare, which TiVo would then put into the mix and eliminate the adult theme. It worked, and the incident ended up inspiring a humorous episode on his HBO show. As an added bonus, Binder now notes that his TiVo "doesn't look at me funny anymore."

Using the Thumbs-Up and Thumbs-Down Options

Any time you watch live or prerecorded television or browse the TiVo Live Guide, Showcases, or TiVolution Magazine, you can choose a rating for whatever you are viewing. You can use the red thumbs-down or green thumbs-up buttons on the TiVo remote to rate the program. You can assign each program either one, two, or three thumbs up or down.

Using the Thumbs-Up and Thumbs-Down options is very simple.

- Just press the appropriate thumb button to give it a thumbs-up or thumbs-down. You can give a program up to three thumbs up or down, to indicate how much you like or dislike something.
- Something you moderately like would get one or two thumbs up, and something you loathe would get three thumbs down.
- Any time you choose to record a program, it will get one thumb up.

TiVo uses these ratings to build up a profile of what genres of programs, actors, and directors that you like to watch and will suggest other programs that you may like through its Suggestions list.

Suggestions on Thumbing Through TiVo

This is not a scientific guide on how often to use your thumb ratings—these are just some best-practice rules that will come in handy if you're going to seriously attempt to train your TiVo system to come up with the best suggestions to suit your viewing habits and choices:

- **One thumb up** Use this for all programs you like.
- **Two thumbs up** Use this for your top 10–20 favorite shows.
- **Three thumbs up** Use this only for your top one or two favorite shows.
- **Thumbs down** Use these as sparingly as possible.

As a rule of thumb (pun intended), it is better to use the Thumbs-Up button to indicate what you like, rather than to thumb down a higher percentage of programs. TiVo seems to weigh a thumbs down more heavily than a thumbs up. Therefore, too many thumbs down may result in you not seeing programming choices you might find interesting. For example, if you give three thumbs down to a program dealing with open-heart surgery, you might miss out on the year's best medical thriller, or a horror film that happens to be set at a local county hospital.

Voices from the Community

Thumbs Up for TiVo

It's unreal how something as simple as the TiVo has been able to completely revolutionize the way I watch and enjoy television. Prior to welcoming this magic box into my life, I was set in my ways—and there were numerous TV programs that I knew nothing about. Now, by simply rating using Thumbs-Up or Thumbs-Down, TiVo is able to figure out additional programs that I may like based on my interests. It's totally expanded my horizons—plus the Thumbs-Up and Thumbs-Down buttons make great noises! Thanks, TiVo!

—Laura Crovo, PR Account Executive at Baltimore's MGH Public Relations, and TiVo enthusiast

NOTE *It normally takes several hours for Suggestions to first appear after you enter your first thumbs up or down ratings. Be patient, they'll show up eventually.*

Changing Your Ratings

If you change your mind about the rating you want to give a particular program, you can change the rating under most circumstances. If the show appears in the TiVo listings, you can pull the ratings up and change them as follows:

1. Go to TiVo Central by pressing the TiVo button on your remote.
2. Select Suggestions.
3. Navigate to your selections by using the arrow keys on the remote.
4. Press the Thumbs-Up or Thumbs-Down buttons to change the ratings.
5. When you are done making your selections, press the right arrow on your remote.

Resetting Your Ratings

If TiVo is not suggesting programs that you enjoy, then you may want to remove you thumbing profile and start thumbing from scratch. It is possible to erase all your thumbs up and down data by using the system menus. To do this, follow these steps:

1. Go to TiVo Central by pressing the TiVo button on your remote.
2. Select Settings.
3. Select Restart or Reset System. You'll see a screen that looks like Figure 3-3.
4. Select the Reset Thumb Ratings & Suggestions option by using the arrow keys on the remote. Press Select to choose the option.
5. When you are done making your selections, press the right arrow on your remote.

CAUTION *It's best not to use the Clear and Delete Everything option. Yes, this will erase your thumb ratings and suggestions, but it will also clear out a lot of other data that you may have already entered. This is overkill, so unless you really need to, don't use this option. Also, keep in mind that TiVo will not react instantly to the erasure of all your thumbing records. A proper reorganization within the TiVo system will not be performed until after the box's next daily call.*

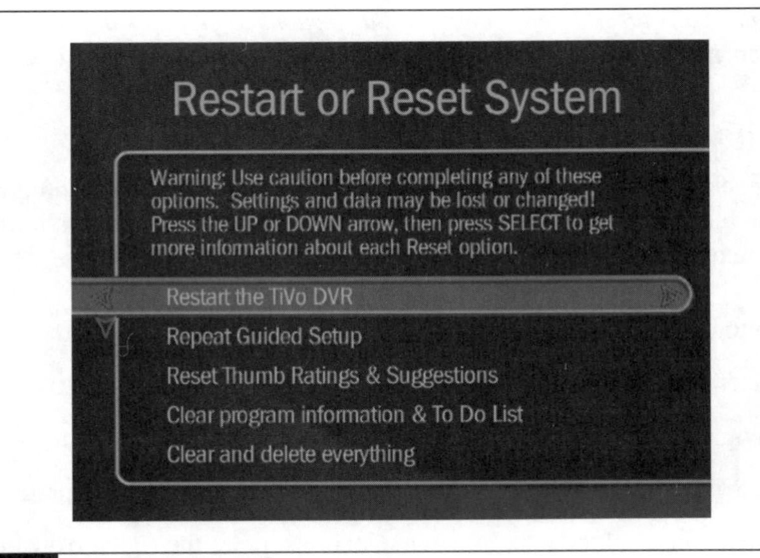

FIGURE 3-3 Restart or Reset System

Understanding the Sharing of Your Personal Information and Viewing Habits

Traditionally, companies involved in the high-stakes, high-value industry of gathering "granular" consumer data for marketing purposes have downplayed the idea of corporate surveillance to minimize consumers' suspicions. TiVo and other collectors of consumer data have gone out of their way to assure customers that their practices do not constitute intrusion, and they adopt practices they claim will minimize privacy risks. Ironically, the first real publicity about the power of TiVo's ability to gather data from its boxes came from its report on the record number of rewinds involving the 2004 Super Bowl halftime show.

NOTE *TiVo data is highly prized by consumer marketing departments at major corporations because of the information's* granularity. *Granularity refers to the level of specificity that the data provides. For example, if you market snack cakes, you're more likely to pay money for data that shows which neighborhoods tune in to the Snack Cake channel, and for which hours, rather than data that just tells you that people in a sample large city tend to watch more snack-cake commercials during the week. Why? Because the information gathered in the first example is more specific, allowing you to figure out where and when you can directly target consumers who will buy your product.*

Data mining is the term for studying recorded consumer choices by using a combination of machine learning, statistical analysis, modeling techniques, and database technology. With these techniques, a company can extract information from databases and discover things that aren't immediately obvious. Data mining generally involves analyzing data to identify patterns and establish relationships, usually between a type of consumer and a type of consumer good. And although it might seem illogical at first glance, it's actually in the data-mining company's best interest to make sure your information is kept private.

IBM, Oracle, and other companies that specialize in data mining allow marketers to filter out only the most precious tidbits of information from a given sample. But to these companies' credit, they have been working on ways to continue receiving this valuable data while still preserving consumer privacy. Although skeptics disagree, it's generally assumed that unless consumers of the service have a high degree of confidence in the privacy controls, they will specifically "muddy up" their profiles and customer data to avoid being pigeonholed by marketers.

For example, take a customer who is suspicious of how they are being monitored, but who still wants information on a particular medicine or surgical procedure. This user might avoid using TiVo to get this information, and instead go to another source like the Internet. Alternatively, they could order additional programming they're not interested in, such as a half-dozen historical dramas, in order to swamp the TiVo box with erroneous data and make the choice they had originally asked for appear insignificant.

One method TiVo uses to improve the level of privacy is to strip the data of any identifying informational markers at the source (your TiVo box). The rest of the information is encrypted before it is sent out to TiVo. This allows information to be passed freely back to the TiVo home office with fewer concerns about the misuse of information from corporate marketing, or hackers looking for people's personal information.

Consumers can also rely on the basic privacy protections that developed organically on the Internet in the late 1990s as e-businesses took off and began handling more and more consumer information over their networks. TiVo has adopted the following commonly accepted procedures:

- Account numbers or device serial numbers that could be used to identify an individual are removed.
- No demographic information is sent to TiVo.
- Information collection practices that provide aggregated (anonymous) data are used. These practices can show the total numbers of people engaged in a particular activity, but they don't indicate which household took which action.

TiVo's Privacy Policies

TiVo is well aware that its use of consumer information has caused a certain amount of consternation among their user base. To reassure its customers, TiVo has been very good about making its privacy policies available online and keeping them up to date. TiVo's Site Privacy Policy describes its information-gathering and dissemination practices for the web site (http://www.tivo.com). TiVo also has a separate privacy policy that governs the use of information that is gathered as part of the TiVo service.

TiVo's Web Site Privacy Policy

TiVo's full policy on information gathered via the web site can be found at http://www.tivo.com (click on the Privacy Policy link at the bottom left). Here are some of the points covered in this policy:

- TiVo won't collect any personal information about you unless you voluntarily submit such information.
- TiVo asks children to obtain the consent of their parent or guardian before providing any personal information.
- TiVo partners with or employs other companies to perform functions such as processing credit card payments and providing marketing assistance. According to TiVo, these parties have access to personal information needed to perform their functions, but are bound by contract to not use it for any other purposes.
- TiVo uses cookies for site-visitor tracking. The information collected is anonymous, and TiVo has no way of accessing any of your personally identifiable information through this system.
- TiVo leaves itself an out when it states that it can sell aggregated (non-personally identifiable) information about its Web site users. TiVo will also release personal information if necessary to comply with the law or to protect the rights, property, or safety of the company, its user base, or others.
- Information provided to TiVo can be used for internal marketing purposes, or can be sold as aggregated (non-personally identifiable) user information.

If TiVo is acquired—and stranger things have happened in the fast-moving technology industries, no matter how healthy the company—customer information will be one of the transferred assets. No guarantee can be made as to how the new owner might use or sell the information.

TiVo gives you the following options to remove your information from TiVo's database so you will not receive future communications:

- You can send an e-mail to *webmaster@tivo.com*
- You can send mail to the following postal address: Webmaster, TiVo Inc., 2160 Gold St., Alviso, CA 95002-2160
- You can call the following toll-free telephone number: 1-877-367-8486

CAUTION *TiVo uses industry-standard information-protection methods to prevent unauthorized access, such as storing information on a computer behind a firewall, placing the computer itself in a secure location, and placing heavy restrictions on internal employee access. However, keep in mind that there is no such thing as perfect security in a world full of hackers.*

What is potentially more troubling is that TiVo's servers are actually operated by a third-party Web host. While this party is not identified in any of TiVo's documentation, it's worth noting that this company's actions are not controlled by TiVo.

TiVo's Privacy Policy for Their Service

TiVo's full policy on information gathering via the TiVo service can be found at http://www.tivo.com (click on the Privacy Policy link at the bottom left). The policy covers the following points:

- While TiVo collects certain types of information from its users, TiVo has no way of knowing what shows you or your household have watched *unless you have already agreed to release this information.*
- TiVo also has no way of knowing what shows you've either recorded, or rated with its Thumbs-Up and Thumbs-Down feature *unless you have already agreed to release this information.*
- TiVo does collect anonymous information from all users—information such as your viewing selections. However, it cannot identify you as an individual or household.
- If you've decided to use the Remote Scheduling feature of TiVo's Home Media Features package, TiVo will collect the Personally Identifiable Viewing Information associated with this feature.

- If you decide to participate in a commercial transaction, such as responding to an advertisement or promotion, TiVo will collect and disclose your commerce information to the commerce partner so that the product can be paid for and shipped to you.
- TiVo also collects diagnostic information from a few randomly sampled TiVo DVRs for quality-control purposes.

NOTE *If you subscribe to the TiVo Plus service and don't want your anonymous viewing information and diagnostic information used in any way, write to TiVo or call them at 1-877-367-8486 to inform them of this. However, if you subscribe to TiVo's Basic service, you are not allowed to opt out of TiVo's collection of these types of information.*

Determining Whether to Opt Out of TiVo's Information-Collection System

Before you pick up the phone, pencil, PDA, or computer keyboard to contact TiVo, consider whether you think TiVo Suggestions and tailored marketing information are an invasion of privacy. It's worth looking at how TiVo has dealt with the subject in the past. From there, you can decide whether you think the personal information you give to TiVo will be held in strict confidence or not.

TiVo's Positive Record on Privacy

It's worth noting that while any of TiVo's users (including you) can opt out of the data-collection system at any time, TiVo claims that relatively few of its users have requested to be removed from the information-gathering system.

This is particularly interesting in light of past controversy. In March 2001, the Denver-based consumer-protection group Privacy Foundation reported that TiVo boxes use their built-in modems to send nightly activity reports back to the TiVo corporate headquarters. Richard Smith, the chief technology officer of Privacy Foundation, claimed that the practice conflicted with TiVo's written promise available in its 2001 documentation. At the time, TiVo's documents contained an explicit promise to its customers that all personal viewing information remained on the home receiver.

The Privacy Foundation report, testimony from Smith at a House Energy and Commerce Committee hearing, and a written request by three members of Congress spurred the Federal Trade Commission (FTC) to action. The FTC convened to determine whether an investigation should be made into the DVR company's privacy policies for possible violations of federal law.

In May that same year, the FTC came to its conclusion and declined to launch a formal investigation of TiVo for possible privacy violations. This proved to many people that TiVo was, absent any evidence to the contrary, dealing with its customers squarely on the release of information.

No One Can Predict the Future

Because it has not been proven otherwise, TiVo seems to have lived up to its promises not to do the television equivalent of flooding a paying customer with spam. On the other hand, the cynic could point out that the low numbers of people opting out might be due to the fact that the opt-out instructions are buried deep within the sleep-inducing contract literature accompanying the box's documentation package.

Despite the FTC's decision not to investigate the privacy policy, some people remain skeptical of the claim that this was because TiVo was above suspicion. It is true that technology companies are prepared to counter government intervention in their business. Indeed, the information technology industry has organized to oppose legislation that would ban any data harvesting not explicitly authorized by the customer.

And finally, while the company's privacy policy forbids the peddling of customer information to advertisers, the manual takes care to note that TiVo's policies, like those of any other company, are subject to change.

How to Opt Out of TiVo's Information-Collection System

If you're one of those concerned about being part of the viewing and diagnostic information tests that TiVo conducts, such as the one during the 2004 Super Bowl, you can call a toll-free number to opt out.

How you opt out depends on the kind of service you have.

- If you subscribe to the TiVo Plus service and want to opt out of TiVo's information collection, simply send a letter to TiVo or call them at 1-877-367-8486.
- If you're receiving the TiVo Basic service, you cannot opt out of TiVo's collection of viewing and diagnostic information. However, you are not required to provide TiVo with any personal contact information.

The Internet is the biggest customer-communication channel in existence, and while most companies aren't doing enough to tap it, TiVo is among the handful of companies that gets it. When you buy a TiVo, you gain a pass into the TiVo online community. These are the die-hard users who chat about the value of TiVo and, most importantly, what the company could be doing better. TiVo is wisely using that information and making constant adjustments and improvements to their service. Should TiVo continue too far down the path of merchandising their customer information, be sure to speak up and be heard: TiVo is listening.

Part II

Try This at Home

Chapter 4

Hook TiVo Up to Your Home Network

How to...

- Determine Whether You Should Hook TiVo Up to Your Home Network
- Understand What Home Networks Are
- Purchase the Right Network Hardware and Software
- Choose Between Wired and Wireless for Your Home Network
- Set Up a Wired (Ethernet) Connection
- Set Up a Wireless Connection
- Transfer Photos and Music with the TiVo Desktop Software
- Go Beyond TiVo Desktop's Features with JavaHMO

Uh-oh, you're thinking, "I'm not a computer geek! How am I going to understand anything in this chapter?" Not to worry. If you're severely technophobic, then take a quick glance at the information under the "Reasons To Hook Up TiVo to Your Home Network" section, and if you want to learn some truly cool stuff, read on. And no, you don't need a master's degree in computer engineering to learn the material here—some basic networking savvy, aided by our instructions, are all you really need.

Reasons to Hook Up TiVo to Your Home Network

If you're an inveterate gadgeteer like many people, or you enjoy being the coolest resident of the 21st century on your block, or you just like challenges involving lots and lots of cable, this is the right section for you. After all, you're a practical person who likes getting the most out of your TiVo, right?

With that in mind, here are a few reasons to hook your TiVo up to your home network:

- **Getting the best out of the Home Media Features** If you want to get the most out of your Home Media Features, which includes using special music features and Multi-Room Viewing, covered in Chapters 5, 6, and 7, you'll want to know how to get your TiVo cooking.

NOTE *Even without hooking TiVo to your home network, you can still use the Home Media Features to connect multiple TiVos to each other to implement the Multi-Room Viewing system.*

- **Reclaiming a phone line** By switching the TiVo over to using a broadband Internet connection, you allow the connection to be made with a DSL or cable modem. This not only allows for faster data transfer rates, it means the TiVo won't need to use your phone line.

CAUTION *Even if you plan to hook TiVo up via a broadband or cable modem connection, it will still need a phone line during the initial Guided Setup described in Chapter 2.*

- **Setting TiVo to record over the Internet** Let's say you're stuck at the office late, and you forgot to set the TiVo to record your favorite sitcom, which is running a once-in-a-lifetime marathon of all your favorite episodes. If you have a new Generation 2 TiVo and the Home Media Features, and you've taken the time to hook up your TiVo to the Internet via your home network, you can log onto your TiVo from any Internet-ready computer and tell it what to record.
- **Surfing the Net throughout your home** By networking your Internet connection throughout your home, you can enjoy a high-speed connection in as many different rooms as you'd like. If you opt for a wireless connection, you'll enjoy even more freedom as you tote your laptop around the house. You can plop down on the couch in front of the TV and watch the sitcom marathon you set TiVo to record and check e-mail at the same time.

NOTE *Lest you think that the major benefit of running a wireless connection is just the "cool" factor, there is a practical benefit. When you go wireless, you won't have to worry about unsightly wires or cables dangling from every Internet port, nor will you have the expense of tearing open walls or pulling up the carpet to install cables throughout your home.*

What Makes Up a Home Network?

A home network is a group of computers, printers, and other devices in your house that are connected together. These connections can be made with cables, which come in several types and configurations, or they can be wireless, and the resulting network allows users on the networked machines to exchange documents, music, pictures, and other sorts of data with each other. The network can be used to print documents on printers in different rooms of the house.

Generally, machines on a network can be set up to share hardware or software, so long as the software is set up to allow the hardware to exchange data. Networks can have anywhere from two to a million nodes or more—the Internet is really just a very big network.

NOTE *A "network node" is simply a device connected to the network, whether it is a computer, printer, or other device.*

Given the budget of most people, though, the size of home networks tends to be limited by physical distance (length of cords) and the amount of data that can be handled by the units on the network. Cabling, for example, can only be so long, because the strength of the signal decreases as the wire gets longer, and wireless networks are similarly limited in distance.

The most popular form of networking connection today is Ethernet, which consists of computers and peripherals connected together in specific ways. Ethernet has become the dominant form of networking because it is relatively inexpensive, moderately reliable, easy to set up and use, and very fast unless overloaded with network traffic.

Choosing to Go Wired or Wireless

You can set up your home network with either a wired or wireless system. Whichever you use will be reasonably fast and reliable, but the two systems have different strengths and weaknesses:

- **Wired Ethernet network** Wired connections and program transfers for Multi-Room Viewing are quicker. The disadvantage is that you have to run wires between all the network devices, and if they're scattered throughout the house, this can be awkward or expensive.
- **Wireless network** Although wireless connections tend to be slower and more expensive (though not always!), there's nothing like the convenience of wireless networking. If you don't want to run Ethernet cables through your home, this may be the best solution. Apart from being slower, though, wireless networks can receive interference from 2.4 GHz (gigahertz) cordless phones or microwave ovens.

Setting Up a Wired Connection

You can plug a USB-to-Ethernet adapter directly into the USB port on your TiVo. This type of adapter allows you to connect to other devices on your Ethernet network. You can purchase adapters at most electronic stores or on the Internet. At the time of writing, the known list of compatible adapters included the following:

Company	Adapter
3Com	3Com USB Ethernet 3C460B
AboCom	UFE1000C USB 10/100 Fast Ethernet
	UFE2000 USB Fast Ethernet
Accton	EN1046 USB Ethernet Adapter
	EN5046 USB 2.0 Adapter
ADMtek	Infineon-ADMtek ADM8511 USB Ethernet
	Infineon-ADMtek AN985B
Allied Telesyn	AT-2000-USB100
Belkin	F5D5050 USB 10/100 Ethernet Adapter
Billionton	Billionton USB-100
	Billionton USBE-128
	Billionton USBKR-100
	Billionton USB2AR
Compaq	iPAQ Networking 10/100 USB
D-Link	D-Link DUB-E100
	D-Link DSB-650TX
	D-Link DSB-500T
Elsa	Elsa Micolink USB2Ethernet
Hawking	Hawking USB NETWORK ADAPTER 11M
Linksys	EtherFast 10/100 LAN Card LNE100TX
	EtherFast 10/100 LAN Card LNEPCI12
	EtherFast 10/100 LAN Card LNEPCI2T
Siemens	SpeedStream USB 10/100 Ethernet
smartBridges	smartNIC 2 PnP Adapter
SMC	SMC1244TX - Fast Ethernet PCI Network Card
	SMC1255TX - Fast Ethernet PCI Network Card with Personal Firewall
SOHOware	SOHOware NUB100 Ethernet

To connect your TiVo to a network hub using Ethernet or Fast Ethernet wired connections, complete the following steps:

1. Plug the USB end of the USB-Ethernet adapter into the USB port on the TiVo.
2. Plug one end of the Ethernet cable into the USB-Ethernet adapter, and plug the other end into your home network's hub or router.

Creating a Network Connection on a Wired DHCP Network

Once your TiVo is connected to the network, press the TiVo button on your remote. A message will be displayed. Review the message (it will usually say that it has detected your network hardware), and press Select on your remote control to go to TiVo Central.

If your home network uses DHCP without a client ID, simply wait 5–10 minutes for the TiVo to automatically complete its network settings, and your job is complete.

NOTE *DHCP automatically, or "dynamically," assigns IP addresses to devices on a network. Most home networks that include a home gateway or router use DHCP.*

If your home network uses a DHCP client ID, or it has trouble making the DHCP network connection, you'll have to assign an IP address to your TiVo. Follow these steps:

1. Press the TiVo button on the remote control to go to TiVo Central.

CAUTION *The IP address is where your TiVo will "live" in your home network. The address consists of four numbers, each being from 1 to 255. On a home network, the first two numbers are typically 192 and 168 (written as 192.168). The last number has to be unique on your network, or it will not work.*

2. Highlight Messages and Setup and press the Select button.
3. Highlight Settings and press the Select button.
4. Highlight Phone & Network and press the Select button.
5. Highlight Edit Phone or Network Settings, and press the Select button.
6. Highlight TCP/IP Settings as in Figure 4-1 and press the Select button.

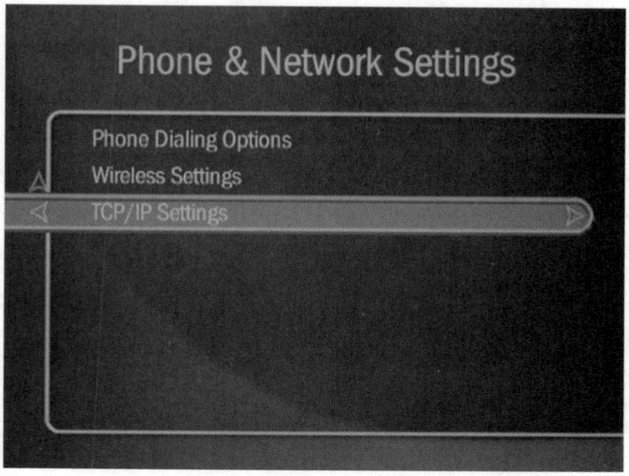

FIGURE 4-1 Highlight TCP/IP Settings

7. Select how your TiVo should get an IP address: obtain it automatically or statically. If you're not sure, select Obtain It Automatically, as shown in Figure 4-2. This selection will work for most people.

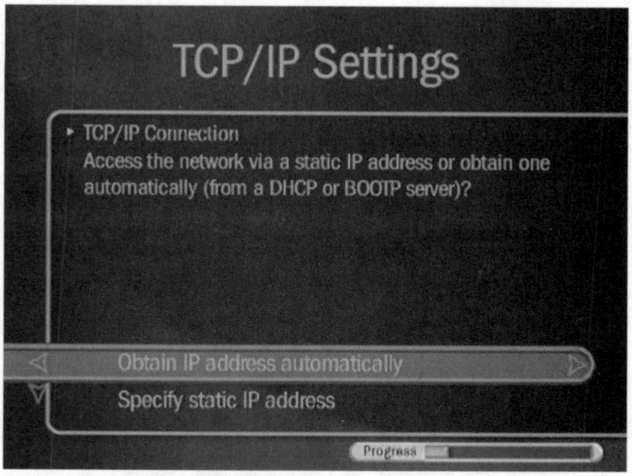

FIGURE 4-2 Selecting the Obtain It Automatically option

8. You'll be prompted for your DHCP Client ID.
   - If your Internet service provider (ISP) did not provide you with a DHCP Client ID, you can select the I Don't Have a DHCP Client ID option. Again, this selection will work for most people.
   - If your ISP did provide you with a DHCP Client ID, enter it using the arrow keys to navigate around the screen's keyboard, and press Select to select each character.
9. Highlight the Done Entering Text option as shown in Figure 4-3 when you are finished, and press the Select button.

NOTE *Client ID names are case sensitive. Use the Thumbs-Up and Thumbs-Down buttons to change between upper- and lowercase on the screen's keyboard.*

10. Review your settings. If they're correct, highlight the Accept These Settings option and press Select. You should now be connected to your home network!

Creating a Network Connection on a Wired Static IP Network

If your home network does not use DHCP, you'll have to assign an IP address to your TiVo. Follow these steps:

FIGURE 4-3 Selecting the Done Entering Text option

1. Press the TiVo button on the remote control to go to TiVo Central.
2. Highlight Messages and Setup and press the Select button.
3. Highlight Settings and press the Select button.
4. Highlight Phone & Network and press the Select button.
5. Highlight Edit Phone or Network Settings and press the Select button.
6. Highlight TCP/IP Settings and press the Select button.
7. Select how your TiVo should get an IP address: obtain it automatically or statically. If you don't have DHCP, select the Specify Static IP Address option as in Figure 4-4.
8. When prompted to enter your IP address as in Figure 4-5, enter 192.168.1. followed by whichever fourth number you choose between 1 and 255, so long as the number is not currently assigned on your network. Press the right arrow button when you are finished.
9. When prompted to enter your network mask, enter 255.255.255.0 (see Figure 4-6). Press the right arrow button when you are finished.

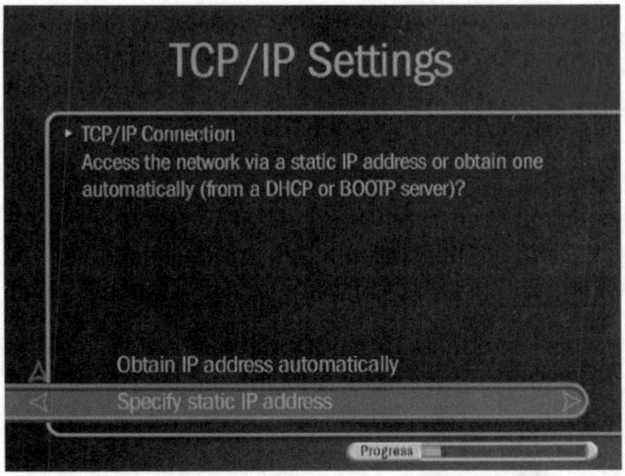

FIGURE 4-4 Selecting the Static IP address Text option

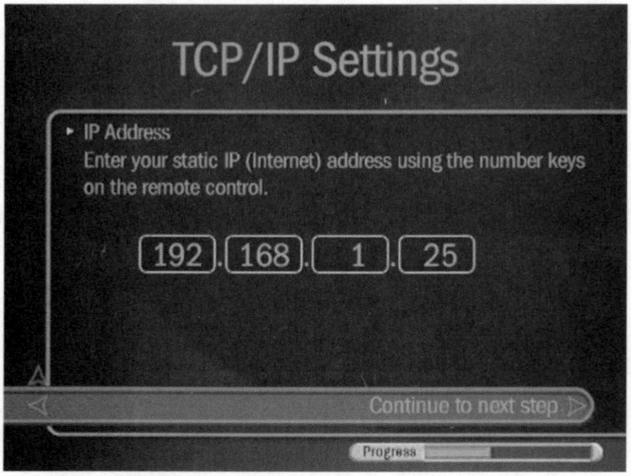

FIGURE 4-5 Entering the IP address

10. When prompted to enter your Gateway (Router) address, enter it. This IP address is usually available from the configuration screens of your gateway or router. If your home network uses an ISP's router, get the IP address from the ISP if they haven't given it to you already.

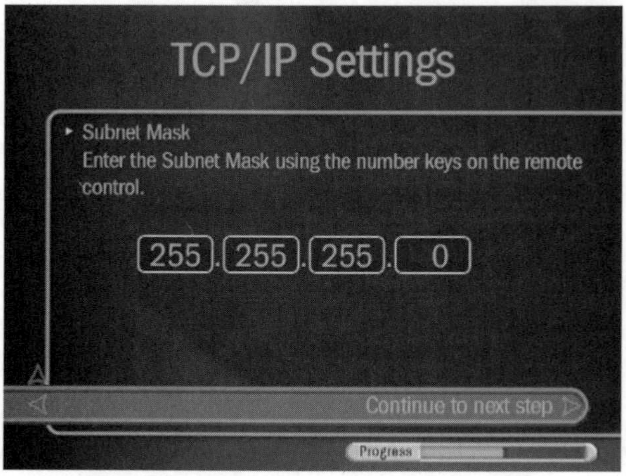

FIGURE 4-6 Entering the Network Mask

11. If your home network includes a broadband connection to the Internet, the ISP will also have given you a DNS (Domain Name Server) address. Enter it when prompted here. If you don't have a broadband connection, skip this step. Press the right arrow button when you are finished.
12. Review your settings and select Accept These Settings if they are correct. You should now be connected to your home network.

Setting Up a Wireless Connection

To set up a wireless connection, you'll need to purchase a network adapter that allows you to connect wirelessly to other devices on your network. You can purchase these adapters at most electronics stores or on the Internet. One of the best vendors for wireless adapters is Linksys, whose models, such as the Linksys WUSB11 Wireless USB Network Adapter, have a good track record of working with TiVo. (Visit their Web site at http://www.linksys.com/ or visit the TiVo community Web page at http://www.tivocommunity.com for consumer feedback on which new models work the best for this task.

To connect a TiVo to a network hub using a wireless connection, follow these steps:

1. Connect one end of the USB cable that came with the wireless network adapter to the port on the adapter.
2. Connect the other end of the USB cable to the USB port on the TiVo.
3. The power light on the adapter may come on before the cable is completely inserted into the USB slot; this is normal. Be sure to firmly plug the cable into the USB slot.
4. Position the adapter by placing it in an open location so the signal can be transmitted easily.

CAUTION *Don't place the wireless adapter near the TiVo's power supply, on a power strip, or near a surge protector, microwave, or cordless telephone. Doing so will cause serious signal interference and corresponding loss of quality.*

When you first connect the wireless adapter to the TiVo, you may see a screen that says "Firmware Update Required." This means that the TiVo has a software update for the wireless adapter you've selected. Select the Proceed, Update Firmware option, and allow the TiVo to complete the update. This should take no longer than 2 or 3 minutes and will complete the physical hookup process.

CAUTION *During the update, we recommend that you refrain from disconnecting either your network adapter or the power to the TiVo unit. Either action might damage the adapter, which you will then need to replace.*

TIP *One other option for you to consider is using an RCA RC930 Caller-ID-Compatible Wireless Modem Jack. This device consists of two units, the base and the extension. The base unit plugs into a phone jack and an electrical outlet, and it transmits the telephone signal on the electrical wiring using the FM band. The extension unit plugs into any electrical outlet in the house and receives the telephone signals from the electrical wiring—you can plug a phone line into the telephone jack on the extension unit just as you would use a regular telephone jack. Any number of extension units can be operated from the original base unit, and it can accommodate modem speeds up to 56 Kbps, which is relatively slow for Web surfing but ideal for your TiVo's connectivity needs.*

Configuring Network Settings for Wireless Home Networks

To configure your network settings, follow these steps:

1. Press the TiVo button on the remote control to go to TiVo Central.
2. Highlight Messages and Setup and press the Select button.
3. Highlight Settings and press the Select button.
4. Highlight Phone & Network and press the Select button.
5. Highlight Edit Phone or Network Settings and press the Select button.
6. Highlight Wireless Settings as in Figure 4-7 and press the Select button.
7. You'll be presented with a Wireless Checklist telling you to find out the following:
   - The name and password of the wireless network
   - Whether the password is alphanumeric or hexadecimal
   - For alphanumeric passwords, the level of network encryption utilized

   All of this information should be listed in the configuration settings of your network's wireless access point or router. When you're ready to proceed, press Select on your remote to continue.

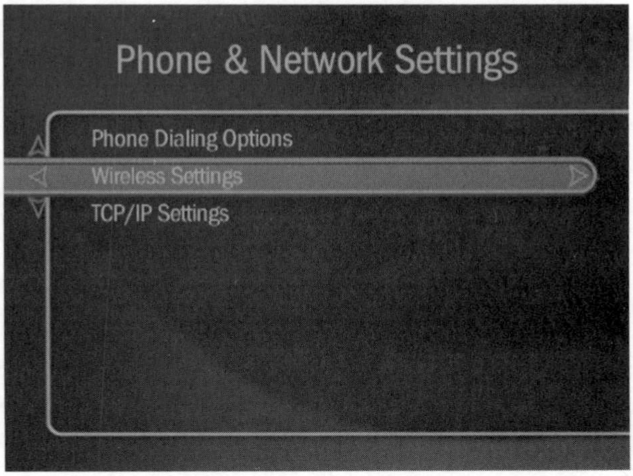

FIGURE 4-7 Selecting the Wireless Settings option

NOTE *If your wireless network doesn't use encryption, you only need to provide the wireless network name—no password format or password is needed in your setup process.*

8. You'll be prompted to enter the name of your wireless network. The screen will list names of nearby available wireless networks that the TiVo has identified.
9. If your network's name is listed, select it. Otherwise, select the Connect to a Closed Wireless Network option as shown in Figure 4-8.

NOTE *If you have a wireless router or access point and did not set a network name, your network is probably using the system default, which is the name supplied by the router's manufacturer.*

10. On the Wireless Network Name screen, enter the name of your network by using the arrow keys to navigate the onscreen keyboard and pressing the Select button to select the highlighted letter. Remember that wireless network names are case sensitive, so use the Thumbs-Up and Thumbs-Down buttons to switch between uppercase and lowercase letters.
11. Select the Done Entering Text option when you are finished, and press the Select button.

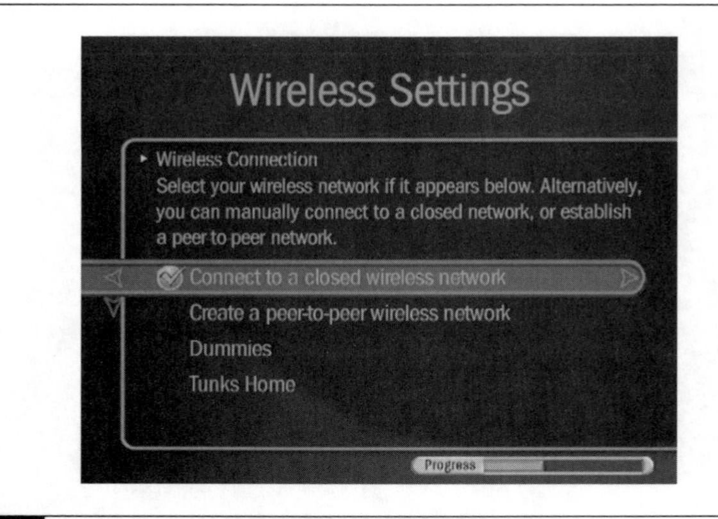

FIGURE 4-8 Selecting the Closed Wireless Network option

12. If your wireless network doesn't use encryption, you'll be asked to Accept These Settings. If they are correct, press Select to accept the settings you've entered.

13. If your network does use encryption (WEP), the Wireless Password Information screen will appear. You'll first be asked to select the password format: hexadecimal or alphanumeric.

NOTE *WEP (Wired Equivalent Privacy) is a form of encryption used by wireless networks to improve security. There are different levels of WEP encryption, ranging from 40-bit to 256-bit encryption. The higher the level of encryption, the better the level of security provided. At the higher levels (104-bit or above), WEP encryption can slow down the transfer of programs between DVRs due to the extra amount of processing time needed for the level of encryption. WEP is usually switched on or off through the configuration settings of a wireless access point or router.*

14. Next, you'll be prompted to enter the password itself. Whether entering a hexadecimal or alphanumeric password, use the arrow keys and thumbs keys to navigate the onscreen keyboard and change case. A hexadecimal password is simply a password that contains a combination of letters from A–F and numbers 0–9. By contrast, an alphanumeric password is a combination of the letters A–Z and numbers 0–9.

NOTE *Hexadecimal passwords are* not *case sensitive, but alphanumeric passwords* are *case sensitive.*

15. If you're using an alphanumeric password, you'll be asked to select a level of wireless encryption. Select 40- or 64-bit encryption or 128-bit encryption, depending on the wireless network's settings.
16. Now you'll be asked to Accept These Settings. If they are correct, press Select to accept the settings you've entered.

The final step in a wireless setup is to finish configuring the network settings. If your network uses DHCP and you do not have a DHCP client ID, you are finished. However, if you do have a DHCP client ID or your network uses static IP addressing, you have one last thing to configure on your TiVo. Follow the steps in whichever of the two following sections is appropriate to your networking situation.

NOTE *There is a relatively good list of recommended accessories for TiVo and reviews of networking equipment at http://www.tivo-direct.com/tivo-accessories.html.*

Creating a Network Connection on a Wireless DHCP Client ID Network

If your home network uses DHCP and has a DHCP client ID or just has trouble making a network connection, you'll have to assign an IP address to your TiVo. Follow these steps:

1. Press the TiVo button on the remote control to go to TiVo Central.
2. Highlight Messages and Setup and press the Select button.
3. Highlight Settings and press the Select button.
4. Highlight Phone & Network and press the Select button.
5. Highlight Edit Phone or Network Settings and press the Select button.
6. Highlight TCP/IP Settings and press the Select button.
7. Select how your TiVo should get an IP address: obtain it automatically or statically. If you're not sure, select the Obtain It Automatically option. This selection will work for most people.

8. You'll next be prompted for your DHCP client ID:
   - If your Internet Service Provider (ISP) did not provide you with a DHCP client ID, then you can select the I Don't Have a DHCP Client ID option. Again, this selection will work for most people.
   - If your ISP did provide you with a DHCP client ID, enter it using the arrow keys to navigate around the onscreen keyboard and pressing Select to select each character.
9. Highlight the Done Entering Text option when you are finished and press the Select button.
10. Review your settings, and select Accept These Settings if they are correct. You should now be connected to your home network.

Creating a Network Connection on a Wireless Static IP Network

If your home network does not use DHCP, you'll have to assign an IP address to your TiVo. Follow these steps:

1. Press the TiVo button on the remote control to go to TiVo Central.
2. Highlight Messages and Setup and press the Select button.
3. Highlight Settings and press the Select button.
4. Highlight Phone & Network and press the Select button.
5. Highlight Edit Phone or Network Settings and press the Select button.
6. Select TCP/IP Settings and press the Select button.
7. Select how your TiVo should get an IP address: obtain it automatically or statically. If you don't have DHCP, select the Specify Static IP Address option.
8. When prompted to enter your IP address, enter 192.168.1. followed by whichever fourth number you choose between 1 and 255, so long as the number is not currently assigned on your network. Press the right arrow button when you are finished.
9. When prompted to enter your network mask, enter 255.255.255.0. Press the right arrow button when you are finished.

10. When prompted to enter your Gateway (Router) address, enter it. This IP address is usually available from the configuration screens of your gateway or router. If your home network uses an ISP's router, get the IP address from the ISP if they haven't given it to you already.
11. If your home network includes a broadband connection to the Internet, the ISP will also have given you a DNS (Domain Name Server) address. Enter it when prompted here. If you don't have a broadband connection, you can skip this step. Press the right arrow button when you are finished.
12. Review your settings and select Accept These Settings if they are correct. You should now be connected to your home network.

Using TiVo's Desktop Software

Now that you've created a home network, you need something that uses it. That's where TiVo's free TiVo Desktop software comes into play. After activating TiVo's Home Media Features, you can transfer photos and music files from your computer to your TiVo units. Using TiVo Desktop, you can publish your photos and music files so that TiVo sees them and can display and play them.

CAUTION *If you don't subscribe to the Home Media Features, you will still be able to run and install the TiVo Desktop program. However, you won't be able to take advantage of the Desktop's best feature—the ability to make digital music and photos available to your TiVo Series2 DVR!*

Before you begin downloading and installing the TiVo Desktop, make sure that your home system meets the minimum requirements.

For PC users:

- A networked TiVo Series2 DVR with the TiVo Home Media Features premium feature package
- Microsoft Windows 98, ME, 2000, or XP
- 233 MHz Pentium II class processor or better
- At least 64MB RAM
- At least 25MB free disk space

For Macintosh users:

- A networked TiVo Series2 DVR with the TiVo Home Media Features premium feature package
- Mac OS X v10.2 or later
- iTunes 3.0.1 or later required for music
- iPhoto 2 or later required for photos
- 400 MHz G3 processor or better
- At least 256MB RAM

NOTE *You cannot view photos and also listen to music with a TiVo at the same time. However, this feature may be available in a future software release.*

Voices from the Community

A Fan of the Desktop Photo Feature Speaks Up

One of my favorite features from the TiVo Desktop, once I signed up for the Home Media Features, is the way you can publish photos to the television. The photo slideshows are easy to launch and a real breeze to navigate. You can just let a slideshow run automatically, and once it's going, you can use the pause button to freeze on a photo as long as you want, and the fast forward and reverse buttons let you jump between photos.

Personally, I like to set my own pace by using the channel up and down buttons to navigate between other photos in the set, so I can skip directly to the slides that show off all the fun stuff we've been doing. Whenever I have guests to the house, they've come away impressed to see king-size photos on the TV off my digital camera. Now, once I get that wide-screen HDTV that I've been eyeing, and hook my TiVo up to that, we're really talking!

—James Marchetti, Regional Platform Manager, Barclay's Global Investments, and big TiVo fan

Downloading TiVo Desktop

The TiVo Desktop software can be downloaded for free from TiVo's Web site. Go to http://www.tivo.com and click on the I Have TiVo! link in the menu at the left. Then click on the Home Media Features link. On this page you should see a link for TiVo Desktop Software, which takes you to the TiVo Desktop page.

Under Download TiVo Desktop, you'll see separate links for the Windows and Mac versions of the software. Click on the one you need as shown in Figure 4-9.

After saving and downloading the file, you can begin installing it (we'll install the Windows version of the software here). Follow these steps:

1. Double-click on the file you downloaded.
2. The Welcome to the TiVo Desktop Installation Wizard screen appears. Click Next.

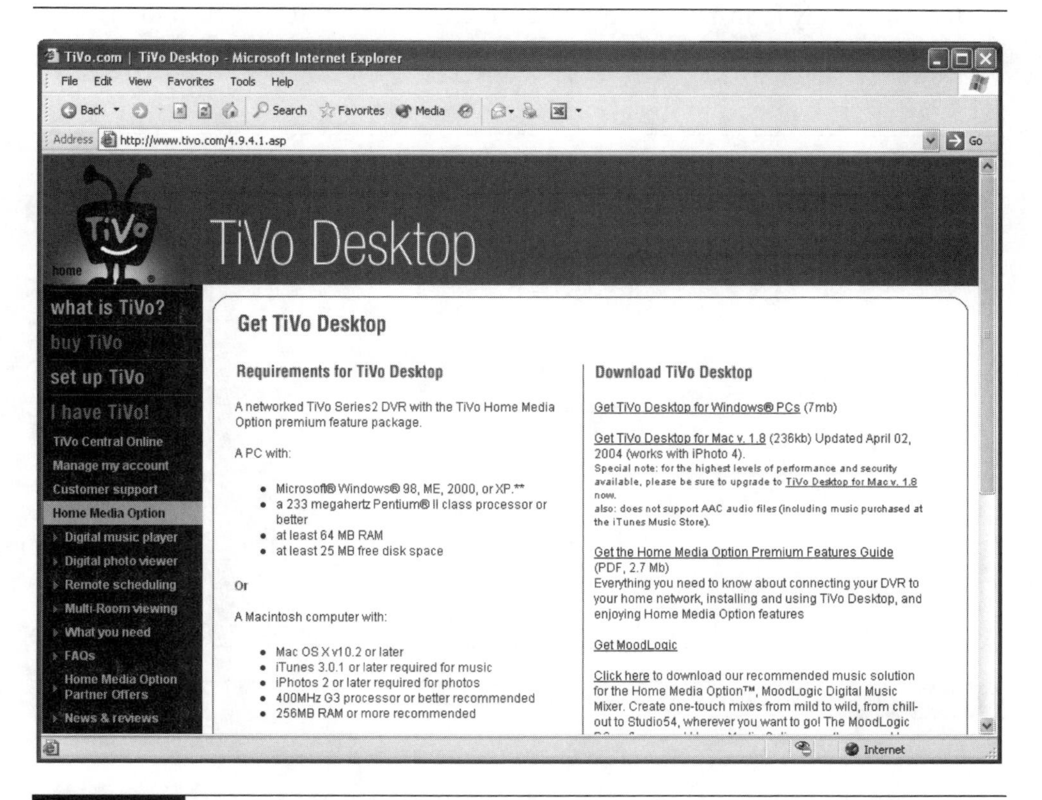

FIGURE 4-9 Downloading the TiVo Desktop software from the TiVo web site

3. The License Agreement screen appears. Select the I Accept the License Agreement option, and click Next.
4. The TiVo Home Media Features screen appears, as shown in Figure 4-10. Select Yes, I Have Purchased Home Media Features, if it's not already selected. Click Next.
5. A screen describing what you need to use music and photo files appears. Click Next.
6. A Readme Information screen appears. Click Next.
7. The User Information screen appears, as in Figure 4-11. Enter your name and organization and click Next.
8. The Destination Folder screen appears. If the destination folder is okay, click Next. Otherwise, click Browse and navigate to the folder you'd like to use.
9. The Ready to Install the Application screen appears. Click Next to install the software.

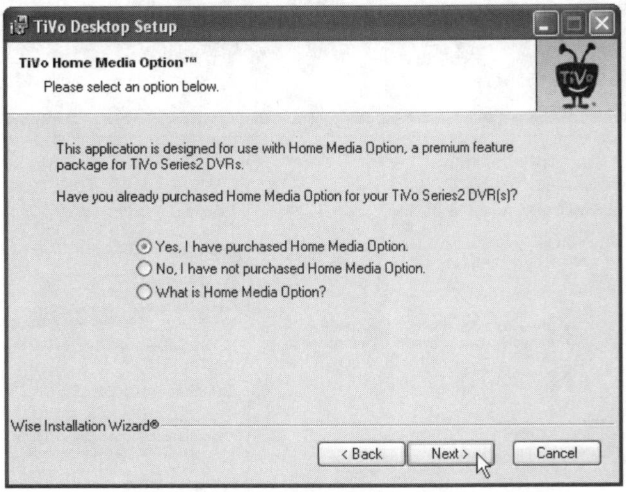

FIGURE 4-10 TiVo Home Media Features screen

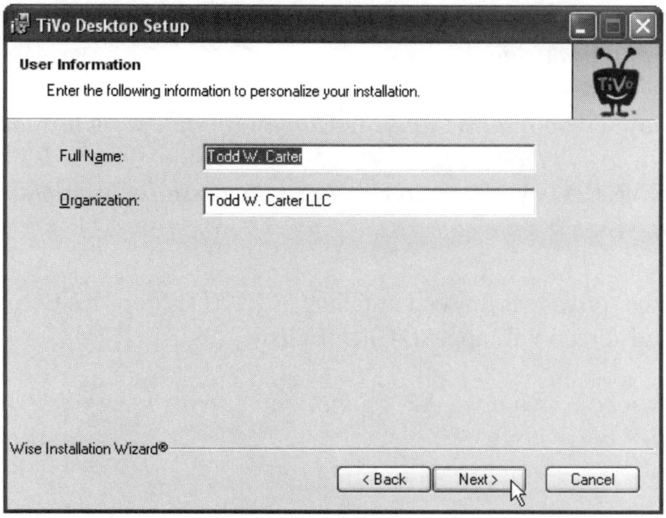

FIGURE 4-11 Entering your name and organization

Did you know?

Software by Any Other Name

You may see two other program names while using the TiVo Desktop software package. That's because TiVo Desktop actually consists of two related software programs:

- TiVo Publisher, which publishes and unpublishes your music and photo files.
- TiVo Server, which operates in the background on your computer while TiVo Desktop runs. It communicates with your HMO-enabled TiVos, sending requested music and photo files from your computer to your TiVo.

NOTE *Using TiVo Desktop to publish photos and music files (which makes them accessible from your TiVo) is covered in Chapter 5 and the two-colored centerfold in this book.*

NOTE *You may see an Internet Sharing Configuration dialog box, asking you if it's okay to edit the Internet Connection Protection settings. It's safe to answer Yes for this. A TiVo Desktop Has Been Successfully Installed screen will appear when it's done.*

10. When the software is finished installing, a TiVo Desktop Has Been Successfully Installed screen will appear. Click Finish.

NOTE *A TiVo icon in Windows XP's notification area shows you that the software installed correctly.*

How to ... Open Your Firewall Ports

If your TiVos cannot connect to your computer, it's possible your computer's firewall software is the culprit. In order to communicate with TiVo, the TiVo Publisher software sends out a beacon on your home network. Basically, a beacon is a short burst of data that says, "Hey, I'm here and I'm ready when you are."

The TiVo Desktop software sends these beacons out from your computer on certain ports, and if your firewall software is blocking these ports, the beacons can't reach the TiVos. Make sure the following ports are open on your computer:

- TCP port 2190
- UDP port 2190
- TCP ports 8080 through 8089

You can open these ports with your firewall software. In many cases, the software senses the need to keep these ports open, asking whether you would like to add a rule that keeps them open.

Did you know?

An Alternative to TiVo Desktop

All popular technologies seem to attract some savvy users who take the manufacturer's platform and make it better by adding features and tweaks. That's the case with JavaHMO, which performs the same tasks as TiVo Desktop, and more.

You can download this software, which is not supported by TiVo, from the developer's Web site at http://javahmo.sourceforge.net/. In the Downloads section, which you can access from the menu at the left, you can select and download the Windows, Mac OS X, or Linux versions.

That's the first thing you'll notice about JavaHMO: it works on three operating systems, while TiVo Desktop only works on Windows and the Mac OS. That's important if you favor Linux and still want access to TiVo's Home Media Features. As its name suggests, JavaHMO runs under Java, a programming language that operates on just about every operating system that exists. The following illustration shows TiVo displaying the six services that JavaHMO adds to the Music & Photos screen.

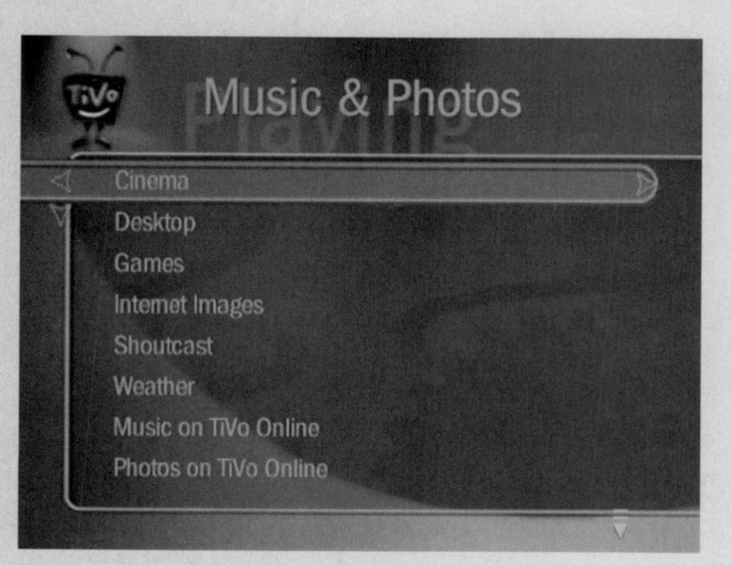

JavaHMO has many features that go beyond TiVo's software, and they allow you to perform the following functions:

- View images in the following formats: BMP, GIF, FlashPix, JPEG, PNG, PNM, TIFF, and WBMP.
- Rotate the images you send from your computer to the TV screen.
- Play MP3s like .m3u and .pls files with a random/shuffle play feature.
- View MP3 file tag information and organize the MP3s by their ID3 tags.
- Play MP3s direct from streaming radio stations on the Internet. (This is a popular option for several reasons, including the ability to pick stations that have little or no advertising.)

- Automatically download the Shoutcast playlists from your favorite streaming Internet radio stations.
- Use the streaming proxy server to improve upon TiVo's support for online streaming stations.
- View live local weather. JavaHMO makes weather information available for viewing: current conditions, five-day forecasts, and radar images. TiVo displays a current weather screen.

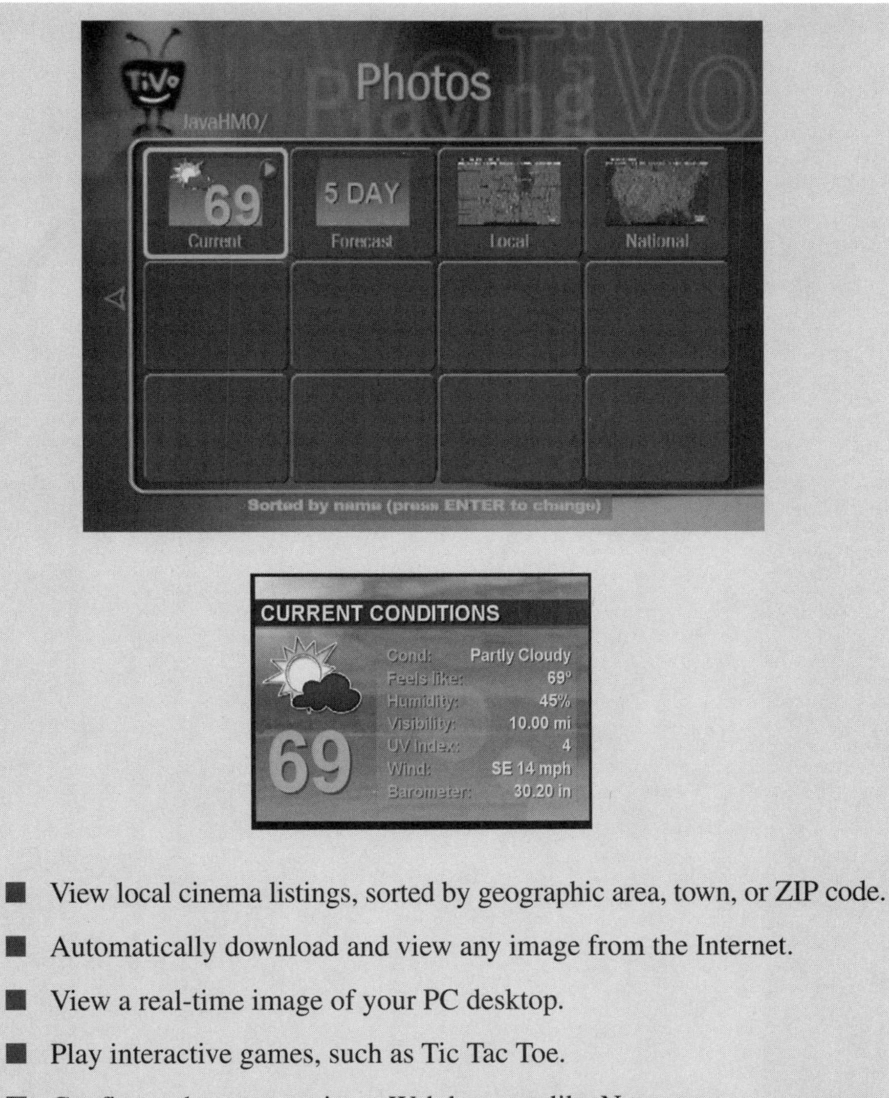

- View local cinema listings, sorted by geographic area, town, or ZIP code.
- Automatically download and view any image from the Internet.
- View a real-time image of your PC desktop.
- Play interactive games, such as Tic Tac Toe.
- Configure the server using a Web browser like Netscape.

NOTE *JavaHMO only works for TiVo's Series 2 boxes. In addition, you need to have purchased the Home Media Features to take advantage of these features.*

Chapter 5

Listen to Music on Your TiVo

How to...

- Organize Your Music Files
- Publish Your Music Files
- Play Your Music Files Through TiVo

Your TiVo can do more than just record classic movies airing at 3 a.m. for your prime-time enjoyment. If you have a TiVo Series2 you can also display photos and listen to music through your TiVo.

Playing music files on your TiVo requires you to do two things:

- Organize your music files into folders and subfolders for best performance.
- Select and publish these files using the TiVo Desktop software. Publishing a music file makes it available to your TiVo.

When you're done publishing your music files, you can play them on your TiVo through your TV set or home theater receiver.

Organizing Your Music Files

Organizing anything, including music or photo files on your computer, is personal. One person's organizational techniques are another person's idea of obsessive-compulsive behavior.

You could create folders for artists or music genres and then file your music files in subfolders within those folders. So Dave Brubeck's "Take Five" song from the *Time Out* album could be stored in a Time Out subfolder in a Dave Brubeck Quartet

Did you know?

Music to Your Ears

Playing music or viewing photos on your TiVo does not reduce the amount of space available for recording TV programs. That's because your music and photo files are stored on your computer, not TiVo's hard drive.

Did you know?

Helping Your Music Load Faster

If you have many music files and store them all in one folder, TiVo may take longer before it begins playing a file. It's smart to find a way to organize your files that breaks them down into smaller groups.

subfolder in a Jazz folder. Or you could file the album's tracks in a Time Out subfolder in a Dave Brubeck Quartet folder.

If you have music from bands in genres you otherwise rarely listen to, you could just set up a folder with the band's name and the album title: The Smiths—Meat Is Murder. For other kinds of music, you may want to just create a folder for the genre, such as Trance, and store everything in it. Figure 5-1 illustrates this sort of organization.

The music organization shown in Figure 5-1 is just one way of organizing a music collection. You may want to organize it by album titles instead of by artist, or by the year the music was released. The important thing is to choose an organizing system that works for you, allowing you to find the music you want when you want it.

Publishing Your Music

Before you can listen to music on your TiVo, you first need to select folders or songs and publish them using the TiVo Desktop software.

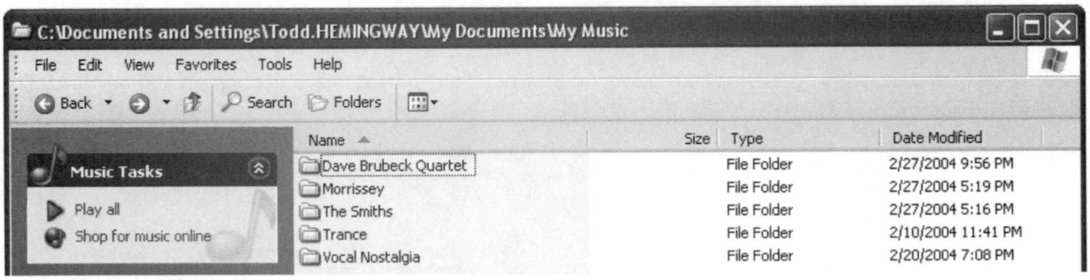

FIGURE 5-1 Organizing music by artist and genre on the computer

NOTE *For information about installing the TiVo Desktop software, see Chapter 4.*

Follow these steps to publish your music:

1. In TiVo Publisher, select the folder that contains your music files.
2. In the Preview pane to the right of the folder list, a list of folders within your selected folder appears. Select the folders or songs you want to publish in one of two ways:
   - Right-click on the folders or songs you want to publish and select Publish, as shown in Figure 5-2.
   - Drag and drop the folders and songs into the bottom pane of TiVo Publisher.

FIGURE 5-2 Publishing music folders

How to ... Convert Music Files and Play Playlists

TiVo can only play music files in MP3 format, so if you want to play any files that are in other formats, such as WMA and WAV, you'll need to convert them to MP3 format. You can find several good conversion programs on shareware software sites like http://www.download.com.

TiVo normally plays tracks in alphabetical order (unless you've turned on the shuffle command), but it can also play music playlists in the M3U, PLS, and ASK file formats. Playlists make it easier to control how your TiVo plays songs. In Windows Media Player, Winamp, and other music players, you can arrange the order of album tracks and save that track list as a playlist.

NOTE *If Publish is unavailable on the pop-up menu, it means the folder is either empty or does not contain any files in the MP3 format. TiVo only supports MP3 files.*

Listening to Your Music

After you select music files and publish them with TiVo Desktop, the next step is listening to them on your TV or home theater system. You can control everything with your TiVo, using your remote control to select songs, play them, and fast forward and rewind them.

NOTE *Remember that the computer where TiVo Desktop is installed must be turned on and TiVo Server must be running for TiVo to play the music files (the music files are on the computer, after all). In TiVo Publisher, you can ensure TiVo Server is active by selecting Start/Resume from the Server menu.*

Here's how you navigate TiVo's screens to select and play music files:

1. From the TiVo Central screen, select Music & Photos, as shown in Figure 5-3.
2. On the Music & Photos screen, select your music server by pressing Select or the right-arrow button. In the example shown in Figure 5-4, the server is "Todd's Music on HOMEOFFICEPC."

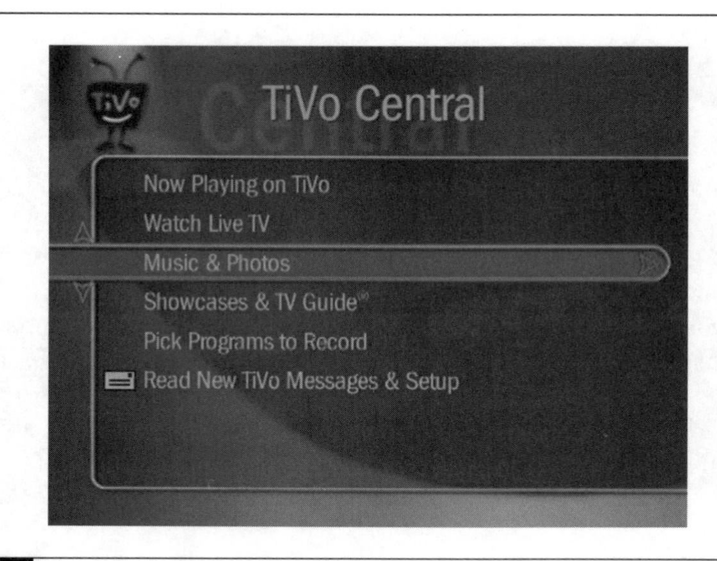

FIGURE 5-3 Selecting Music & Photos

3. Press the up and down arrow buttons to navigate through the list of music folders and files. You can press Select or the right arrow button to view the contents of the highlighted music folder, as shown in Figure 5-5.

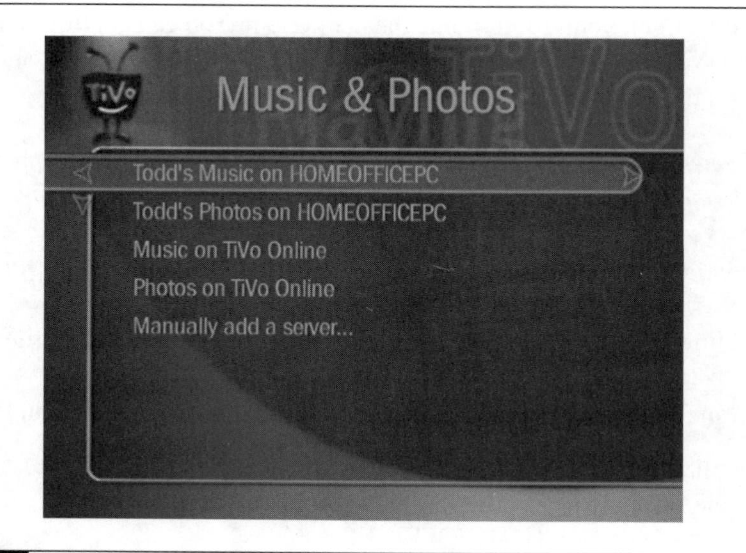

FIGURE 5-4 Selecting a music server

FIGURE 5-5 Selecting a music folder

NOTE *At any time while browsing your music files, you can press Play. The contents of a folder or a single song, depending on what is highlighted at the time, will begin playing.*

4. Press Select to see the highlighted file's contents, as shown in Figure 5-6.

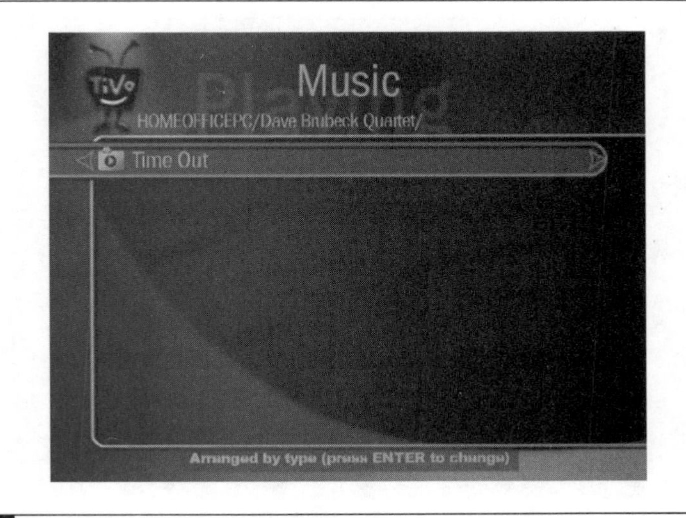

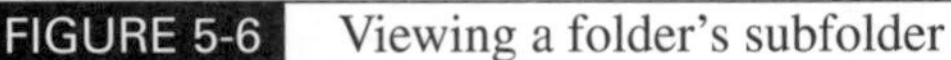

FIGURE 5-6 Viewing a folder's subfolder

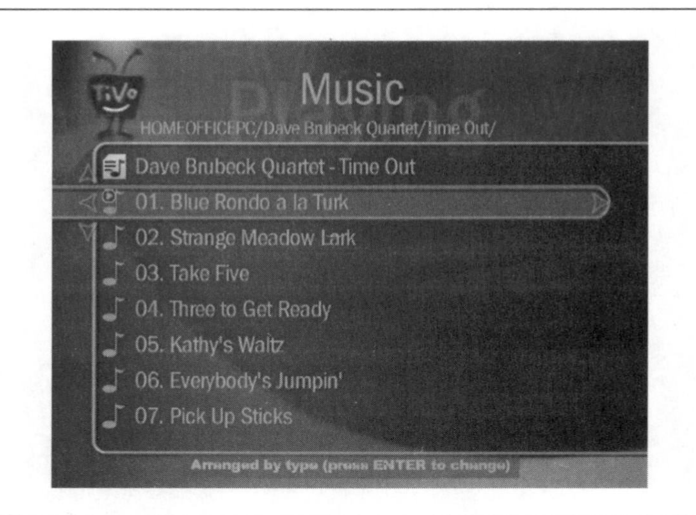

FIGURE 5-7 Selecting a song

5. Highlight a song or playlist, as shown in Figure 5-7, and then press Select to display the Song screen.
6. From the Song screen, you can play the song, change music play options, or return to the previous screen, as shown in Figure 5-8. TiVo is shown playing a song in Figure 5-9.

FIGURE 5-8 You can play songs from the Song screen.

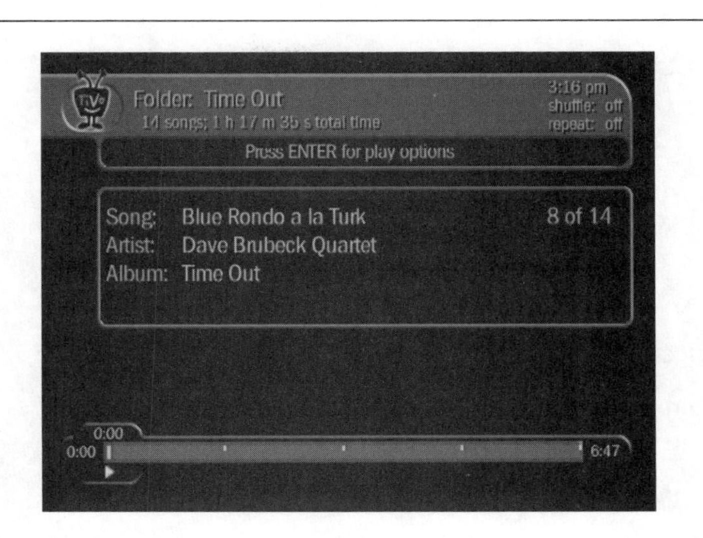

FIGURE 5-9 Playing a song

Music Play Options

You can change some of TiVo's music settings on the Music Play Options screen, shown in Figure 5-10. You reach the screen by highlighting a song, pressing Select, navigating to Options, and pressing Select again.

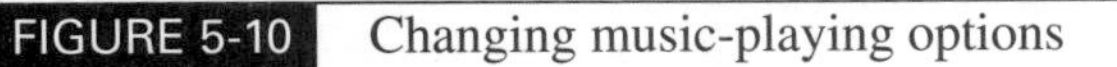

FIGURE 5-10 Changing music-playing options

How to ... Navigate with TiVo's Remote

You'll use TiVo's remote control buttons when playing music. Here's a list of the buttons and their music functions:

- **Play** Plays a song or playlist in a folder.
- **Select** Opens a folder or displays song options.
- **Chan Up** Goes to a song's beginning if pressed once. Press it twice to go to the previous song.
- **Chan Down** Goes to next song.
- **Fast Forward** Scans forward through a song.
- **Rewind** Scans backwards through a song.
- **Replay** Rewinds a song eight seconds.
- **Right** Displays a banner while a song is playing. The banner displays at the top of the screen and shows the folder or playlist name, the number of songs in the folder or playlist, and the total time of all songs. In addition, the banner shows the status of the Shuffle and Repeat features.
- **Clear** Removes banner and progress bar, which TiVo displays on the bottom of the screen to show graphically how much of a song has played.

From the Music Play Options screen, you can change several settings:

- **Shuffle** When this is turned on, TiVo plays songs in a playlist or folder in a random order. Otherwise, songs in folders are played alphabetically and songs in playlists are played in the same order they were created.
- **Repeat** You can choose to repeat all songs, one song, or to turn off the feature.
- **Include Subfolders** When this is turned on, TiVo plays all songs and playlists in the selected folder and in the folder's subfolders. Otherwise, only songs in the selected folder are played.

Music Browsing Options

While you're browsing music on your TiVo, you can press Enter on your remote control to reach the Music Browsing Options screen, as shown in Figure 5-11. You can change the way TiVo lists music files by turning Arrange by Type on and off:

- **On** TiVo displays folders at the top, followed by playlists and then all songs.
- **Off** TiVo displays folders, playlists, and songs all sorted in alphabetical order.

How to ...

How to Troubleshoot TiVo Desktop

If you've published music within TiVo Desktop, but the files are not appearing on TiVo's Music & Photos screen, check the following points for a (hopefully) quick solution:

- Is the TiVo Desktop software running? If not, launch the application.
- If TiVo Desktop is running, but the music files are still not appearing on your TiVo, go to TiVo Desktop and choose Start/Resume from the Server menu. The software should now be communicating with your TiVo.

If the problem remains, advanced users may want to try the following steps:

1. Go to Control Panel | Administrative Tools | Services.
2. Double-click on Services.
3. Scroll down to TiVo Connect Beacon.
4. Confirm that the Status column displays "Started."
5. If the Status column is blank, indicating that the service is stopped, right-click on TiVo Connect Beacon service, and select Start.

Your TiVo Desktop software should now be communicating with your TiVo units. Your published music should appear on TiVo's Music & Photos screen.

If you're still having problems, you can try TiVo's support site at http://customersupport.tivo.com.

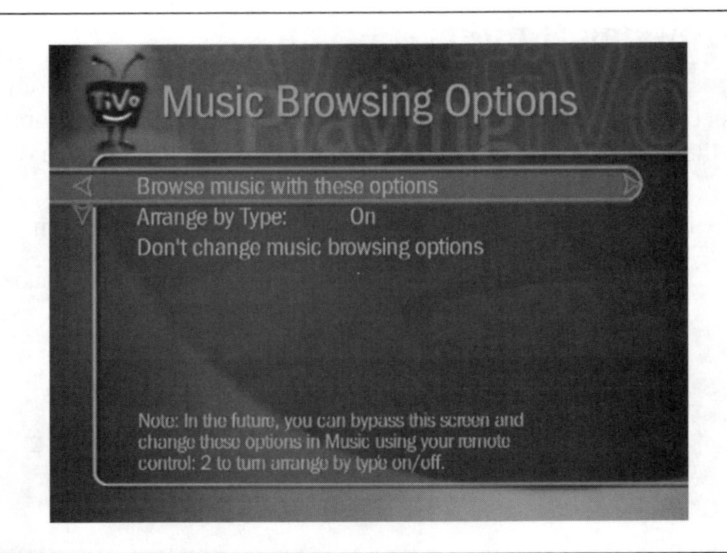

FIGURE 5-11 Setting music browsing options

Using MoodLogic

Are you tired of listening to your songs in the same old order, an order dictated by musicians and engineers with no input from you? Would you like your music automatically organized by tempo and genre? If so, MoodLogic might be worth trying.

Given a list of songs, the MoodLogic software can create new mixes of your entire music catalog. You choose what kind of "mood" you're in, and the software, using its knowledge of what other users like to listen to, produces a new mix of the songs you select. TiVo and the MoodLogic software maker are working together, which means TiVo users can easily play these mixes through TiVo.

MoodLogic runs on your Windows PC. In addition to the main software that creates the mixes, there's a separate piece of code that runs separately and keeps your PC in contact with your TiVo. Just like with the TiVo Desktop software, MoodLogic must be running on your PC or your mixes won't appear on the TiVo screen.

NOTE *You can find specific information about using MoodLogic with TiVo at http://www.moodlogic.com/tivo.html.*

What's the downside? For one, you need to purchase the software after a trial period ($39.95 at the time this book was written). Also, after this trial period, MoodLogic creates mixes only if you "activate" your songs, which means that the

How to ...

How to Listen to Internet Streams on Your TiVo

While it is not officially supported by TiVo, it is possible to play music and other streaming audio on your TiVo. We've had success with this simple technique, though it is reputed to be hit and miss, presumably depending on the characteristics of the Internet stream.

As you know, TiVo can play MP3 files and can use several types of playlists. It is possible to trick TiVo into thinking that the Internet stream is simply a song on a playlist. Here's what you need:

- A text editor—Windows XP's Notepad is fine.
- The IP (Internet protocol) address of the streaming audio. An IP address looks like this: 192.168.1.1. It is four numbers ranging from 0 to 255, each separated from the next by a period.
- A port number, which will be something like 8001.

When you have that information, just combine the IP address with the port number, and enter that information into a text file whose name ends with the .m3u extension. The examples given above would result in an IP address and port number combination of "192.168.1.1:8001". Just type that line into Notepad and save it as stream.m3u, for example.

In TiVo Desktop, you would publish the file just as you would a music file. Then, when you select this file and press Play from the TiVo screen on your TV, you can listen to this live stream through your DVR and home stereo.

Unfortunately, this technique only works when an Internet audio stream is accessible by an IP number. It does not seem to work when you only have a domain name, like digitallyimported.com, and the location of the stream on the service's server (/mp3/trance96k.pls).

JavaHMO, which is discussed in Chapter 4, can use that format to play Internet streams. You may want to consider installing and using JavaHMO if you're interested in playing live music and other audio streams from the Internet.

software checks your music files against those in its database so it can categorize them by genre and mood. After the trial ends, you're charged "activation credits" past the first 10,000 songs.

The first thing you should do is visit MoodLogic's web site at http://www.moodlogic.com and download the software. Click on the TiVo Owners button and then click the Download button. Save the program file somewhere on your PC that you can remember.

NOTE *Be sure to download the TiVo version of MoodLogic! The regular version does not work with TiVo.*

Now run the installation program, which will take you through the setup steps. Once the installation is complete, run MoodLogic. The first time you do, the software will ask you to create an account and enter your username and password. It'll then go through a multistep process to get the service up and running for you.

Your first choice is for MoodLogic to either automatically scan your computer for music files, as shown in Figure 5-12, or you can manually select a directory where you keep your music files.

After it identifies your music files, the software activates each of them. This can be time-consuming: MoodLogic estimates it takes eight minutes per 100 songs. Figure 5-13 shows MoodLogic after it has organized music files by genre and artist.

Now the fun part: listening to your music through TiVo. First, confirm that the MoodLogic server is running on your computer. If it is, you'll see a MoodLogic icon in Windows XP's task area:

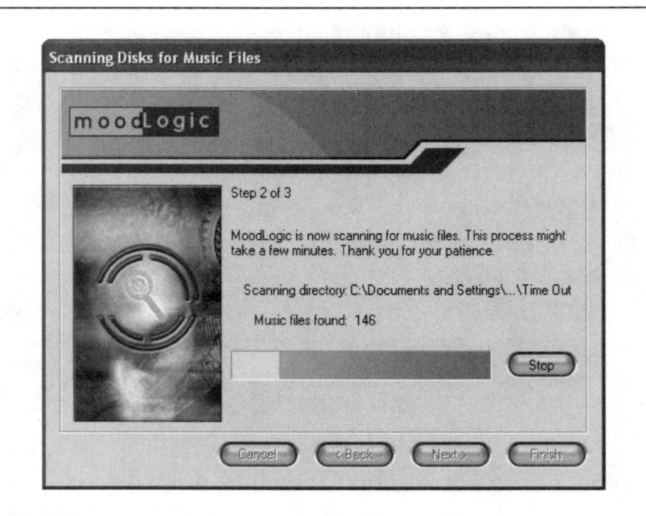

FIGURE 5-12 MoodLogic can scan a PC for music files.

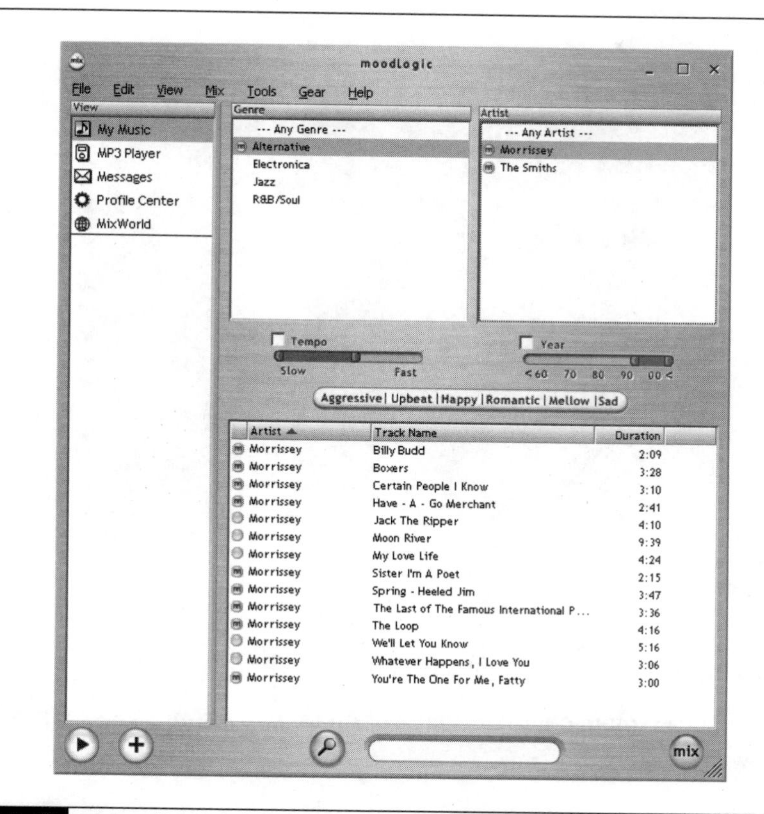

FIGURE 5-13 MoodLogic's My Music area categorizes songs by genre.

Next, to view MoodLogic files on TiVo, follow these steps:

1. From the TiVo Central screen, scroll down to Music & Photos and press the Select button or the right-arrow button. The Music & Photos screen appears.
2. Highlight the MoodLogic entry and press Select or the right-arrow button. A list of MoodLogic folders appears, as shown in Figure 5-14.
3. Select a folder. You can choose from My Mixes, Artist & Album, Genre & Artist, Genre & Mood, Genre & Tempo, Genre & Decade, and Surprise Me! The contents of the folder you selected appear, as shown in Figure 5-15. (In this example, we chose Genre & Artist.)

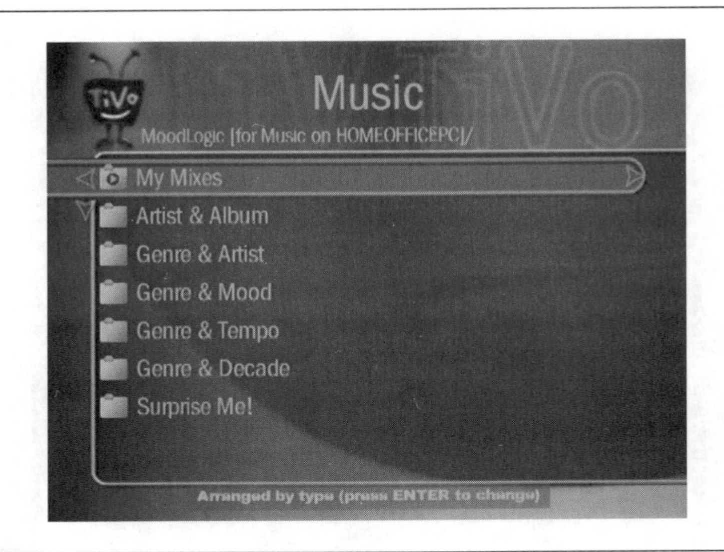

FIGURE 5-14 TiVo displays MoodLogic's folders.

4. Select a subfolder. The contents appear, as shown in Figure 5-16.
5. Press your remote control's Play button to play a song. The song plays, as shown in Figure 5-17.

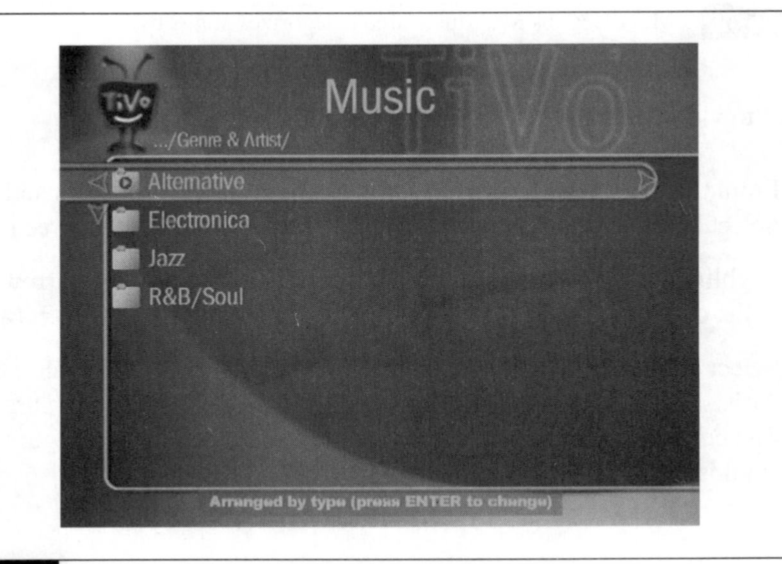

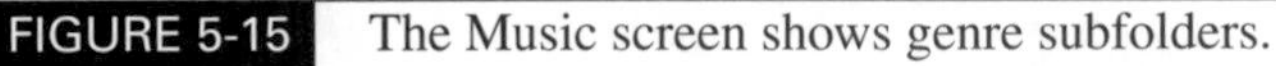

FIGURE 5-15 The Music screen shows genre subfolders.

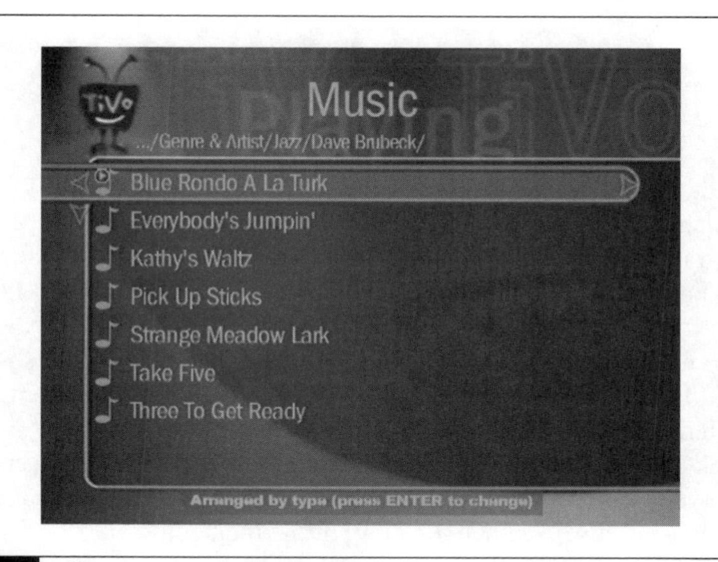

FIGURE 5-16 Viewing an artist's songs in the Jazz genre folder

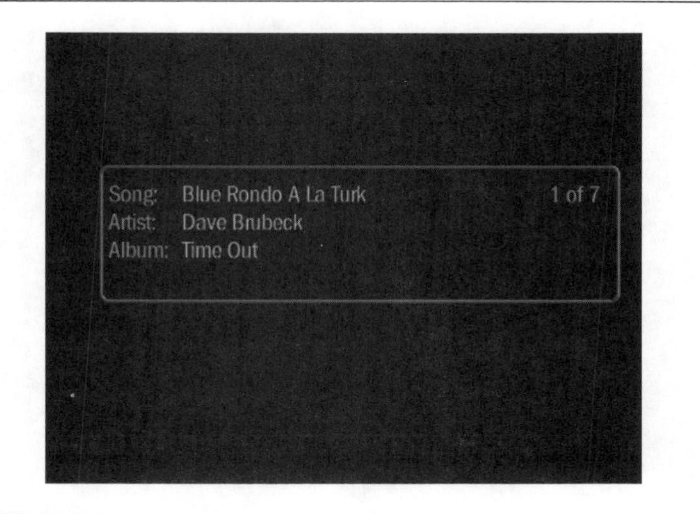

FIGURE 5-17 TiVo playing a song

Voices from the Community

A Yellow Submarine It's Not, But TiVo Helps Nonetheless

Jill Spiegal, an author and speaker, says she uses her TiVo on a daily basis for work, which involves weaving pop-culture examples into her speeches, books, and appearances. She uses her TiVo to help collect music-related information.

"I am busy during the day, so I TiVo all the entertainment news shows and then watch them at night to catch up on the latest trends with celebrities.

"I also have my own call-in talk radio show where I play Beatles music as bumpers. I have The Beatles on my TiVo WishList to collect Beatle insights to share with my listeners. And to watch for my own pleasure.

"I think TiVo is one of the all-time greatest inventions! I can play scenes I love over and over. Sometimes I'll watch a short clip 10 times to study someone's body language or just to re-live the glory of the moment."

Chapter 6

Remotely Schedule Programs

How to…

- Use TiVo's Remote Scheduling Feature
- Understand the Limitations of TiVo's Remote Scheduling Feature
- Search TV Listings on TiVo Central Online
- Get a Season Pass for a TV Show Remotely
- Browse TV Listings by Channel
- Schedule a TV Program for Recording
- Use AOL's Customized Version of Remote Scheduling

If you own a Series2 TiVo, you can use a Web browser to tell your TiVo units to record TV programs. You can use this feature anywhere you have Internet access, whether you're at work, across town, or vacationing on the other side of the planet. Just make sure to keep your TiVo plugged in and your broadband or dial-in phone connection alive while you're away!

Remote Scheduling Over the Internet

Depending on the way your TiVo unit connects with TiVo's servers, you can remotely schedule a recording for a show that begins in as little as an hour. If luck is on your side, and you're using a broadband connection, you may be able to schedule a recording with just 15 minutes left before airtime.

You can't be as spontaneous if you're using a phone-line connection, as TiVo recommends you schedule recordings up to 36 hours in advance. While TiVos using broadband connections check back with the mother ship every 15 minutes or so, units depending on dial-in connections may only update themselves every 24 hours.

TiVo's Web site, shown in Figure 6-1, provides several ways to find the TV programs you want to record remotely. You can use a basic or advanced search feature or browse for programs by channel. Once you find the program you want to record, you schedule it by selecting the recording options. If you provide an e-mail address, TiVo will send you an e-mail confirmation that the request was received and another one after your TiVo unit makes contact with TiVo's computers. This second e-mail will indicate whether the request was successful or whether there was a conflict.

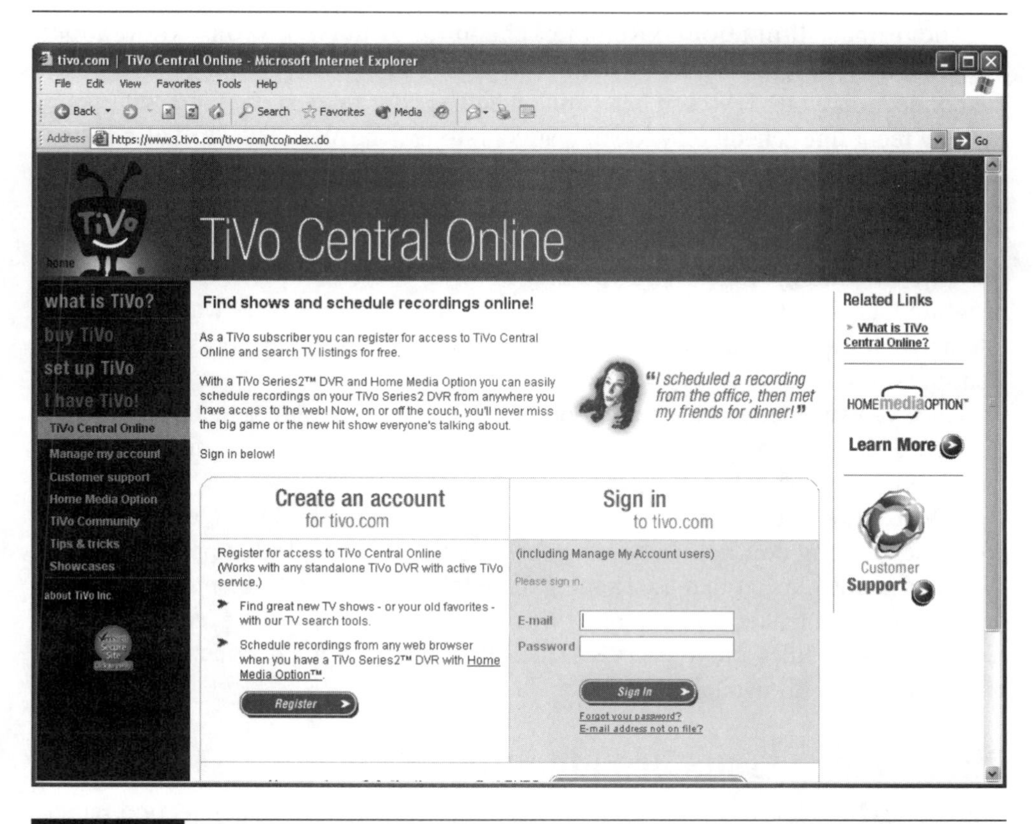

FIGURE 6-1 Remote Scheduling is controlled through TiVo Central Online.

NOTE *If you haven't already registered on TiVo's Web site, you'll need to do so before using TiVo Central Online and its Remote Scheduling feature.*

Limitations of Remote Scheduling

There are some limitations to the Remote Scheduling feature. Most importantly, it's not interactive, at least not in real time. When you request the scheduling of a program, TiVo's central computers wait for your unit to contact them. When that happens, it's determined whether the program can actually be scheduled or not.

If you provide your e-mail address when you make a Remote Scheduling request, you'll receive an e-mail confirming that your request was received on the Web site (which you'll already know when you make your request). You'll receive another e-mail either confirming the recording or explaining why it cannot be scheduled.

The primary limitation to Remote Scheduling is that you won't know about a scheduling conflict until after your unit connects to TiVo headquarters. After it makes the connection, your TiVo can determine whether your request conflicts with your existing recording schedule by looking at your To Do and Wish lists. It'll also factor in the priority you gave the Remote Scheduling request.

If there are scheduling conflicts, your unit will tell TiVo's computers, which will then inform you via e-mail (if you provided your e-mail address when you made the Remote Scheduling request). So you may not learn about conflicts until after a TV program has already started. At the time of writing, there was no way for you to interact with your TiVo unit in real time to resolve these scheduling conflicts via TiVo Central Online.

Once you've received an e-mail informing you that a program cannot be scheduled for recording, however, you can return to the TiVo Web site and try again—assuming there's enough time before the program airs.

TiVo cautions that the Remote Scheduling feature is not foolproof, even if there are no scheduling conflicts between your request and your To Do or Wish lists: "TiVo does not warrant that the Remote Scheduling service will be uninterrupted or error-free and makes no guarantees that any particular recording request will be received or scheduled on your DVR. The Remote Scheduling service is provided on an 'as is' and 'as available' basis."

Phone Line versus Broadband

With Series2 models, you can connect to TiVo over a broadband connection rather than using a phone line. This frees up your phone line, while also making the connections to TiVo much quicker.

If you are connecting to TiVo via a broadband connection, there is another advantage: your unit will communicate more frequently with TiVo headquarters, so it will pick up scheduling requests from the Remote Scheduling feature much more quickly.

In fact, your unit will contact TiVo several times an hour, compared to once every 30 hours or so when using a phone-line connection. This means that you can schedule programs remotely within an hour of airtime. If you're using a phone-line connection, you may not be able to schedule programs later than 36 hours before they are broadcast.

Given that the Home Media Features requires a broadband connection for its other features, such as transferring photos and music files, there's really no reason not to switch from a phone-line connection to a broadband connection for communicating with TiVo's computers. Doing so will give you more flexibility when scheduling programs remotely.

Searching TV Listings on TiVo Central Online

Before you can record a program remotely, you need to find the program in the TV listings. Your first step is to aim your Web browser at TiVo Central Online (http://www3.tivo.com/tivo-com/tco). You also can reach TiVo Central Online by going to TiVo's main Web site (http://www.tivo.com) and clicking on the I Have TiVo! link in the menu at the left, as shown in Figure 6-2; from there, click on the TiVo Central Online link, as shown in Figure 6-3.

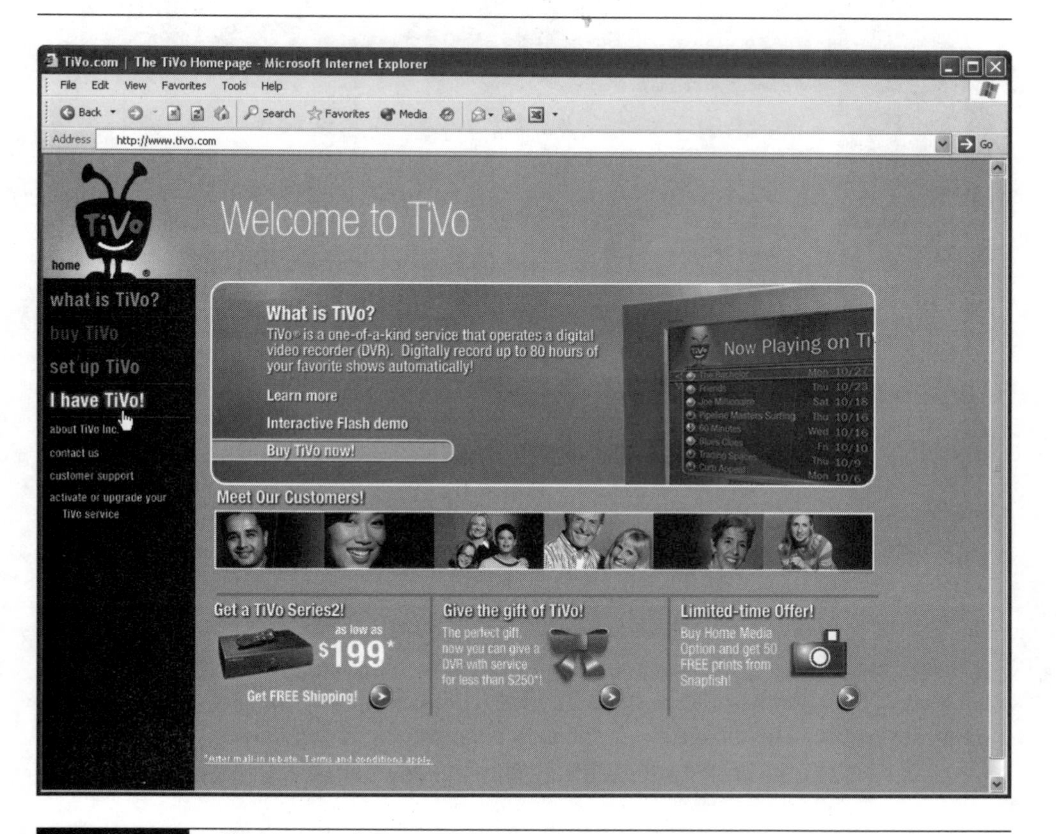

FIGURE 6-2 Clicking the I Have TiVo! link on TiVo's main web page

Select the Right TiVo

If you have more than one TiVo Series2 unit, you'll need to select the unit that should record the program you're scheduling. TiVo units that are active are listed on a pull-down menu on the search page, as shown here.

Assuming you have already registered for access to TiVo Central Online, simply enter your e-mail address and password on the TiVo Central Online Web site, and click the Sign In button. That takes you to the web page shown in Figure 6-3, where you can search for TV listings or browse for programs by channel.

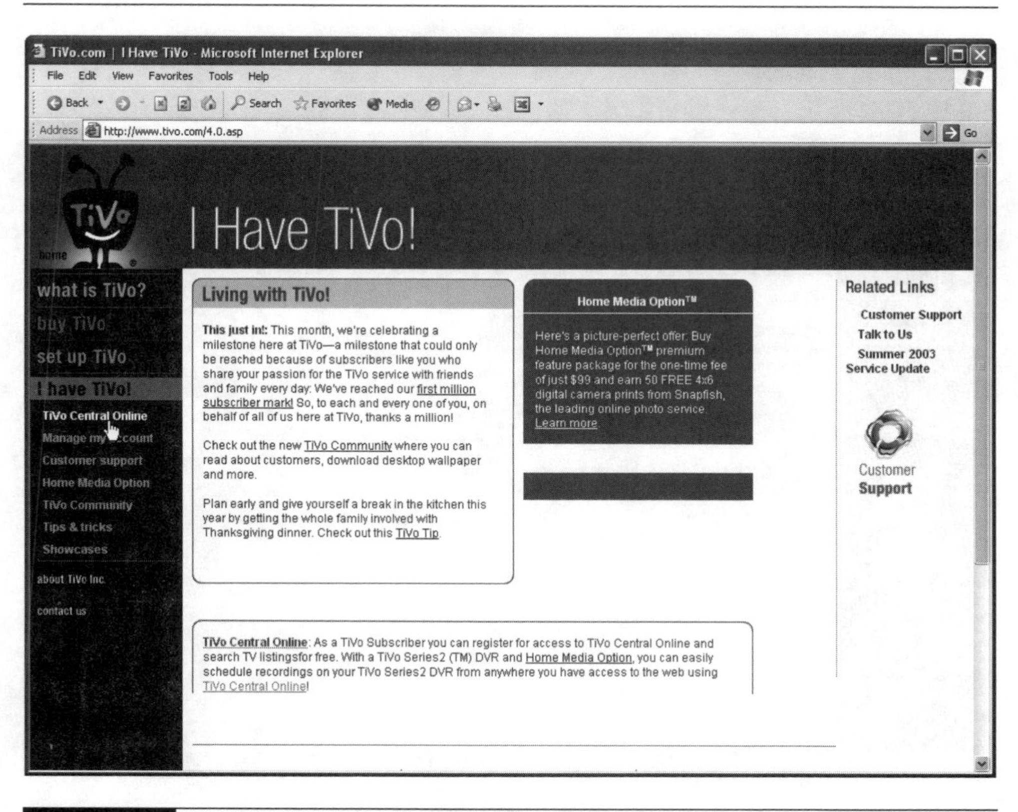

FIGURE 6-3 Choosing TiVo Central Online

There are two ways to search for program listings: using the basic search or the advanced search. You can also browse for listings by channel, which we'll discuss later in this chapter.

Doing a Basic Search

The basic search feature allows you to search for a program by title, keyword, or actor/director. You simply type in a search term, specify where to search (title, keyword, actor/director), and then click Go, as shown in Figure 6-4.

For example, you can search for the *Columbo* TV show by typing "Columbo" and selecting Title as the type of search. Or you can search for it by typing "Peter Falk" or "Falk" and selecting Actor/Director as the search type.

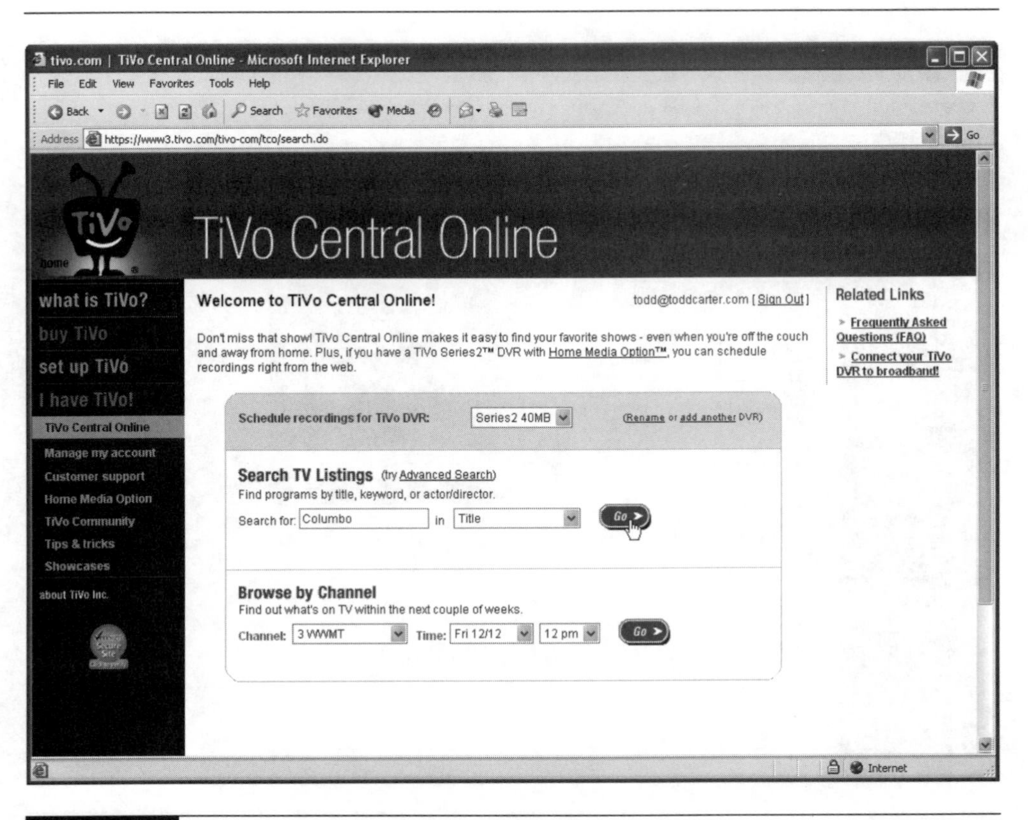

FIGURE 6-4 Search TV Listings

Whichever search type you choose, TiVo Central Online will display a Search Results page if there is a match, as shown in Figure 6-5. (If there are no matches, the Web site will notify you that it found "0 results.") Listings within the next hour are displayed in red, indicating there's likely not enough time to remotely schedule them, even if your TiVo unit is connecting to TiVo headquarters with a broadband connection. You still can try, however.

You can click on a program listing to advance to the Program Info page, where you can remotely schedule the program for recording. We'll cover that a little later in the "Scheduling TV Shows" section of this chapter.

Doing an Advanced Search

The advanced search feature includes the same search options as the basic search feature, but it also allows you to select a category and subcategory, as well as indicating

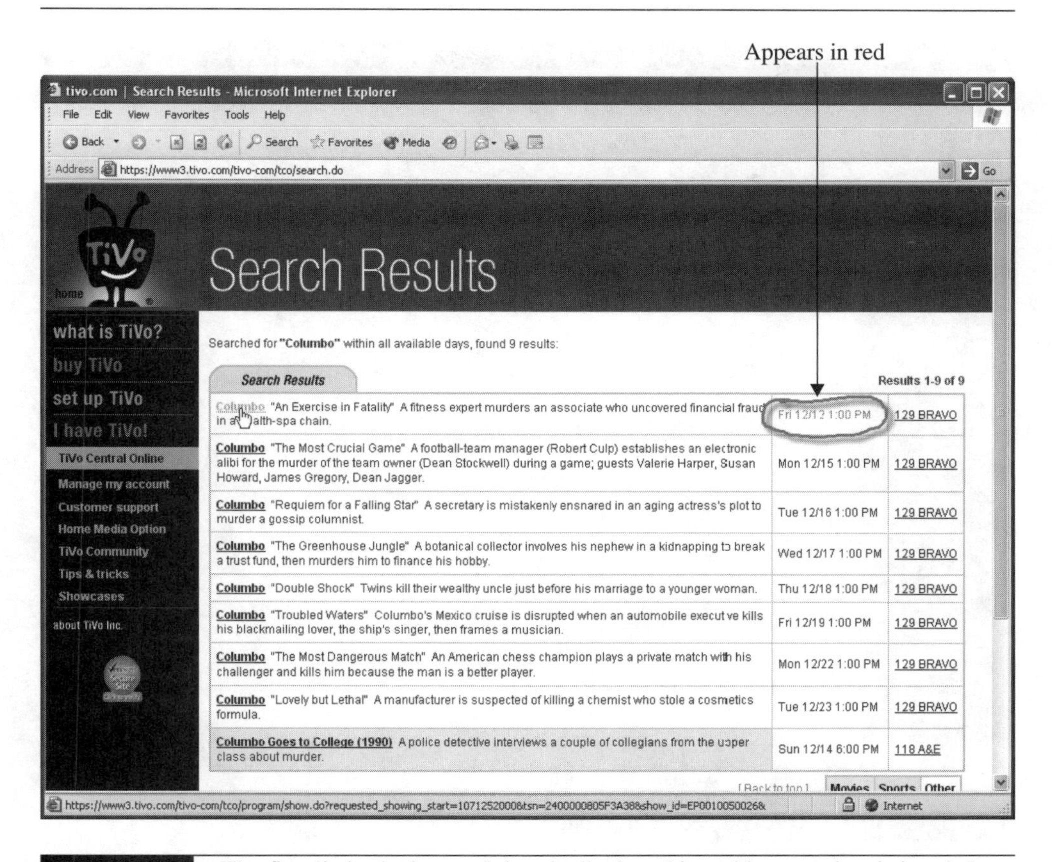

FIGURE 6-5 The first listing's date and time is displayed in red because its airtime is less than an hour away.

the time period of the program listings you want to search: all available days, the next seven days, or the next three days. The Advanced Search page is shown in Figure 6-6.

The available categories include action and adventure, arts, children, comedy, daytime, documentary, drama, educational, interests, languages, movies, mystery and suspense, news and business, sci-fi and fantasy, science and nature, sports, and talk shows. You can also choose to search all categories.

The subcategories available depend on the category you choose. For instance, in the drama category the following subcategories are available: action and adventure, animated, comedy, crime drama, docudrama, historical drama, law, medical, mystery, romance, and suspense.

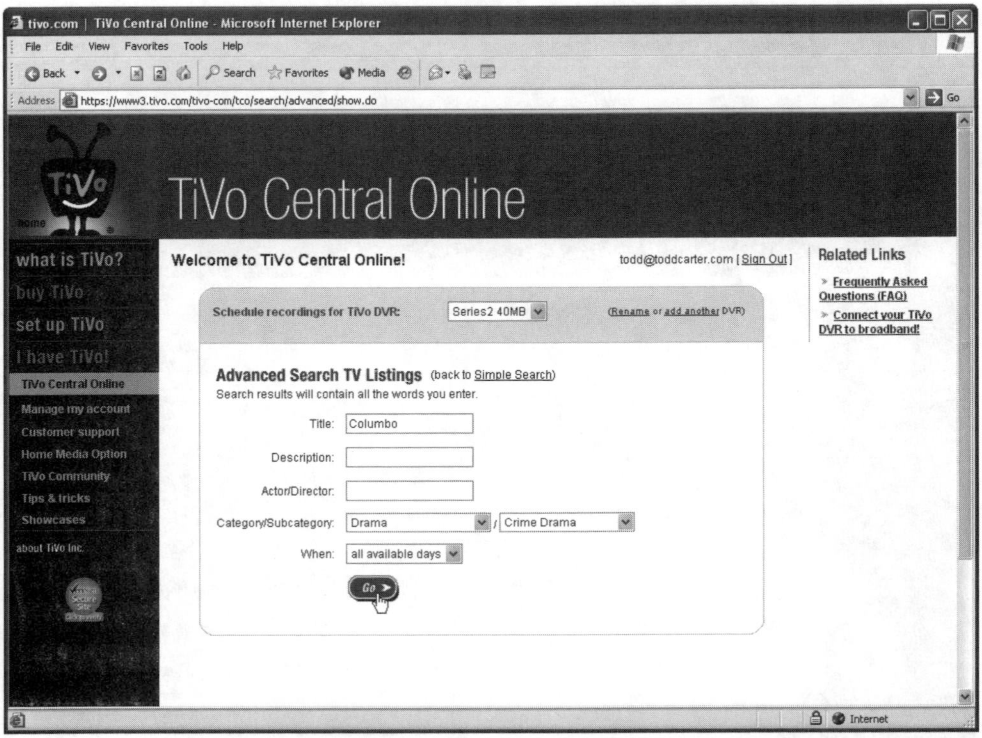

FIGURE 6-6 The Advanced Search page

Once you've entered your search criteria, click the Go button to go to the Search Results page. If you want to clear your search criteria and start over, simply click your Web browser's Refresh button.

Depending on the search, you may find that an advanced search produces fewer listings on the Search Results page, shown in Figure 6-7. Clicking a program in the search results will take you to the Program Info page.

Browsing TV Listings by Channel

If you'd prefer, you can browse for programs by channel, rather than using the search function. You can view TV listings for about 10 days or so into the future. From the

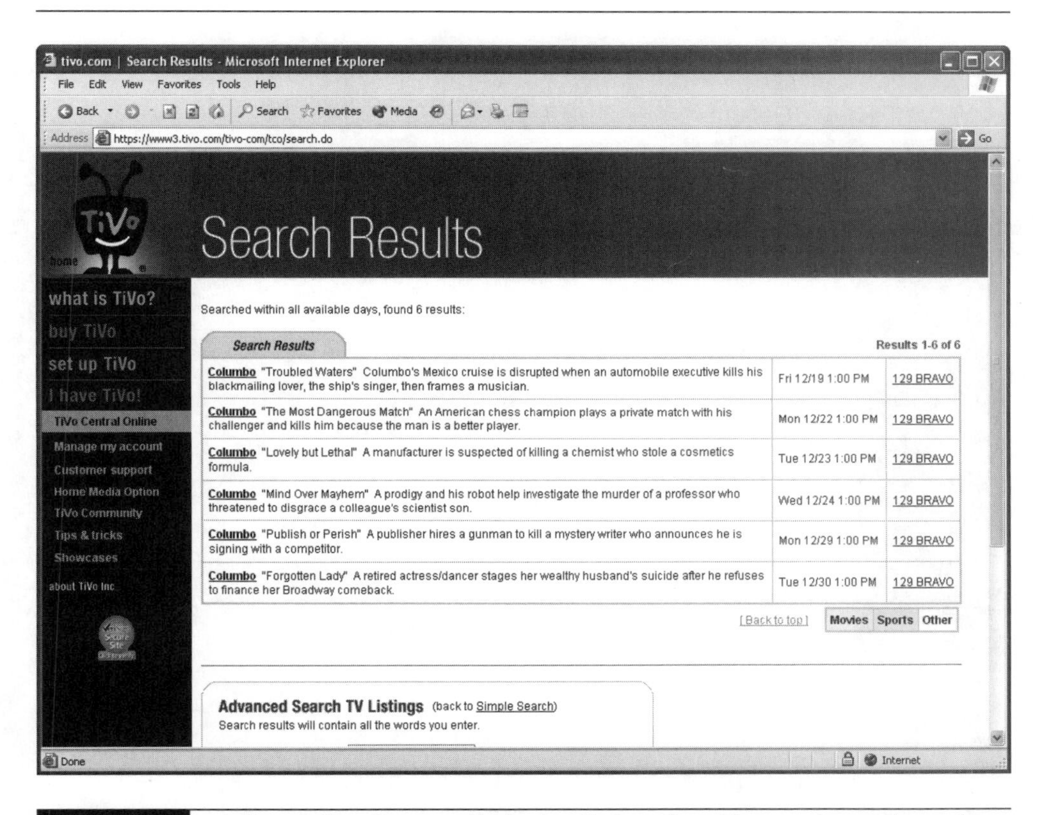

FIGURE 6-7 A Search Results page displaying listings found by an advanced search

Did you know?

Making Sure You Receive Channels

Whether you're searching the TV listings or browsing by channel, you obviously can only schedule a program for recording on a channel that you receive. While your TiVo unit knows what channels you receive (because you entered that information in your TiVo during the setup process), TiVo Central Online does not.

When searching or browsing, make sure the program you want to record is airing on a channel you actually receive. If you don't subscribe to Showtime, for instance, your TiVo cannot record a program on that channel, even though you can browse Showtime listings on TiVo Central Online.

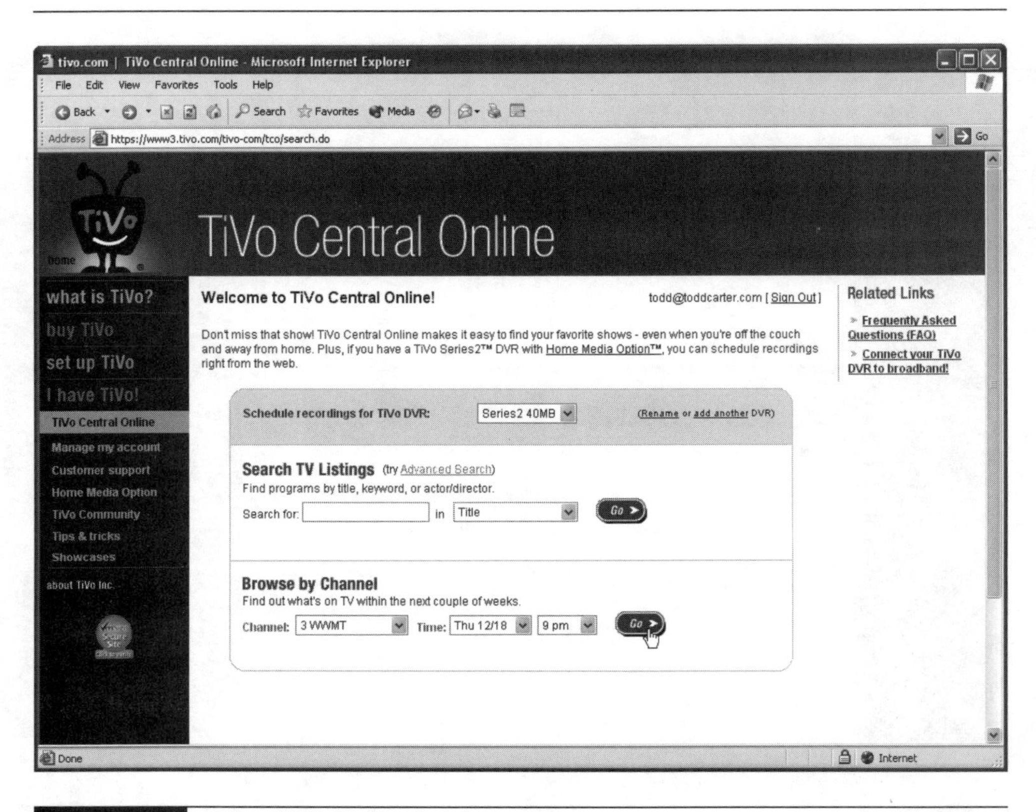

FIGURE 6-8 View TiVo listings by channel.

basic search page, simply select a channel from the pull-down menu, choose a date and time, and then click Go, as shown in Figure 6-8.

A channel schedule is displayed, listing TV programs airing in the next 24 hours or so, as shown in Figure 6-9. Again, a program's date and time will appear in red if the broadcast occurs within the next hour. You can choose a program by clicking on its title; this takes you to the Program Info page.

Scheduling TV Shows

After finding a program by searching or browsing and then selecting it, you'll find yourself on the Program Info page for that program, as shown in Figure 6-10.

FIGURE 6-9 Browse by Channel page shows a Channel Schedule

All you need to do now is schedule the program for recording. There are two basic steps in the process:

1. Click the Record This Episode button.
2. Set your recording options.

If you provide an e-mail address, you'll also receive e-mail notifications indicating the success of the recording.

Scheduling the Recording

If the program is scheduled to air within the hour, you'll see a symbol on the Program Info page (see Figure 6-10)—a red exclamation mark inside a red triangle—and a caution that reads, "May not record."

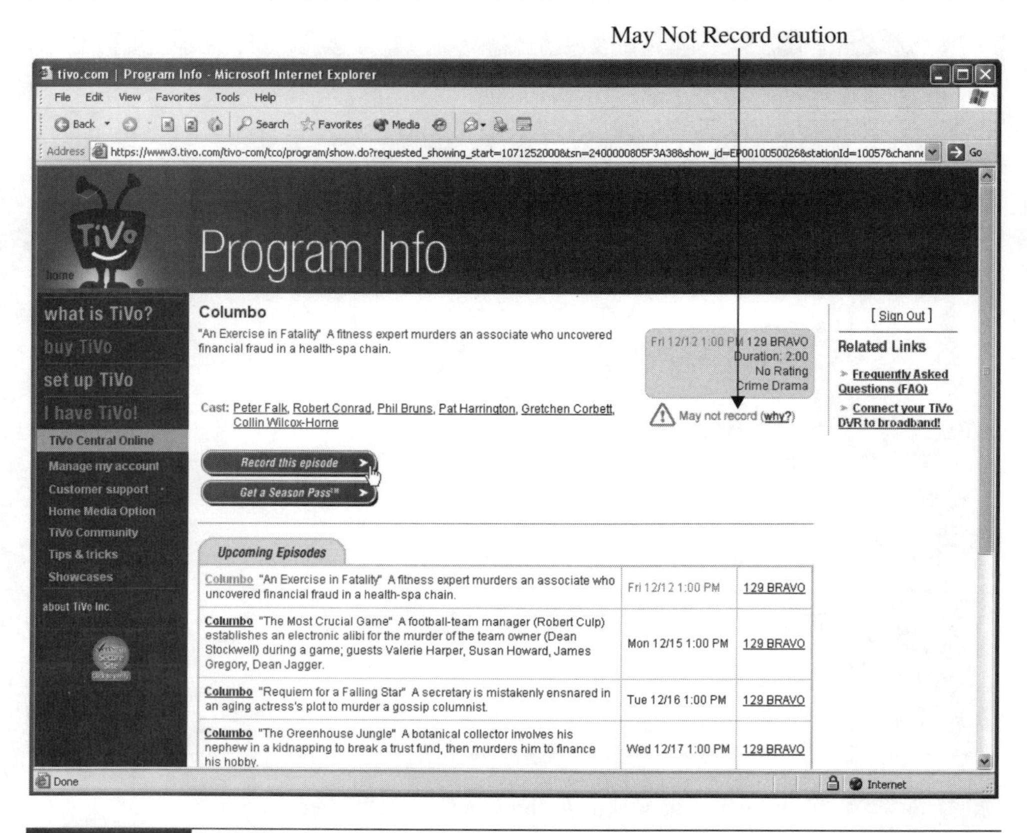

FIGURE 6-10 Program Info page for the selected "Columbo" airing

How to ... Recording Upcoming Episodes

The Program Info Web page will also display an Upcoming Episodes list, and you can remotely schedule these other episodes by clicking on the program listings and going through the same remote scheduling process for each one. Alternatively, you can get a Season Pass for a program, as discussed in the "Getting a Season Pass Remotely" section later in this chapter.

Note that the Upcoming Episodes list may include shows on channels you don't receive.

If you're satisfied that the program displayed is the one you want to record, click the Record This Episode button to instruct the TiVo server to tell your TiVo unit to schedule the show for recording.

Confirming Recording Options

On the Schedule It page, shown in Figure 6-11, you need to specify your recording options:

- Which unit you want to record the program (if you own multiple TiVo units)
- The priority of your request
- The quality of the recording

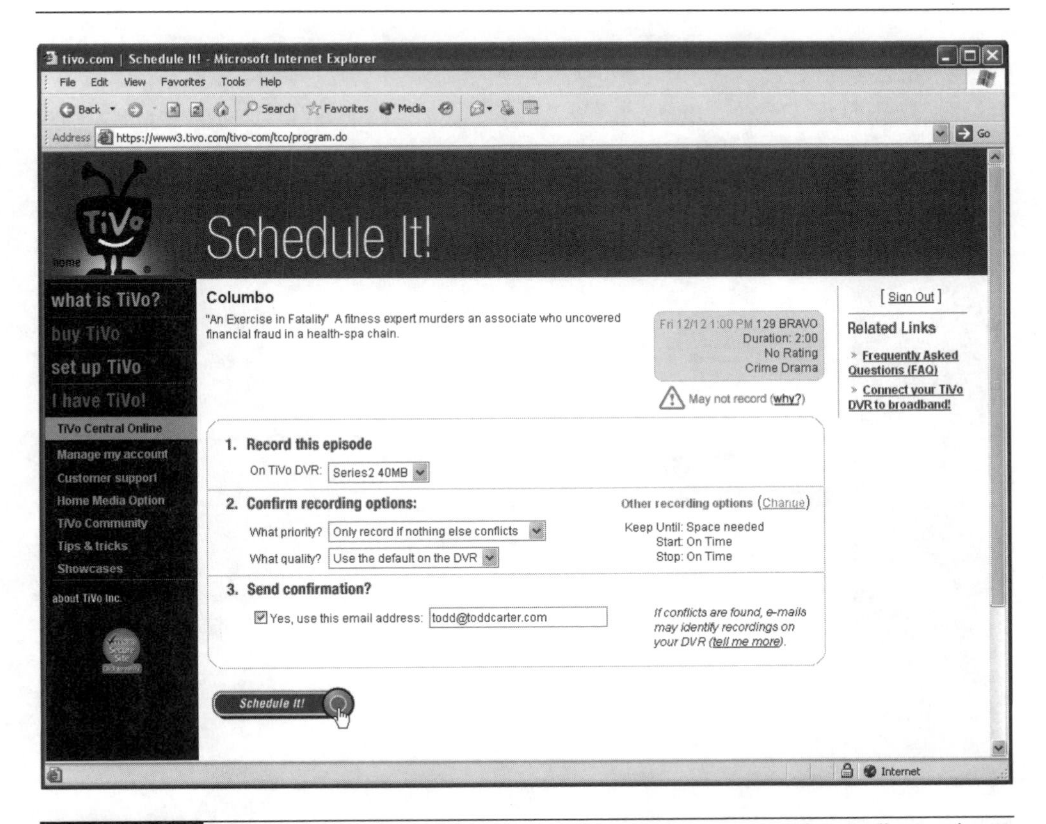

FIGURE 6-11 The Schedule It page allows you to choose from several recording options.

The first of these options is simple enough. The other two only need a little explanation.

Priority

You can choose from two priority options:

- Only record if nothing else conflicts.
- Cancel other programs if necessary.

If you absolutely want to make sure the program is recorded, choose the Cancel Other Programs if Necessary option. Otherwise, if you don't want to risk not recording another program you already scheduled, choose the first option.

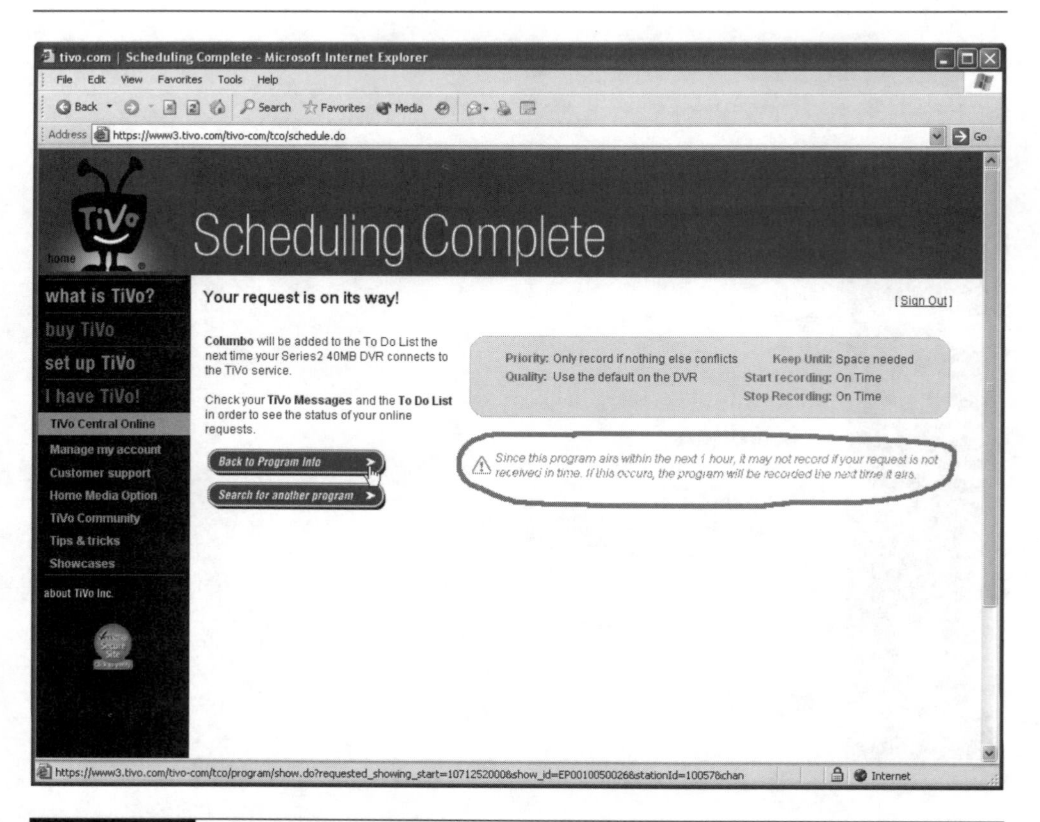

FIGURE 6-12 The Scheduling Complete page

If there is a conflict, and you provide your e-mail address on this page, TiVo will let you know that another program already was scheduled for recording and that this latest request was not fulfilled.

Quality

Remote scheduling will use your TiVo's default recording quality setting unless you change it here. Your choices are Default, Best, High, Medium, and Basic. As discussed elsewhere, if you're recording an action movie or a sports event, and you have enough space available on your unit, you likely will want to choose Best or High (or Default, if one of these settings is the default on your TiVo unit).

Other Recording Options

When remotely scheduling a recording, the default setting is to keep the program until space is needed, and to start and stop the program on time. If you want to change these settings, click the Change link beside Other Recording Options. Three pull-down menus will appear, letting you change one or more of these settings.

If you're recording a sports event, for example, you may want to select a stop time that extends up to three hours beyond the scheduled ending of the broadcast. But remember that this could create a scheduling conflict if your TiVo unit is already planning to record a program that begins sometime within that extended period.

Receiving E-mail Confirmations

You can have TiVo send you an e-mail when the scheduling request is received and after TiVo communicates with your unit. On the Schedule It page, check the Send Confirmation check box and enter a valid e-mail address.

Once you've submitted your remote scheduling request by clicking the Schedule It! button, you'll see the Scheduling Complete page. You can now return to the Program Info page or search for another program to record, as shown in Figure 6-12.

Once you have scheduled your recording, TiVo will inform you via e-mail (if you chose that option) that your request on the web site was received, as shown in Figure 6-13.

After your DVR connects to TiVo, you'll receive another e-mail that either confirms your recording request, as shown in Figure 6-13, or explains why an event cannot be recorded. For example, if a higher priority program is already scheduled to be recorded during the same time, the e-mail will explain that. The e-mail may

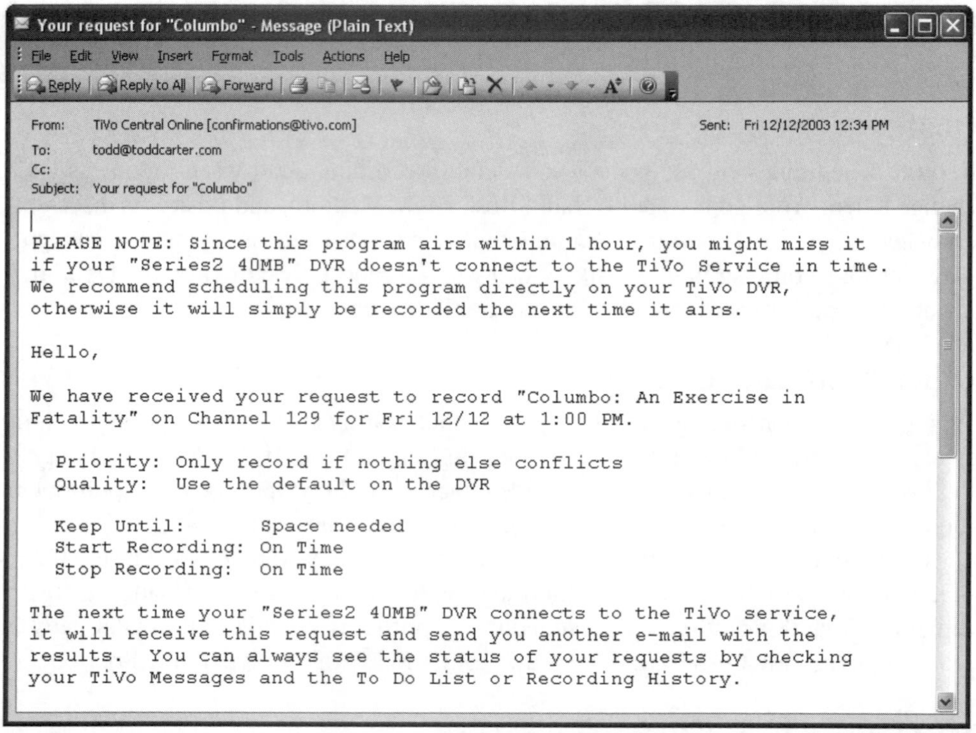

Your request for "Columbo" - Message (Plain Text)

File Edit View Insert Format Tools Actions Help

Reply | Reply to All | Forward

From: TiVo Central Online [confirmations@tivo.com] Sent: Fri 12/12/2003 12:34 PM
To: todd@toddcarter.com
Cc:
Subject: Your request for "Columbo"

```
PLEASE NOTE: Since this program airs within 1 hour, you might miss it
if your "Series2 40MB" DVR doesn't connect to the TiVo Service in time.
We recommend scheduling this program directly on your TiVo DVR,
otherwise it will simply be recorded the next time it airs.

Hello,

We have received your request to record "Columbo: An Exercise in
Fatality" on Channel 129 for Fri 12/12 at 1:00 PM.

  Priority: Only record if nothing else conflicts
  Quality:  Use the default on the DVR

  Keep Until:      Space needed
  Start Recording: On Time
  Stop Recording:  On Time

The next time your "Series2 40MB" DVR connects to the TiVo service,
it will receive this request and send you another e-mail with the
results.  You can always see the status of your requests by checking
your TiVo Messages and the To Do List or Recording History.
```

FIGURE 6-13 An e-mail confirming your request; note the cautionary message that appears when airtime is less than an hour away.

also confirm your request, while alerting you that it or another program may be recorded over sooner than expected because your unit is approaching its storage capacity, as shown in Figure 6-15.

After TiVo has communicated with your DVR and confirmed the program will be scheduled, you'll also receive a note in your TiVo Messages inbox, as shown in Figure 6-15.

Getting a Season Pass Remotely

In addition to recording a program once with the Remote Scheduling feature, you can get a Season Pass so that the program is recorded each time it airs. From the Program Info page (see Figure 6-9), click the Get a Season Pass button rather than Record This Episode.

Getting a Season Pass remotely is similar to using this feature directly on your TiVo unit:

1. Choose the TiVo unit you want to handle the Season Pass (if you own more than one unit).
2. Choose a channel from those broadcasting the program.
3. As with scheduling a single program, select the priority and quality settings for the Season Pass.
4. Click the Change link for Other Recording Options to select the maximum number of episodes you want to keep, as well as whether the Season Pass should record both repeats and first-run episodes, first-runs only, or all.
5. Check the Send Confirmation check box if you want to receive e-mail confirmations for the Season Pass.
6. Click the Schedule It! button.

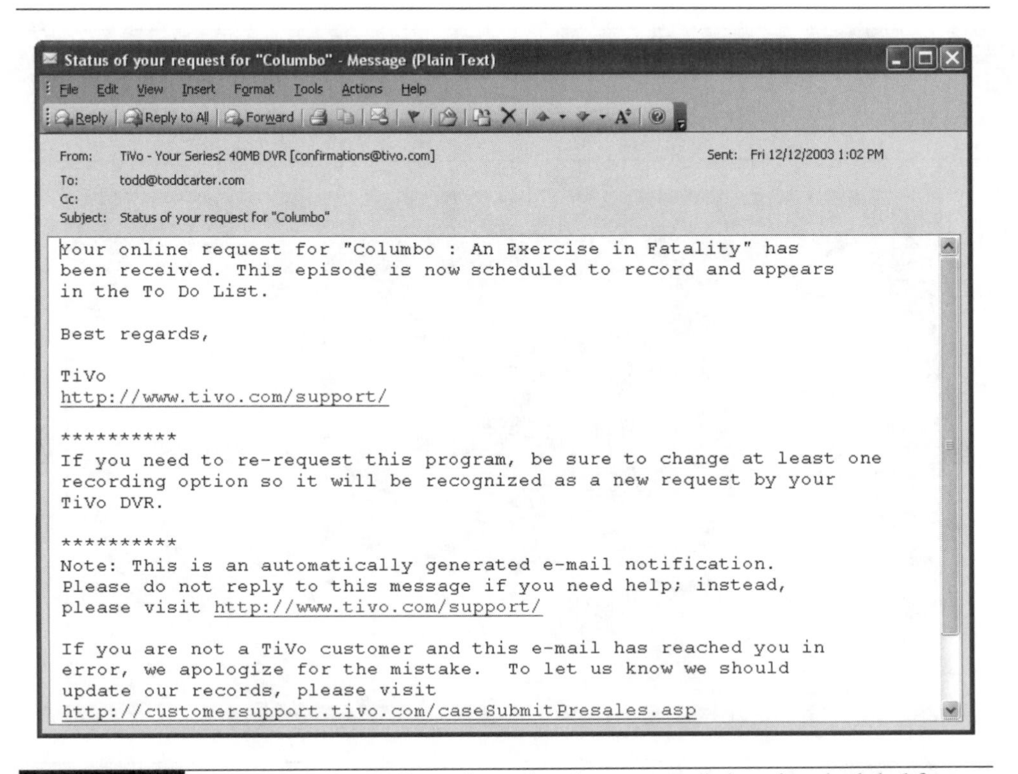

FIGURE 6-14 This second e-mail confirms that the requested show is scheduled for recording.

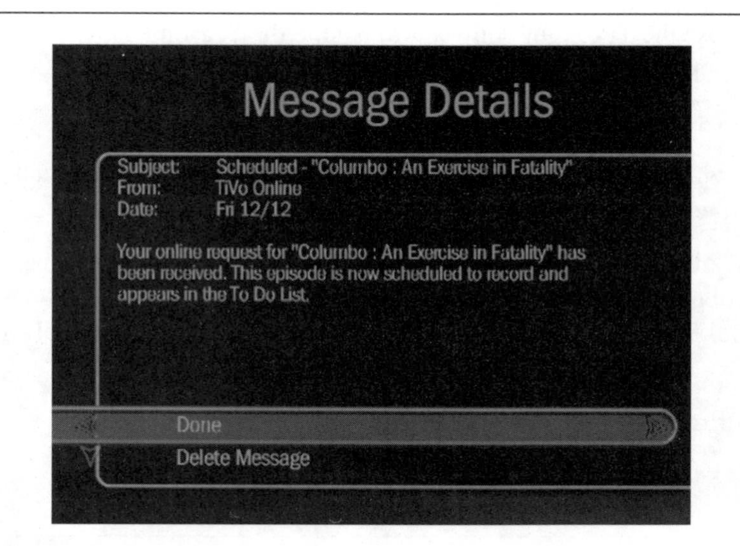

FIGURE 6-15 This note will appear in your TiVo Messages inbox.

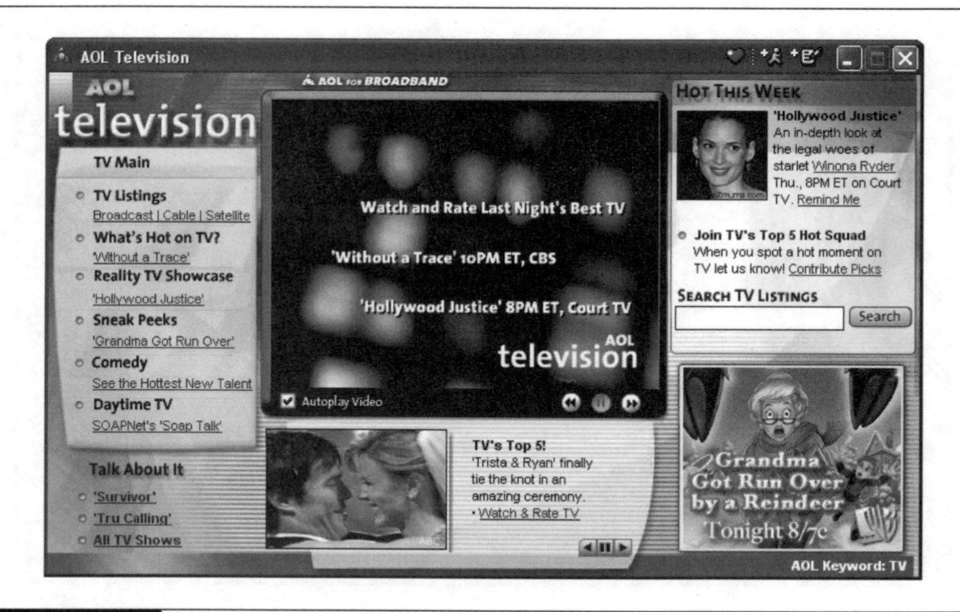

FIGURE 6-16 AOL Television window

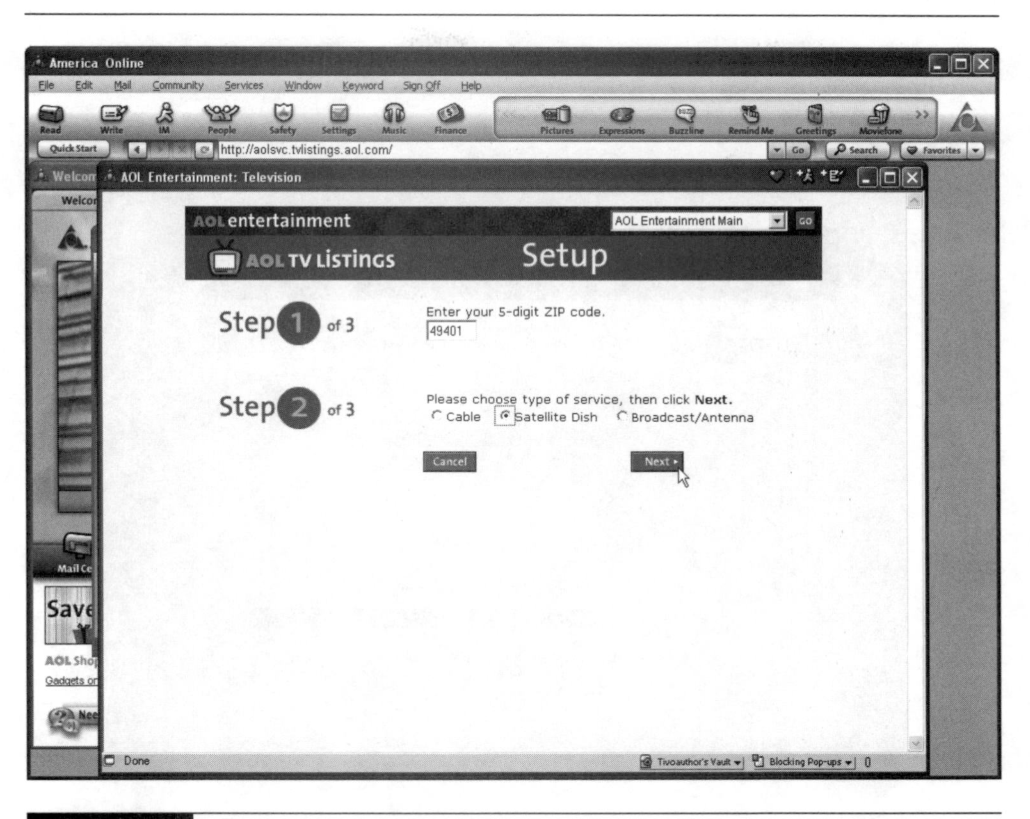

FIGURE 6-17 Enter your zip code and choose your type of service.

If you requested e-mail confirmations, you'll receive e-mails from TiVo like the ones you get when you schedule a single episode.

Using AOL's Customized Version of Remote Scheduling

AOL and TiVo have partnered to create a customized version of Remote Scheduling for AOL members.

The AOL process for scheduling a program is similar to the process on TiVo's web site, described in the previous sections. Here's how to schedule a recording:

1. Using AOL's Keyword feature, type "TV" into the search form and click Go. You'll see the AOL Television window, as shown in Figure 6-16.

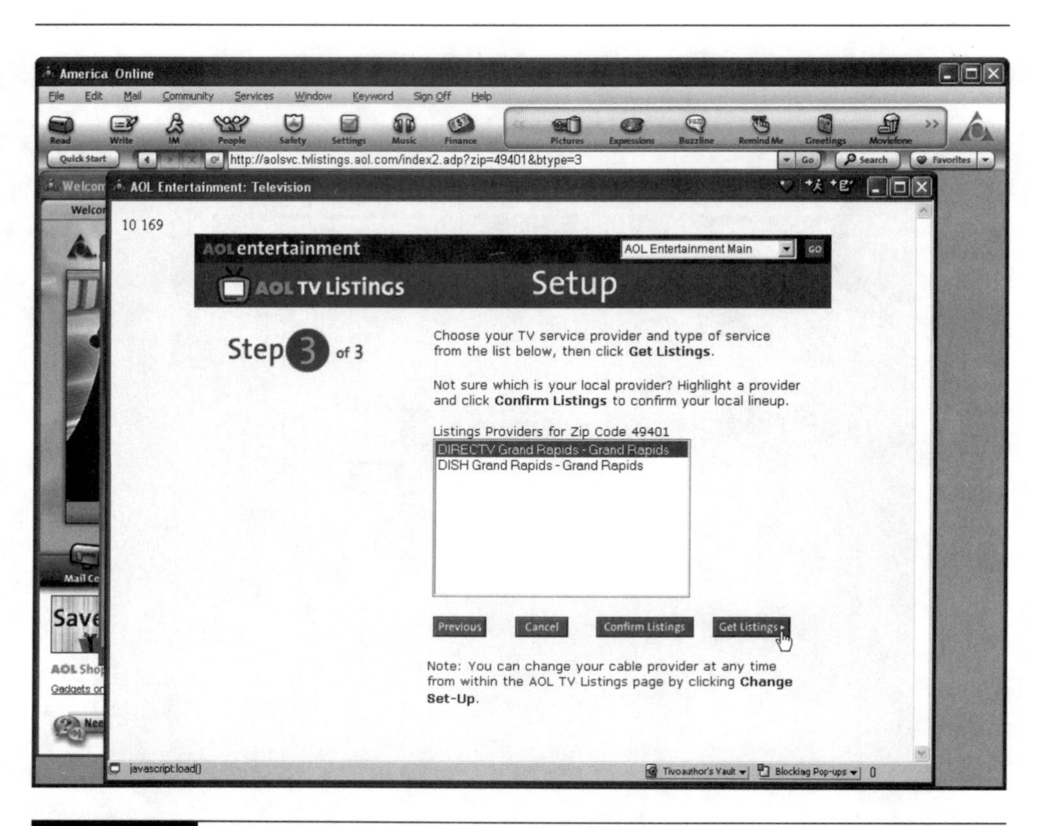

FIGURE 6-18 Select your TV service provider.

2. Click on the TV Listings link in the menu at the left.
3. If you haven't already viewed these listings, you will see a TV listings setup screen (if you have viewed the listings before, go ahead to step 5). Enter your ZIP code and choose the type of service you have, as shown in Figure 6-17. Click Next.
4. In the final setup window, select your specific TV service provider, as shown in Figure 6-18. When you're done, click the Get Listings button.
5. AOL's TV listings will appear, showing current and upcoming programming (see Figure 6-19).

FIGURE 6-19 You can select a program to record from the AOL TV Listings screen.

6. You can change the day, time, and category of the TV listings displayed by using the pull-down menus at the top and clicking the View button when you're done.
7. To make a TiVo scheduling request, click on the title of the program you want to record. This will take you to the specific program information for that show, as shown in Figure 6-20.
8. Click the Record on My TiVo DVR link to schedule the program for recording.
9. If it's your first time using the AOL TiVo feature, you'll see an Update DVR List window identifying the TiVos you can select for recording, as shown in Figure 6-21. If you want to add additional DVRs, click the Add DVRs button. Otherwise, click the close button (the X) in the upper-right corner to close the window.

Record on my TiVo DVR.

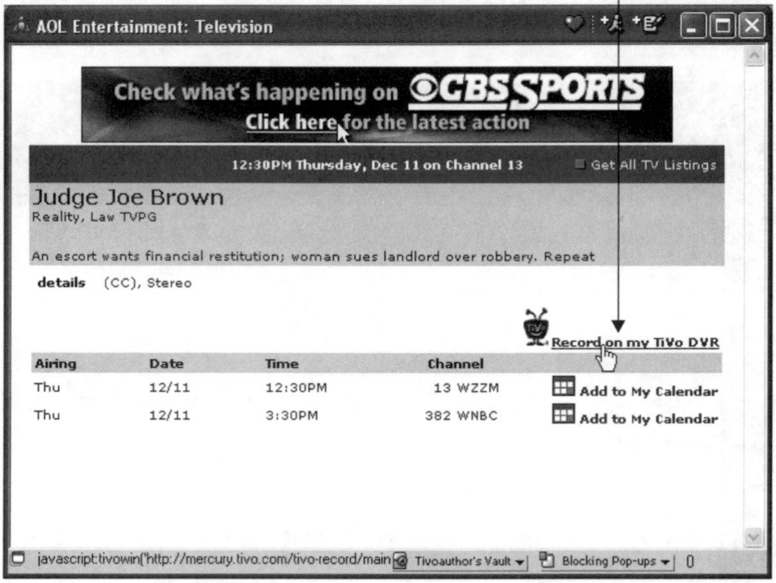

FIGURE 6-20 Make a TiVo scheduling request.

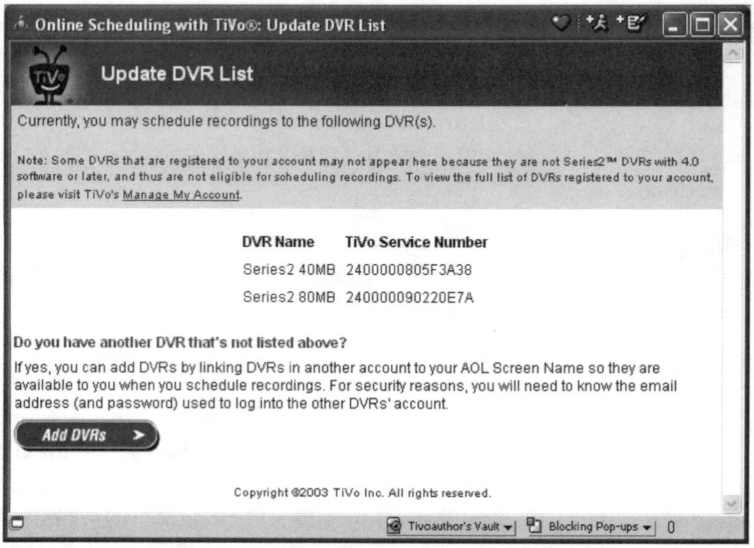

FIGURE 6-21 You can add TiVos that are permitted to record programs scheduled through AOL.

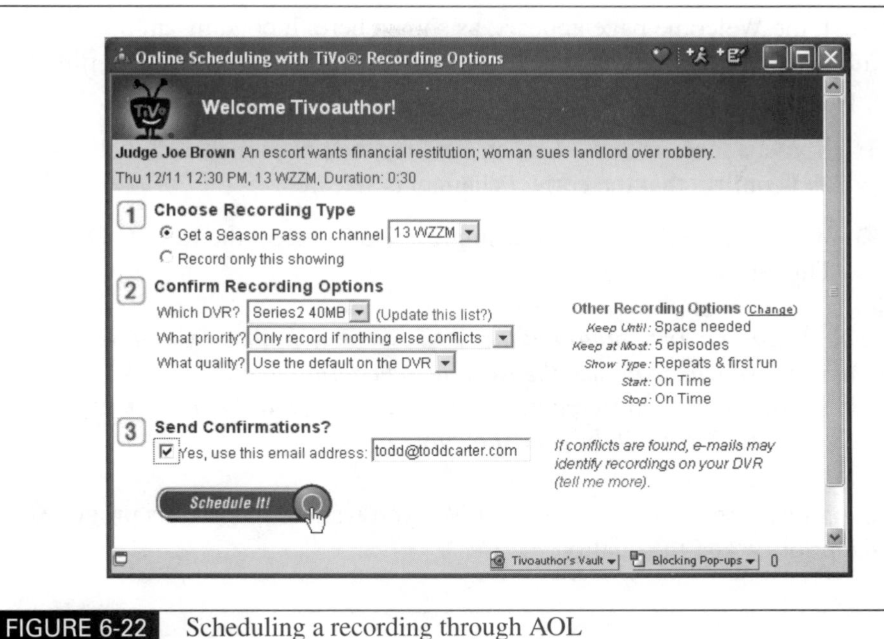

FIGURE 6-22 Scheduling a recording through AOL

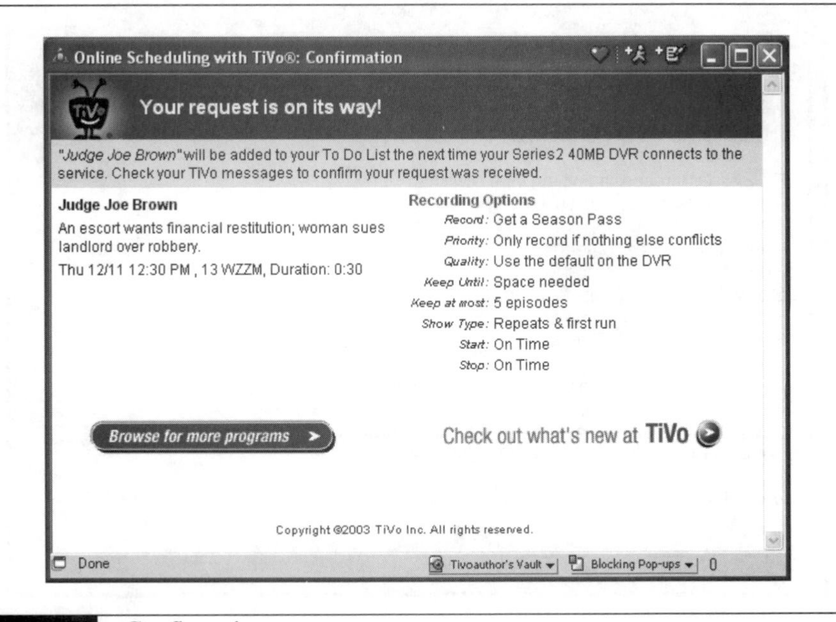

FIGURE 6-23 Confirmation screen

10. Next, the Welcome page appears, as shown here. It contains all the information necessary to schedule a recording. First, choose a recording type:

- If you want to create a Season Pass for the program, select that option and confirm that the correct channel is selected.
- If you only want to record a single episode, select the Record Only This Showing option.

TIP *The Recording Type option in the AOL system defaults to Get a Season Pass, unlike when you use the Remote Scheduling feature on TiVo's web site. If you only want to record a single episode, you'll need to change the selection to Record Only This Showing.*

11. In part two, select which of your TiVos you want to record this program, the priority, and the quality.

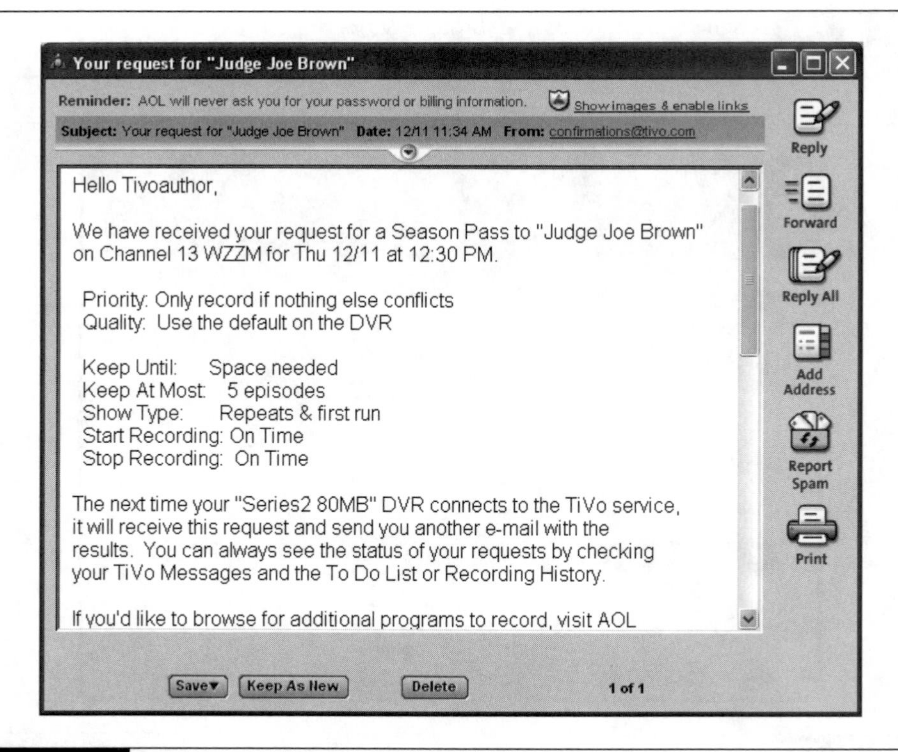

FIGURE 6-24 Confirmation received via e-mail

Message Details

Subject: Scheduled - "Judge Joe Brown"
From: TiVo Online
Date: Thu 12/11

Your online request for a Season Pass to "Judge Joe Brown" has been received. All upcoming episodes of this Season Pass have been scheduled and now appear in the To Do List.

Will record all episodes of:
Judge Joe Brown on channel 13 WZZM

Done
Delete Message
Press CHAN UP/DOWN to scroll

FIGURE 6-25 Confirmation received via your DVR

12. Click the Change link beside Other Recording Options to change how long the recording will be kept, how many episodes should be saved at any one time (if you're requesting a Season Pass), and whether your TiVo should record only first-run episodes or also include repeats. You can also adjust the recording's start and stop times.

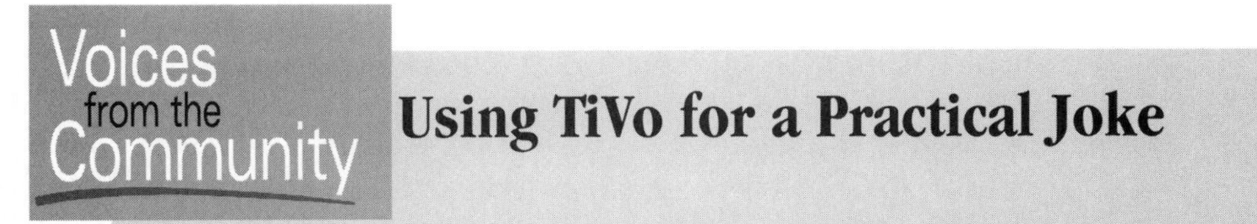

Using TiVo for a Practical Joke

Peter Shankman has found one use for TiVo's remote-scheduling feature: "I travel constantly on business. I have a TiVo that connects wirelessly to a broadband Internet connection. When I'm on the road, I'll occasionally tell TiVo, via Web access, to start taping some Cinemax movies that come on at midnight. The only reason I do this is that it freaks the heck out of my girlfriend. She is sitting at home, watching Letterman, and then a message from TiVo appears on her TV screen: "'TiVo needs to change the channel to record a movie at 12:15 A.M.'"

13. If you want to receive an e-mail confirmation of the scheduled recording, check the Send Confirmations check box and enter your e-mail address.
14. Click the Schedule It! button when you've finished. You'll now see the confirmation screen, shown in Figure 6-23.

If you requested confirmations via e-mail, you'll receive a message in your AOL inbox similar to the one shown in Figure 6-24.

After your DVR checks in with the TiVo server and schedules the request, you'll also see a confirmation in your TiVo Messages inbox on your DVR, as shown in Figure 6-25.

Chapter 7 Transferring Programs Between TiVos

How to...

- Set Up Two TiVos on One Home Network
- Use the Multi-Room Viewing Feature to Transfer Programs Between TiVos
- Select Programs for Transfer
- Estimate Transfer Times
- Play a Program Recorded on Another TiVo
- Directly Connect Two TiVos
- Troubleshoot the Multi-Room Viewing Feature

TiVo's Multi-Room Viewing Feature

In addition to viewing photos on your TV, listening to music, and remotely scheduling the TV programs you want your TiVo to record, the Home Media Features (HMF) includes a Multi-Room Viewing feature. It lets you record a program on one TiVo Series2 unit and view it on another Series2 TiVo.

If you have a TiVo in your bedroom and another one in your family room, you can record different programs on each unit and then transfer the programs between units. You can freely move the units to different rooms, as long as all of the TiVos using the Multi-Room Viewing feature can access the same home network.

And while it's called the Multi-Room Viewing feature, the TiVos can be located in the same room. In fact, you don't even need a home network if the TiVos are located close together, as you can use a special cable to connect the two units.

Whether your TiVo units are in separate rooms or stacked atop one another, the Multi-Room Viewing feature provides you with more storage space for your recorded programs, because you're combining the capacity of these separate units. And although you're combining the capacity, you're keeping each TiVo's program suggestions, thumb ratings, and other personal preferences separate.

Setting Up Two TiVos for Multi-Room Viewing

You can only transfer programs between Series2 units. If you have an original TiVo model and the newer Series2 unit, you cannot transfer programs between them because Multi-Room Viewing is available only on Series2 models.

If you have two Series2 TiVos, all you need to do is activate the Multi-Room Viewing option on TiVo's Web site. To do this, follow these steps:

1. Sign in to TiVo Central Online, and click the Manage My Account option in the menu at the left.
2. You'll see a list of options at the left of the Web page, including DVR Preferences. Click this option, which displays the DVR Preferences page.
3. Under Multi-Room Viewing Permissions, shown in Figure 7-1, check the boxes next to the TiVos you want to activate for Multi-Room Viewing. In this example, there are two TiVo units.
4. Click the Save Preferences button when you've made your changes.

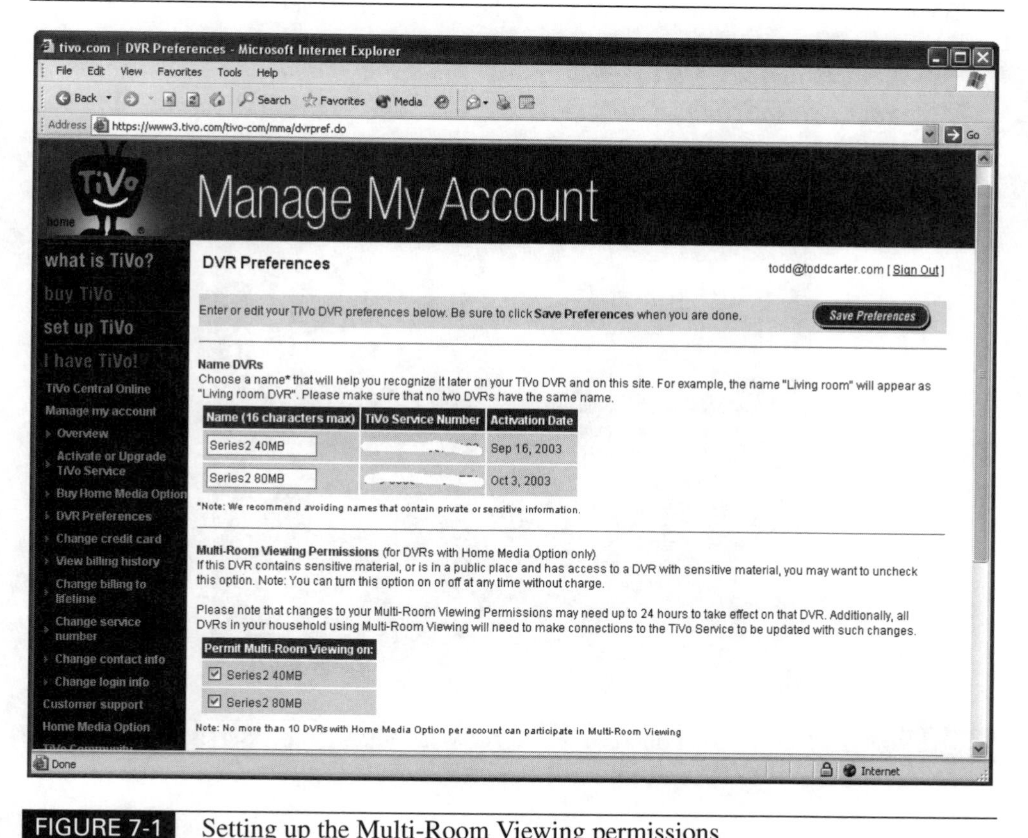

FIGURE 7-1 Setting up the Multi-Room Viewing permissions

Transferring Programs Between Two TiVos

Now that your TiVos are set up for multi-room viewing, we'll discuss what you need to do to transfer programs from another TiVo. And we'll talk a little bit about how you can estimate the time it'll take to transfer these programs, based on things like the quality of the recording and the type of network you're using.

Setting Up Programs for Transfer

If everything is working correctly, your local TiVo should list all the units it can transfer programs to or from under the Now Playing on TiVo screen. For example, the Now Playing on TiVo screen in Figure 7-2 shows one remote TiVo unit, which I've named "Series2 80MB."

If you have multiple TiVos, each of them should be listed on the Now Playing on TiVo screen by the name you assigned when registering them on TiVo's Web site.

TIP *A newly-activated TiVo appears under the Now Playing on TiVo screens of other networked units after its next connection to the TiVo service. You can wait for its scheduled connection or force a manual connection.*

FIGURE 7-2 Now Playing on TiVo screen shows a remote TiVo, "Series2 80MB"

If there are recorded programs on your TiVo, the Multi-Room Viewing units will be listed at the bottom of the Now Playing on TiVo list. In this example, there are no programs recorded on the local unit. You can pick the remote TiVo unit you want to transfer a program from by simply selecting it, like you would a locally recorded program.

After pressing the Select button, your local TiVo displays a list of programs recorded on the remote TiVo, as shown in Figure 7-3. Any of these programs can be transferred to a remote TiVo using the Multi-Room Viewing feature.

In this example, we're going to select the second program listed, an episode of *Columbo*. Scroll down to the program you want to transfer, and press Select. This takes you to the Program screen listing details for the remote recording. You can select either Watch on This TV or Don't Do Anything. For this example, we'll select the first option, which causes a Please Wait… This Might Take a Minute message to show up in the middle of the screen.

NOTE *Selecting Don't Do Anything simply returns you to the previous screen, which lists the programs stored on that remote TiVo.*

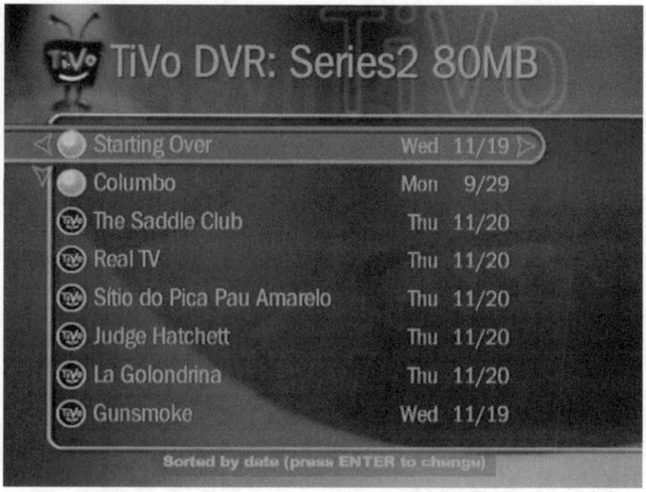

FIGURE 7-3 The remote TiVo displays its recorded programs, any of which can be transferred.

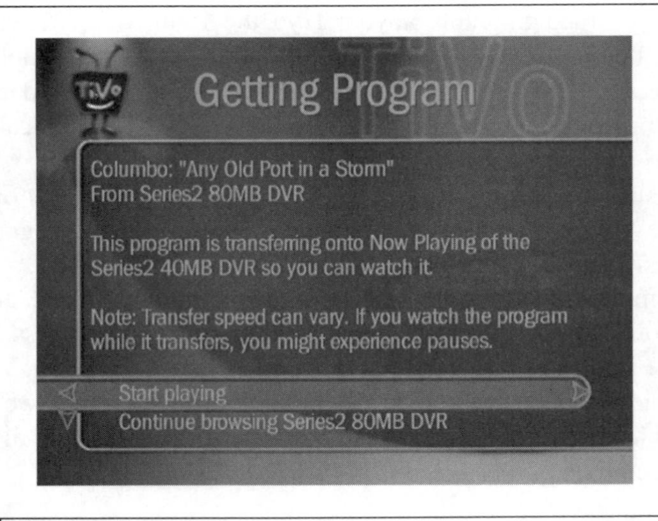

FIGURE 7-4 The remote TiVo displays the Getting Program screen.

If this is the only transfer taking place, you should see the Getting Program screen, as shown in Figure 7-4. It has two options: Start Playing and Continue Browsing. You may want to continue browsing through the listings of the remote TiVo if you have a slow network, or if the program you selected for transfer is recorded at a high

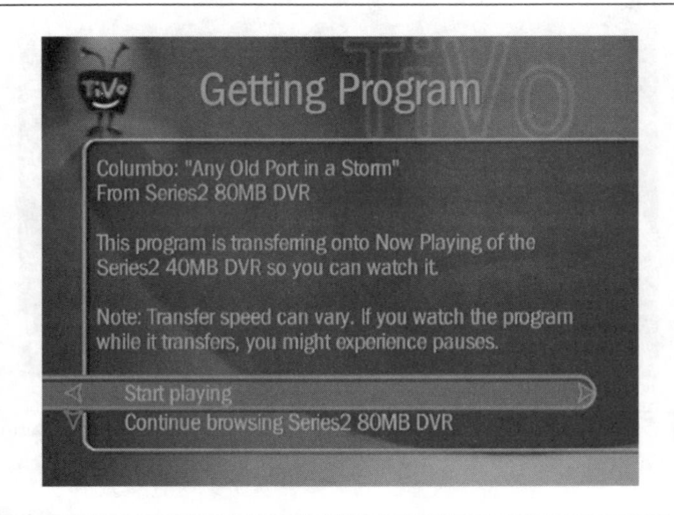

FIGURE 7-5 Choose Start Playing if you want to watch the program as it streams.

Did you know?

It's OK to send and receive at same time

You cannot transfer a program while the remote TiVo is recording it, but you can otherwise both send *and* receive programs to and from a single TiVo at the same time. However, TiVo recommends that you not simultaneously transfer recordings from one DVR to two or more receiving DVRs.

quality level. Otherwise you can start watching the program as it continues to be transferred.

If you have already selected other programs for transfer, you'll see a Will Get Later screen, as shown in Figure 7-6. This informs you that the program you've selected has been added to TiVo's To Do list. The program will be transferred to your local TiVo after the previously requested transfers are completed. You can press Select and then return later to see if the program has transferred to your local TiVo.

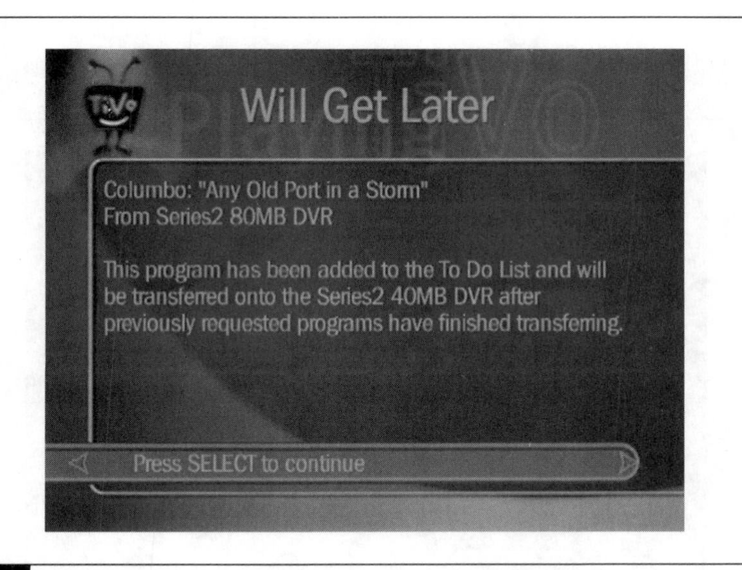

FIGURE 7-6 TiVo stacks your transfer requests in its To Do List

Estimating Transfer Times for Programs

The amount of time it takes to transfer a program between two TiVos depends on several factors:

- The length of the program you're transferring
- The level of quality it was recorded at
- The type of network (wired Ethernet, wireless, or a combination of the two)

The higher the bandwidth available on your network, the closer to "real time" you can watch a streaming program. If there is little bandwidth available, either because you're operating a relatively slow wireless network or because other data is moving over the network at the same time, you'll find that the program you're streaming will pause when your viewing catches up to the transferred part of the program (see Figure 7-7).

The program takes longer to transfer than it does to view, so you'll need to wait for the receiving TiVo to collect more data before it can continue the program on your TV. You'll need to press Play to continue viewing the program—possibly repeating this many times depending on the speed of your network. Of course, you can just wait until a program is done copying and then watch it.

FIGURE 7-7 TiVo pauses a program during a remote transfer.

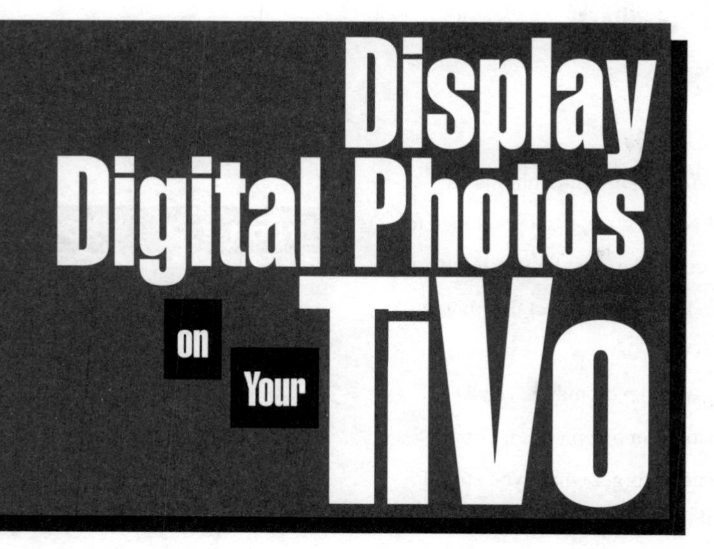

How to...

- Share Your Photo Library with Your TiVo
- Navigate Your Photo Directory on Your TiVo
- Display Your Photo Library on Your TiVo
- Remove Photos After Publishing Them
- Change Slide Show Options on Your TiVo
- Use Adobe Photoshop Album's TiVo Feature
- Use Picasa's TiVo Feature

Are you tired of your family and friends crowding around a small computer monitor when showing them the latest photos of your kids? Wouldn't Junior's smile look better on a big TV screen instead?

That's exactly what TiVo's Home Media Features does. By running the TiVo Desktop software on your PC, you can publish photos to your TiVo Series2 unit, which in turn lets you view those digital photos on your TV set. Once the photos are published, which makes them available for viewing on your TiVo, you can create slide shows of your photos. Just start the automatic slide show, and everyone can sit back and view the photos, one by one, on your TV screen.

Third-party software providers are also adding features so you can quickly publish photos to your DVR. If you already use one of these software packages to organize photos on your PC, watching snapshots on your TV just got easier.

Getting Started with Digital Photos

Before you can view your digital photos on your TiVo and your TV set, it's a good idea to organize them on your computer. Just like with music files, we'll use the TiVo Desktop to select the photos we want to publish to TiVo.

If you're really organized, you may have all of your digital photos stored in one place on your computer. If you're prone to forget where you stored your files, collecting all of your digital photos together is a good idea.

You can use any folder to store photos—one you make yourself or one created by the operating system, such as the My Pictures subfolder in the My Documents folder on Windows machines. If it's more convenient, you can use several folders, although this will initially take longer to navigate to them and select the photos you want your TiVo to display. You can also use specialized photo-management software to ease this organization task (Adobe Photoshop Album and Picasa are discussed later in this section).

When you view your photos on your TV, the photos are resized to a resolution of about 640×480 so that they fit on the TV screen. Lower-resolution photos, however, are not resized to look larger, so they may not fill your entire TV screen. You can use photo editing software to enlarge these smaller photos, but their quality may suffer in the process.

NOTE

TiVo currently works with six different graphic formats: BMP, TIFF, DIB, GIF, JPEG, and PNG. That gives you quite a bit of flexibility.

Despite the size of your TV screen, you should think of the photos you view through TiVo as being like the smaller snapshots you pick up from the drugstore. Figuratively speaking, TiVo can't provide the same amount of detail as those nice, big enlargements you frame and proudly display on your fireplace mantle.

TIP

You can download a 30-page guide on digital photography from TiVo's Web site. It includes information about choosing a digital camera and tips for taking quality photos. You can find it at http://www.tivo.com/pdfs/whitepaper/wp_digitalphotography.pdf. You'll need Adobe's free Acrobat Reader to view the document on your computer.

Displaying Photos with TiVo Desktop

Once your photos are organized—or you at least know where they reside on your hard drive—you can publish the ones you'd like to view on your TV. The TiVo Desktop software makes it easy to select the photos and publish them. If you haven't already installed the TiVo Desktop software, see Chapter 4 for information on how to do so.

NOTE

Chapters 4 and 5, which cover the publishing of music files to your TiVo, include more specific information about using the TiVo Desktop software.

Publishing Photos to Your TiVo

If you've got TiVo Desktop installed, follow these steps to publish your photos:

1. Launch TiVo Desktop, either from the Start menu or by right-clicking on the TiVo Publisher icon in your task tray and selecting Open TiVo Publisher.
2. Select the Photos tab, as shown in Figure 1.

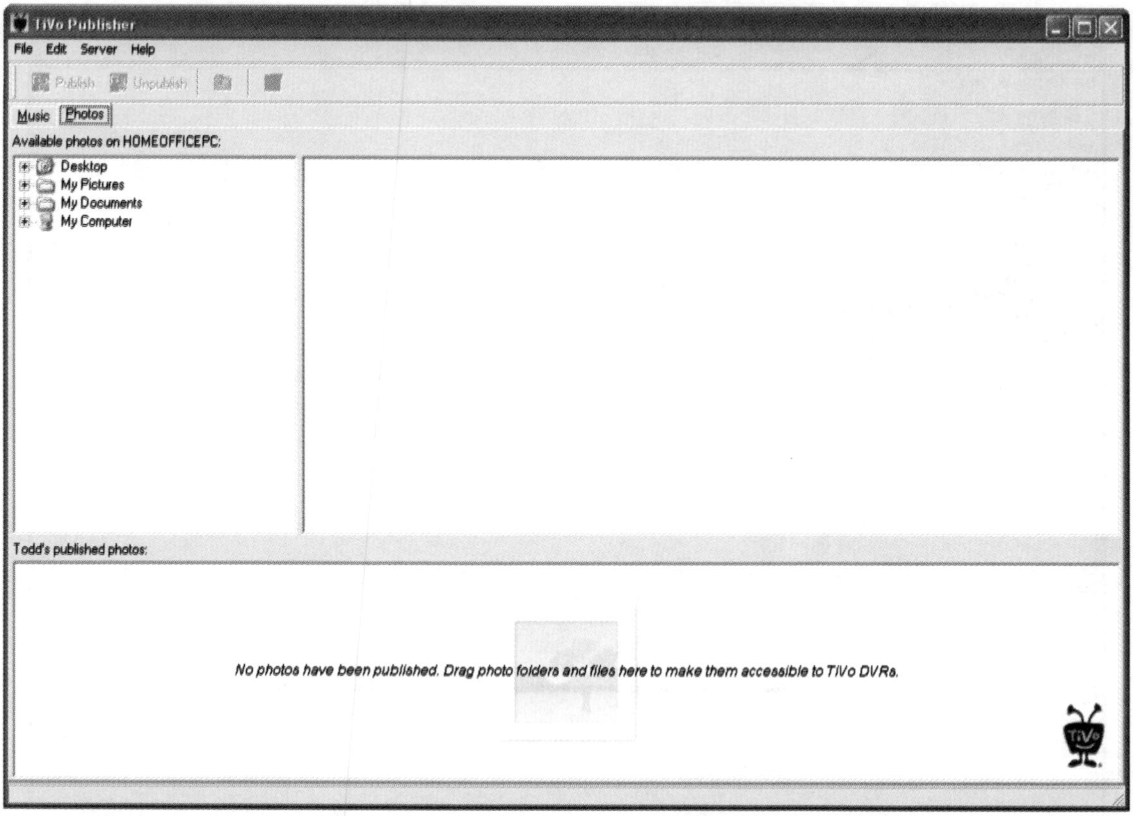

Figure 1. TiVo Desktop's photo publishing screen

3. Use the left pane of TiVo Desktop to navigate to the folder or subfolder containing the photos you want to publish. You can select a folder in the navigation pane to display its contents in the main window, as shown in Figure 2.
4. To publish the photos, do one of the following:
   - To publish a whole folder, right-click on it and select Publish, as shown in Figure 3.
   - To publish only specific photos from a folder, select those photos in the main window and drag and drop them to the bottom pane, as shown in Figure 4.
   - To publish an individual photo, right-click on the photo and select Publish.

When you publish a folder or specific photos, they will appear in the published photos pane toward the bottom of the TiVo Desktop screen as shown in Figure 5.

Removing Photos After Publishing Them

After publishing your photos in TiVo Desktop, you may decide you no longer want them viewable on your TiVo. You can easily remove, or "unpublish," your photos with the TiVo Desktop application.

To unpublish photos listed in the published photos pane of TiVo Desktop, select one or more files and then right-click on them. In the pop-up menu, select Unpublish as shown in Figure 6.

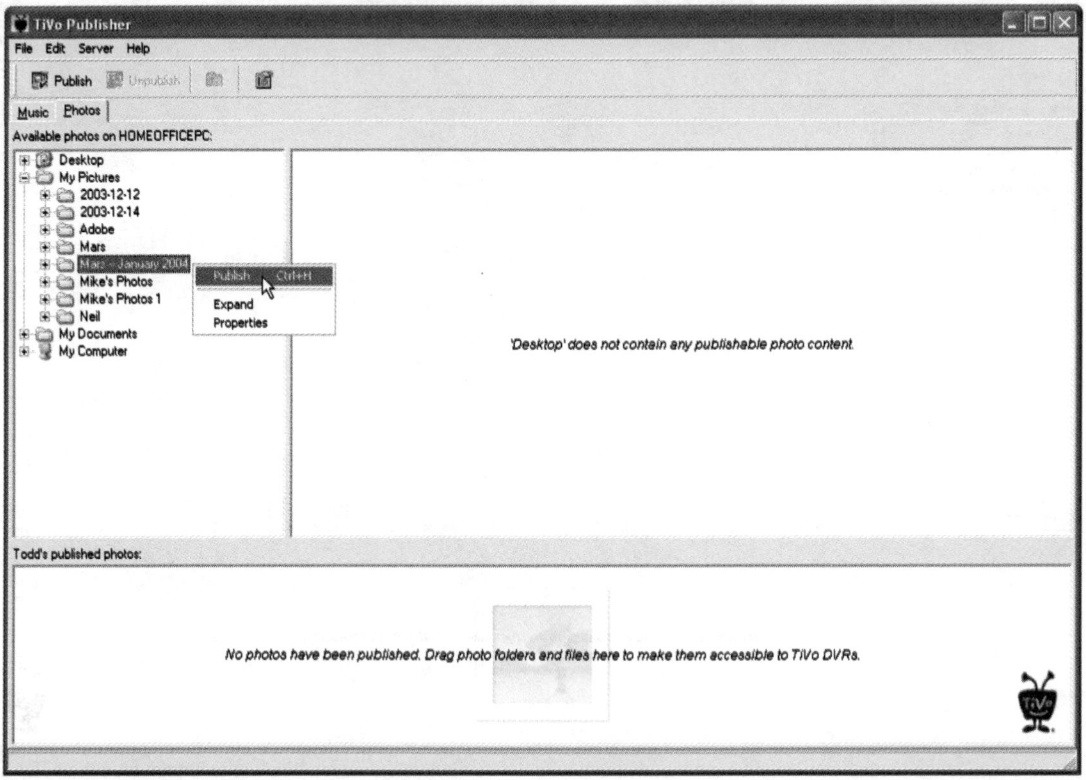

Figure 2. TiVo Desktop displays photos

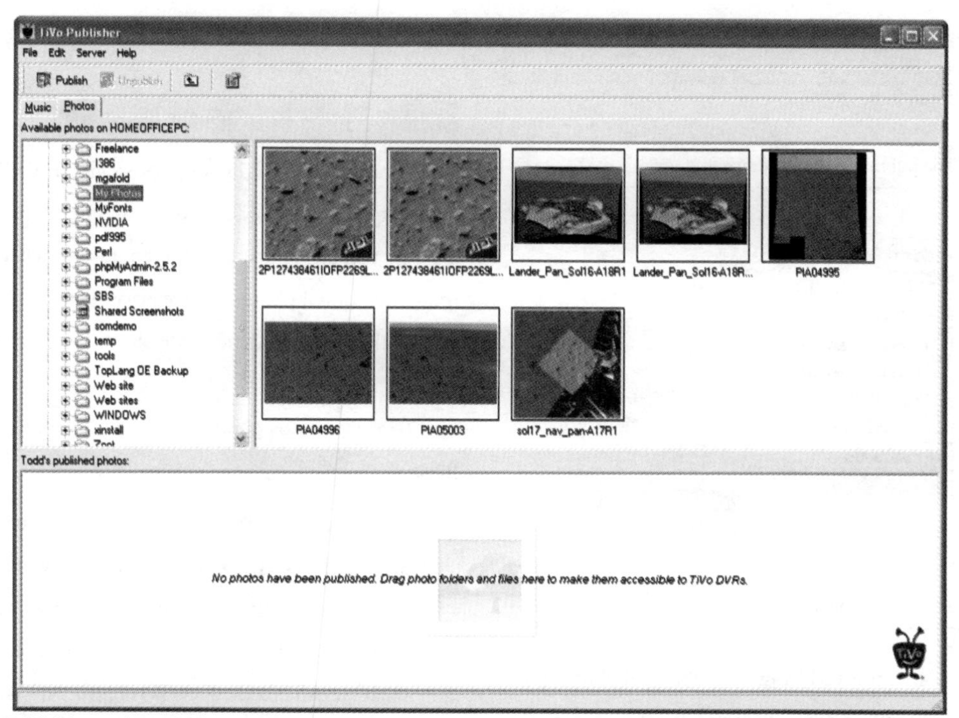

Figure 3. Publishing a photo folder

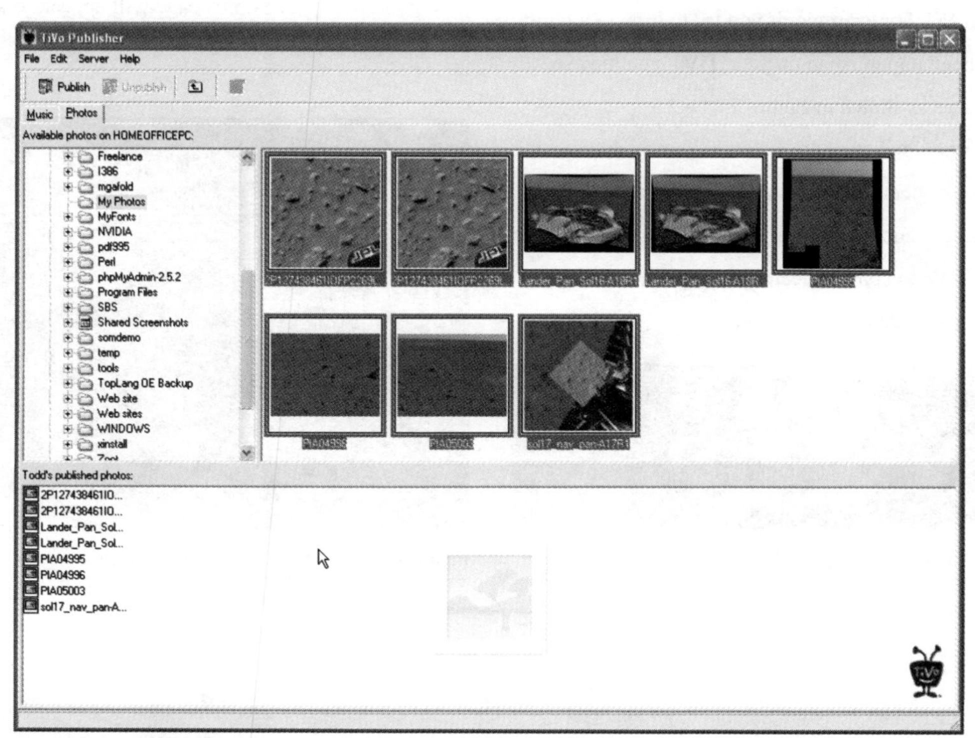

Figure 4. Publishing selected photos

Figure 5. The published photos pane

Figure 6. Select Unpublish.

The photos are removed from the published photos pane of TiVo Desktop, but some information about the photos may remain on your TiVo until the next communication with your PC. For example, if you try to browse a photo that has been unpublished on your TiVo, you may see a black square as shown in Figure 7.

The unpublished photo will continue to show on the TiVo Photos menu until you move back to the Music & Photos screen and then return to the Photos screen. This prompts TiVo to communicate with the TiVo Desktop software and update its file listings.

Figure 7. Browsing an unpublished photo

Setting Photo Options on Your TiVo

Once you've published your photos with TiVo Desktop, it's pretty easy to browse through these pictures on your TV set. You can adjust how the photos are listed and rotate individual photos if necessary. This is all in preparation for displaying slide shows of your photos on your TiVo.

To view photos published to your TiVo, follow these steps:

1. Select Music & Photos from the TiVo Central screen. If TiVo Desktop is running on your PC, you'll see a listing for your photo server, as shown in Figure 8. If you don't see a listing, check to make sure TiVo Desktop is running on your PC.
2. Highlight the menu item for your photo server and press Select. A Photos page will appear, as shown in Figure 9.

Figure 8. The TiVo Music & Photos screen

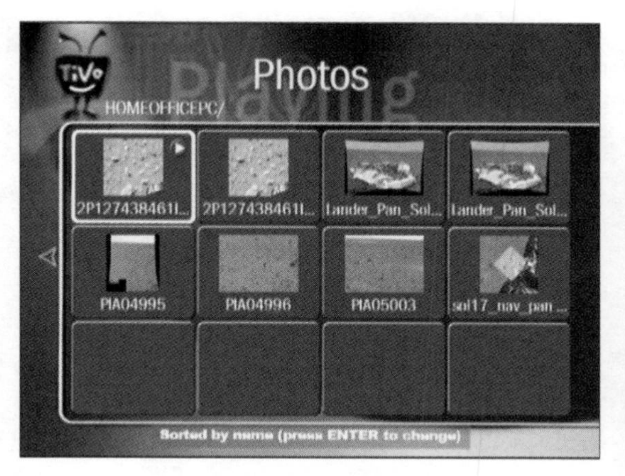

Figure 9. The Photos screen shows thumbnail images of your photos.

3. Navigate among the small, "thumbnail" images using the arrow keys on your remote control. If you want to view a photo, highlight it and press the Select button. You'll see a screen similar to the one shown in Figure 10.

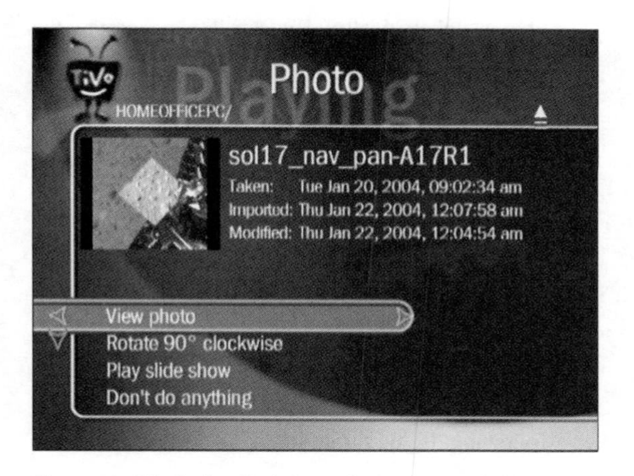

Figure 10. Displaying file information and controls for each photo

4. Select the View Photo option to display the image on your TV set, as shown in Figure 11.

Figure 11. Displaying a full-screen photo on your TV screen

Rotating a Photo

Sometimes photos are scanned upside down or on their sides. Instead of tipping over your TV set, TiVo has a simpler solution: simply rotate your photos. TiVo lets you rotate your photos 90 degrees, and by repeating this, you rotate the photo any way you want. Figure 12 shows the photo after it has been rotated 90 degrees clockwise.

To rotate a photo, follow these steps:

1. In the Photos screen, highlight a photo and press Select.
2. On the Photo screen, highlight the Rotate 90° Clockwise option, and press the right arrow key on your remote control as many times as you want to rotate your photo. If the photo is upside-down, for example, press this option twice. If you change

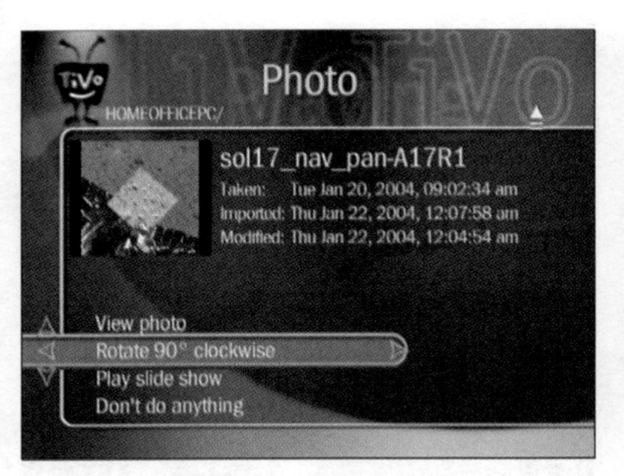

Figure 12. In an upside-down world, the rotating feature is a lifesaver.

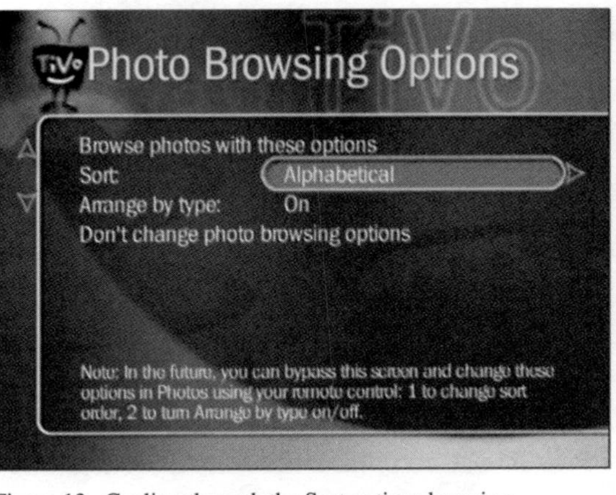

Figure 13. Cycling through the Sort options by using the right arrow key

your mind and want to keep the photo as it was, press the right arrow until the image returns to its original position.

Other HMF-enabled TiVos on your network will show the photo in its new orientation following their next contact with the TiVo Desktop software.

Adjusting Photo Browser Options

You can change the way that photos are organized by TiVo to match your personal preferences. There are two options you can change: the sort order, and whether you want the photos arranged by type.

To change these options, follow these steps:

1. Go to the Photos screen, and press the Enter button on your TiVo remote control. You'll see the Photo Browsing Options page, as shown in Figure 13.
2. To change the sort order, scroll to the Sort line, and use the right arrow button on the remote control to select either Alphabetically (sorts both folders and photos alphabetically), By Date Taken (sorts by the date the photo was taken), or By Date Modified (sorts by the date the photo was last modified on your PC).
3. To separate the folders from the individual photos, scroll to the Arrange by Type line and use the right arrow button to turn it On or Off. If you turn this option on, TiVo displays folders at the top of the list on the Photos page, with individual photos following. If you turn it off, TiVo sorts the folders and photos together, in the sort order you chose in the previous step. These options affect the order that TiVo displays photos during slide shows.

TIP

There's a shortcut for changing these options. From the Photos screen, press the 1 button on your remote control to change the sort order, and press 2 to turn Arrange by Type on or off. If you don't like what you see, you can press the same number button again to return to the previous option.

4. When you're finished setting the options, scroll up to Browse Photos with These Options, and press the Select button (or the left arrow key).

Playing Photos on Your TiVo

Once your photos are published and available to your TiVo, you can play a "slide show" that automatically displays pictures one after another on your TV set.

To begin a slide show, simply highlight your photo server on the Music & Photos screen, and press the Play button on your remote control. The slide show begins, as shown in Figure 14.

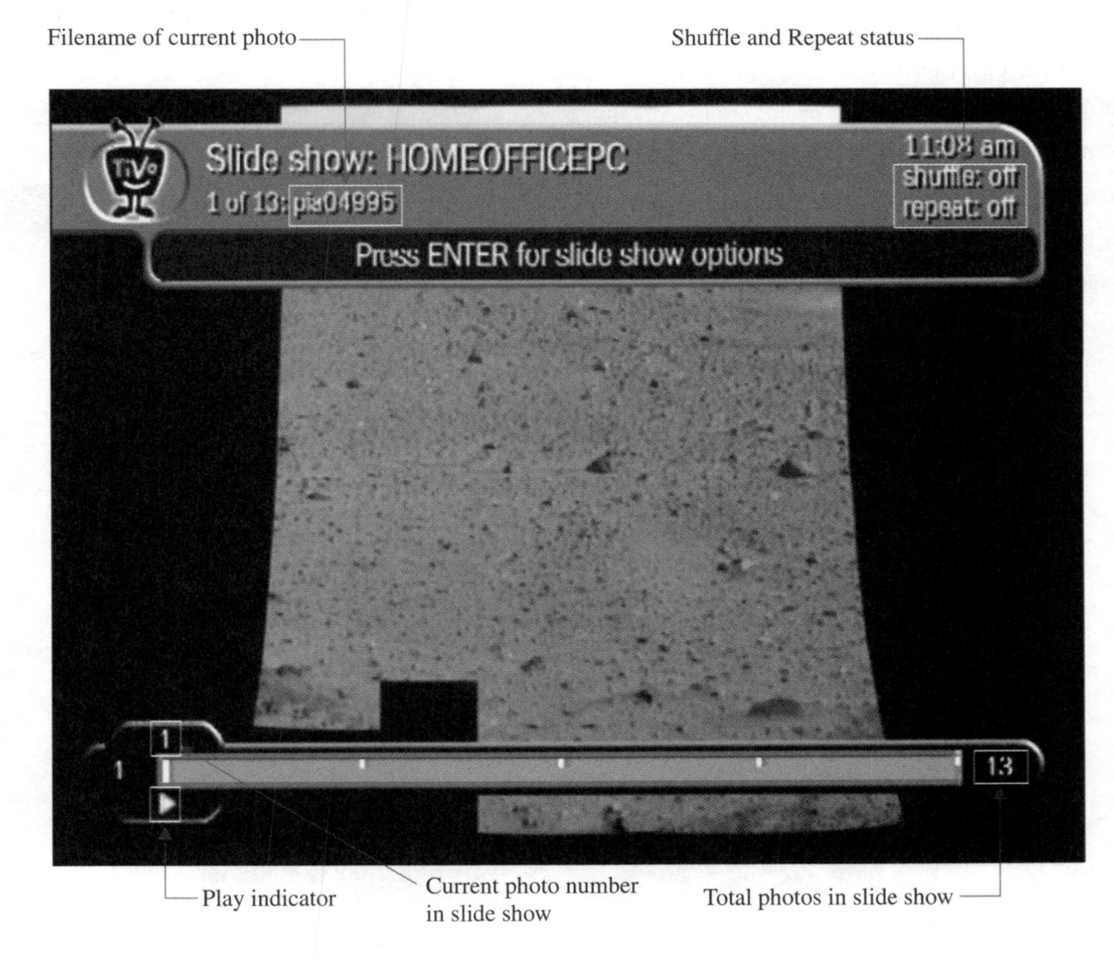

Figure 14. TiVo displays photos one after another during a slide show.

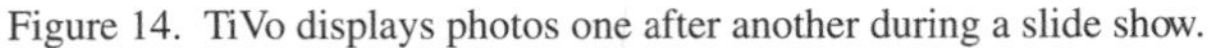

You can use many of the same controls for viewing a slide show as you use for watching TV. In addition to Play, you can use Pause, Fast Forward, and Rewind. You can press the Advance button to skip ahead to the next tick mark on the onscreen status bar while fast forwarding or rewinding. The Instant Reply button lets you jump back one photo.

TiVo headquarters sometimes sends sample photos and music files to your DVR. The photos appear on the Music & Photos screen as Photos on TiVo Online. At the time of writing, there were two TiVo Online subfolders: AroundTheWorld, containing travel photos with superimposed TiVo logos, and Mars Photos, which is separate from the Mars-related photos used as examples in this chapter.

You can view these sample photos just like you would your own. You may want to use them as examples to experiment with until your own photos are published and ready for viewing. Many of the photos are interesting viewing in their own right.

Setting Slide Show Options

You can adjust the amount of time each photo stays on your TV screen before the next picture is displayed. You can also use the Shuffle option to randomly order the photos, and the Repeat option to replay the slide show. If you have subfolders in your photo folders, you can tell TiVo whether or not to include the photos in these subfolders in the slide show.

To change your slide show options, follow these steps:

1. On the Photos screen, scroll down to Play Slide Show, and press the right arrow so that Options appears.
2. Press the right arrow to highlight Options, and press Select or the right arrow. The Slide Show Options screen will be displayed, as shown in Figure 15.

NOTE

All of these options can be changed at any time.

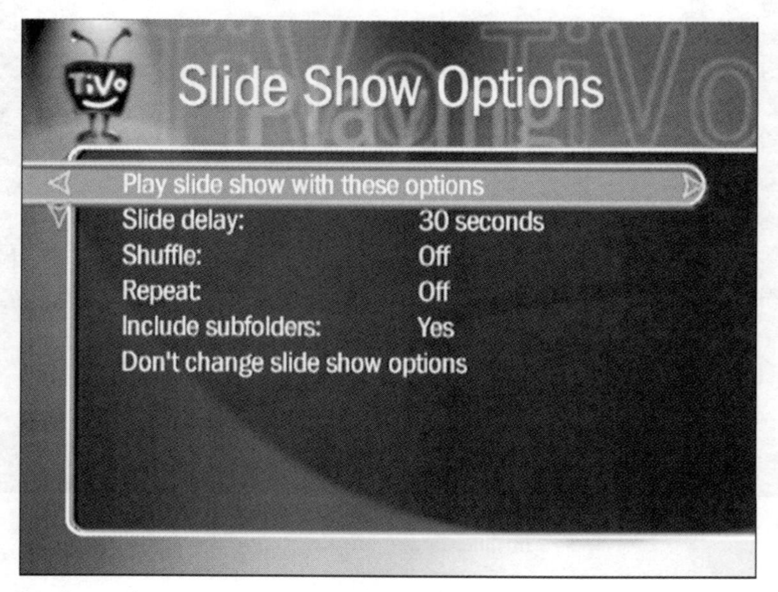

Figure 15. Changing slide show options

3. To change the delay between photos, use your remote control to cycle through the options: 5, 10, 15, or 30 seconds, or 1, 2, or 5 minutes, as shown in Figure 16. If you're showing photos to others, you may want to choose 15 or 30 seconds, which gives viewers enough time to see each photo while still keeping a healthy pace. Remember, not everyone may be as enthralled with your photos as you are!
4. Turn the Shuffle option On to add variety to a slide show, especially one you've already viewed.

Slide delay: 30 seconds

Figure 16. Changing the delay between photos

5. Turn the Repeat option On if you want to see the entire slide show multiple times.
6. When you're happy with the settings, select Play Slide Show with These Options, which saves your settings and returns you to the previous screen.

Publishing Photos with Third-Party Software

At least two makers of photo album software have added features to their software that make it easy to publish photos to TiVo. Adobe Photoshop Album and Lifescape Inc.'s Picasa help you organize your digital photo collections, let you select the photos you want to publish, and then send those photos to your TiVo with a couple of mouse clicks. You can do this without using your TiVo Desktop software, although you'll need it if you decide to unpublish any of the photos.

Adobe has a free, pared-down "starter edition" of Photoshop Album that does not include the TiVo feature. You can download and use Photoshop Album Starter Edition to see if you like the software before buying the full version, which costs about $50. The starter edition can be found at http://www.adobe.com/products/photoshopalbum/starter.html.

You can also download a trial version of Picasa from http://www.lifescapeinc.com/picasa/. The trial version includes the TiVo feature. If you like the software, you can buy a registered copy for about $30.

Publishing Photos with Adobe Photoshop Album

Before you can publish photos to your DVR with Adobe Photoshop Album, you first need to choose the pictures you want the program to display. The photos are collected and viewed from an area in the program called the *photo well*, as shown in Figure 17.

NOTE

This section focuses on Photoshop Album's TiVo feature only, so we're not covering the steps to get your photos into the photo well. The program has a detailed help function that explains how to use it.

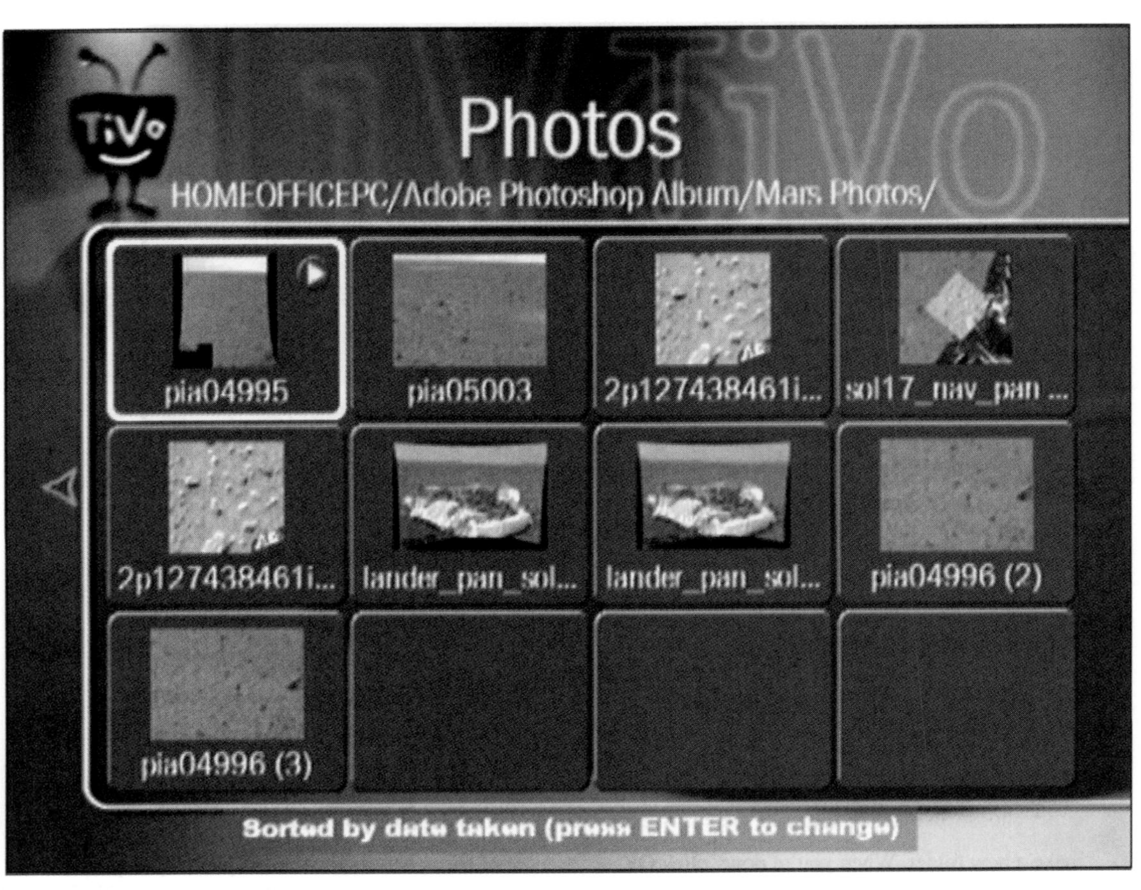

Figure 21. TiVo displays photos published through Photoshop Album.

While you can use TiVo Desktop to unpublish photos that were published with Adobe Photoshop Album, you'll need to manually remove the shortcuts that Photoshop Album creates on your computer's hard drive.

Publishing Photos with Picasa

It's easy to publish your photos to TiVo using Picasa once you've collected your photos together in a Picasa album, as shown in Figure 22.

There are only two steps:

1. Select the photos in the album that you want to publish. To select them all, press CTRL-A.
2. Once you've selected the photos, choose Tools | Export To | TiVo Series2 DVR, as shown in Figure 23.

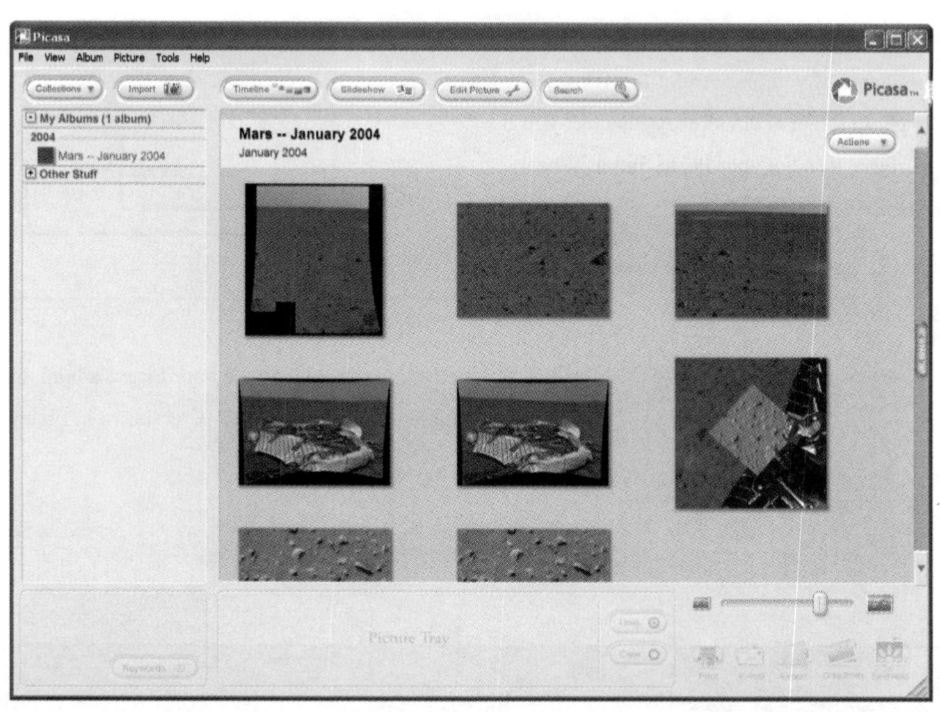

Figure 22. Photos arranged in a Picasa album

Figure 23. Publishing photos to TiVo from Picasa

3. An Export to TiVo dialog box will appear. Change the album name, if you like. And if the TiVo Desktop photo folder is not already selected, you'll need to browse through your PC to find it. When you're done, click OK.
4. If everything went well, you'll see a confirmation dialog box as shown in Figure 24.

Note

TiVo Export Complete!

OK

Figure 24. The confirmation box

You should now see your Picasa album displayed under Photos on your TiVo, as shown in Figure 25.

Figure 25. A Picasa album on TiVo's Photos screen

Even if you wait a few minutes before watching a program being transferred, you may find that it takes less time to watch the program than it does for TiVo to transfer it. If that's the case, it may be difficult—or irritating—to watch the program until it's either substantially or completely transferred. This is especially true for programs recorded at the highest quality levels (which take longer to transfer).

The best solution is to move programs between TiVos before you want to watch them. That way, whether your network is slow or fast, the completed programs will be waiting for you, without the irritating pauses.

If you're watching a program while it's being transferred, and you want to stop playing it or to stop the transfer completely, follow these steps:

1. Press the left-arrow button on your remote control. You'll see the Program screen, shown in Figure 7-8.
2. Select either Resume Playing or Stop Transfer. If you select Resume Playing, the program will continue playing. If you select Stop Transfer, you'll see the Stop Transfer screen shown in Figure 7-9.
3. You can select Stop Transfer, Stop Transfer & Delete From Now Playing, or Don't Stop. If you select Stop Transfer, the partially transferred program is listed on the Now Playing on TiVo screen, shown in Figure 7-10.

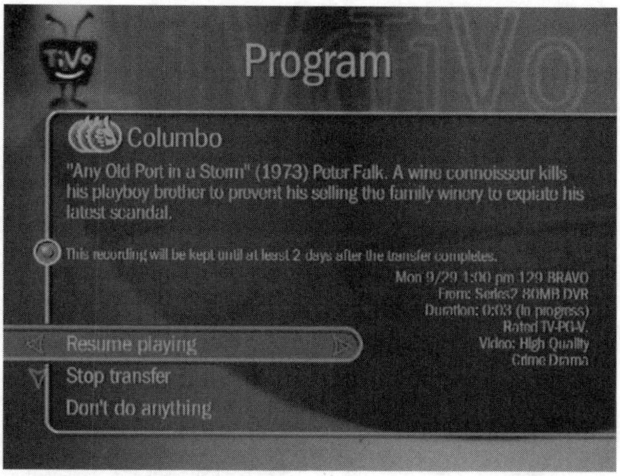

FIGURE 7-8 You can resume playing a streaming program.

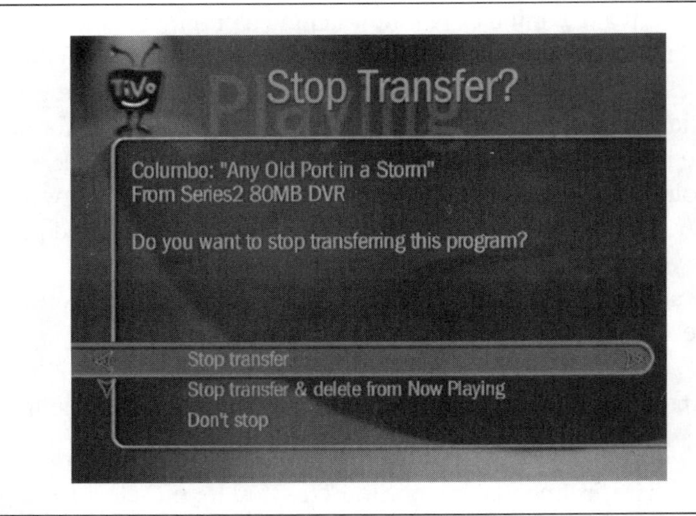

FIGURE 7-9 If you change your mind, you can stop a program transfer.

Network Connections and Speeds

Even if your home is wired with a Fast Ethernet network, which allows transfer speeds up to 100 Mbps, you'll hit a frustrating TiVo engineering bottleneck. Instead of including Ethernet jacks in the Series2 units, TiVo chose a clumsy and much slower

FIGURE 7-10 The transferred program appears under Now Playing on TiVo.

Did you know?

Connecting Your TiVos

Chapter 4 contains more details about connecting your TiVos via wireless and wired networks. Another networking scheme you can use with the Multi-Room Viewing feature is a peer-to-peer wireless network, which is discussed later in this chapter.

alternative: a USB port. So you must use a USB-to-Ethernet adapter to connect the TiVo to your wired Ethernet network. The USB port, as it's currently supported on the TiVo, only allows for 12 Mbps transfer speeds, closely mimicking the speed of the slowest home network—far slower than a Fast Ethernet network.

If you're using a wireless network to transfer programs between TiVos, you'll need a USB wireless adapter for each TiVo. You also can connect one or more TiVos using a wireless adapter, while networking the other TiVos with wired Ethernet connections. In that case, you'll need a wireless access point or similar device to bridge the wireless TiVos with your wired Ethernet network.

The following table will give you an idea of the difference in transfer speeds depending on the type of network you have and the recording quality of the program you're transferring.

	Slow Transfer	Faster Transfer	Fastest Transfer
Network	Wireless	Wired and wireless	Crossover cable
Recording Quality	Best	Medium/high	Medium/high/basic

Playing a Transferred Program

Once you have transferred a program (or have started transferring it), you can start watching it. Programs transferred from a remote DVR to your local unit appear on the Now Playing on TiVo screen just like the programs recorded locally.

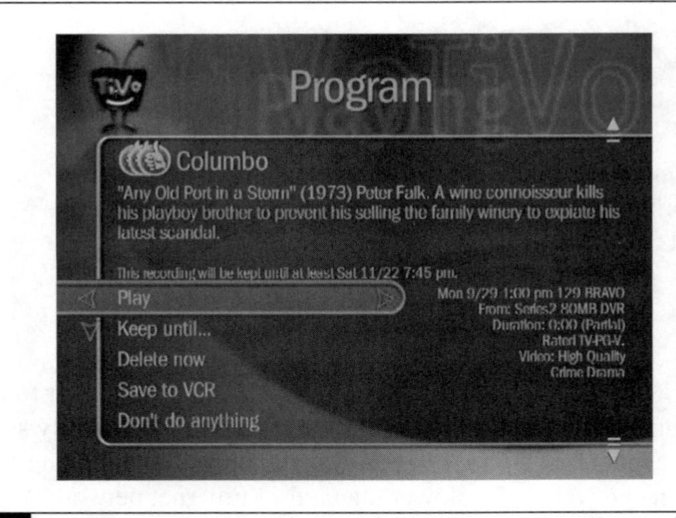

FIGURE 7-11 As you would with other recorded programs, select Play to view a transferred show

Select the program you want to view, and you'll see the Program screen, as in Figure 7-11.

A useful feature lets you resume watching a transferred program where it was paused on the original TiVo. This lets you begin watching a recorded program in the living room, for example, and then pause it and continue viewing it from a TiVo in your bedroom.

If someone partially viewed and then paused a program on the remote TiVo, TiVo will display a slightly different screen after you select a remote recording. It will ask whether you want to Watch From the Beginning or Watch From the Paused Location, as shown in Figure 7-12.

Connecting TiVos Via a Wireless Connection

In Chapter 4, we discussed how you can connect your TiVo to a wireless network. But there's another way to create a wireless network between TiVos that are activated to use the Multi-Room Viewing feature—a peer-to-peer network. It connects one TiVo directly to another TiVo wirelessly, bypassing routers and other wireless networking equipment.

FIGURE 7-12 You can continue watching a paused program on another TiVo

NOTE *If your TiVos are connected via a peer-to-peer network, they aren't connected to the Internet. But they still can call into TiVo corporate headquarters by using a dial-up phone line connection.*

Here's how you create a peer-to-peer network:

1. From TiVo Central, scroll to TiVo Messages & Setup and press Select. The TiVo Messages & Setup screen appears.
2. Scroll to Settings and press Select. The Settings screen appears.
3. Scroll to Phone & Network Setup and press Select. The Phone & Network Setup screen appears.
4. Scroll to Edit Phone or Network Settings and press Select. The Phone & Network Settings screen appears.
5. Scroll to Wireless Settings and press Select. You'll see a Wireless Checklist, showing the information you'll need for the next few steps.
6. Select Create a Peer-to-Peer Wireless Network. The Wireless Settings' Peer-to-Peer Wireless Channel screen appears.

7. Select a channel to use for your network. TiVo suggests you try channel 11 first, and that is the default setting.
8. Press Select to continue to the next step. The Wireless Network Name screen appears.
9. Enter the name, or SSID, of your wireless network—both TiVos must have the same network name. Once that's entered, select Done Entering Text. The Wireless Password Format screen appears.
10. Choose one of the following selections:
    - Enter password with alphanumeric characters
    - Enter hexadecimal password
    - The wireless network doesn't use a password
11. If your network doesn't require a password, go to step 13. Otherwise, if you choose to enter a password, you will see a screen for entering it. Select Done Entering Text when you have done so. The Wireless Encryption screen appears.
12. Select the level of encryption used on your network:
    - 40 or 64 bit
    - 128 bit
13. On the Confirm Settings screen, select Accept These Settings if everything is okay. (Otherwise, select Restore the Previous Settings and make changes to the appropriate settings.) After accepting the settings, you return to the Phone & Network Settings screen. You can move one screen backwards and test the connection. If all is well, you can view the network's signal strength. Otherwise, change the appropriate settings until the test connection works.

If your peer-to-peer connection is working, you'll see the other TiVo listed at the bottom of your TiVo's Now Playing on TiVo screen.

Stacking Your TiVos to Increase Storage Capacity

Would you like to simply increase TiVo's storage capacity, but you don't have any interest in installing multiple units throughout your home? If you have the right setup, you can simply place two or more TiVos in one location and use the Multi-Room Viewing feature to virtually combine the units' storage capacity—albeit in a rather kludgy way.

You'll need two or more Series2 TiVos. You'll also need separate cable TV or satellite TV tuners if you want the TiVos to concurrently record programming from more than one network. And you'll need a way to alternate between the TiVo boxes so you can watch the recordings, either by using a home theater amplifier or by using a simple video switch. (You can manually unplug audio and video cables from one TiVo and plug them into the other box, but that's not very convenient.) The TiVo remote control's DVR 1-2 switch lets you select between two TiVo units, so you can use a single remote to control each DVR separately.

Instead of hooking the TiVos to your home network, you use a *crossover* Cat5 Ethernet cable. Crossover cables are wired differently than regular Cat5 cables, allowing you to directly connect two Ethernet-enabled devices without first going through a network router.

Simply plug a USB-to-Ethernet connector into each TiVo's USB port, and then connect a crossover cable to the two Ethernet jacks. Once this is done, you may need to wait about five minutes for the two units to communicate with each other and establish a connection.

In some cases, you may need to restart one or both TiVos before the TiVos recognize the crossover cable connection. You can do this by restarting the DVRs from the TiVo Messages & Setup screen, or by unplugging the power cords and plugging them back in after 20 seconds or so. You'll need to wait for them to boot up before proceeding.

Even after restarting the DVRs, the Phone & Network Setup screen may show that the last network attempt failed, as shown in Figure 7-13.

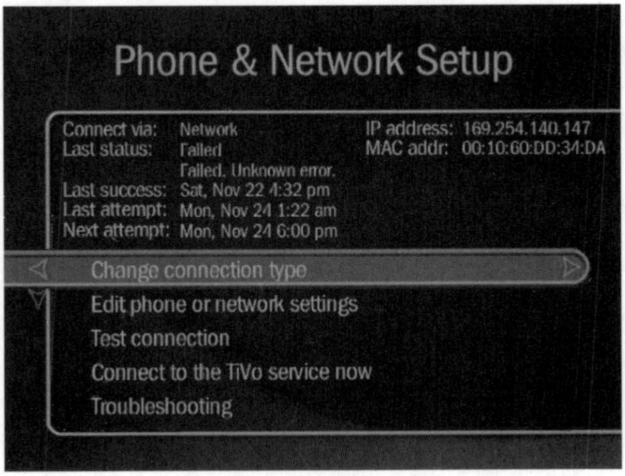

FIGURE 7-13 You likely can ignore network error messages like this while using crossover cable connections.

How to ... Troubleshooting Problems with the Multi-Room Viewing Feature

If you have a problem with your home network connection after you select a remote program for transfer, your local TiVo will display a screen similar to the one shown next. If you see the screen shown here, you should check your network to make sure your cables, routers, and other networking equipment are working correctly.

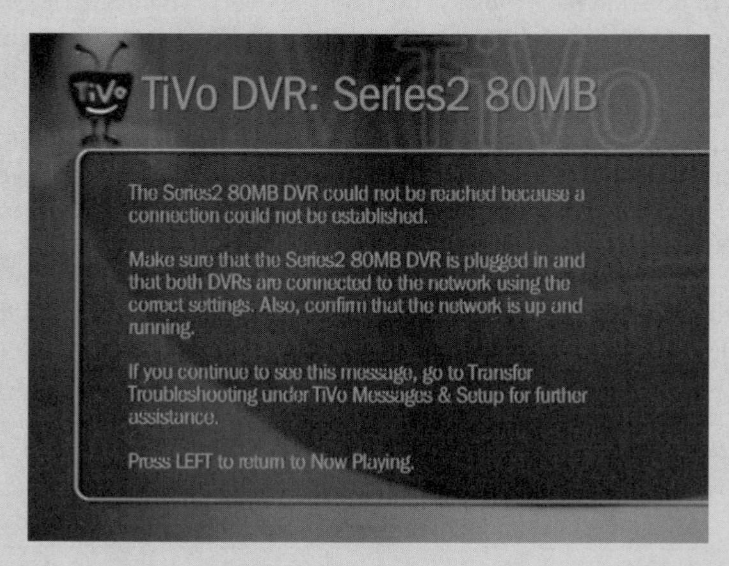

If you're using a wireless network, you should confirm that you have a strong signal. One way to increase your signal strength is to reduce the distance between your wireless equipment. And whether you're using a wired or wireless network, checking for loose cables is always a good bet. If you think you've solved the problem, you can press the left-arrow button on your remote control to return to the previous screen, where you can try the transfer again.

In general, if you're having trouble transferring programs, there are several steps you can take to identify the problem:

- Confirm that you're trying to transfer TV shows among Series2 models. The original TiVo model, which was replaced by the Series2 model, does not have Multi-Room Viewing capabilities.
- All the TiVos must have active service and be under the same person's account.

- Confirm that each Series2 model is running version 4.0 or higher of the TiVo software. You can see which software version a TiVo unit is running by viewing the Settings screen. If any unit is running older versions of the software, connect to the TiVo service so it can be updated.

This is because your TiVo was unable to connect to TiVo's computers, which is to be expected when using the crossover cable–based network. The important thing is that each TiVo can reach the other unit, which you can confirm by checking to see if the other TiVo is listed on the Now Playing on TiVo screen.

If you have two Series2 units with 80MB hard drives, this setup will essentially double your capacity to 160MB. Of course, this is what you're doing if you have TiVos located throughout your home, and you just transfer programs among the units.

By using the crossover cable, though, you don't need a home network, and you won't have the technical hassles that sometimes arise while using a network. Instead, both TiVos have a clean, direct connection to each other.

You can also place two TiVos in separate rooms and connect them with a long crossover cable. But you may not want to look at that cable all the time, and risk tripping over it.

You Can't TiVo Your Cat—Yet

Jennifer Massaro, a public relations manager, would like to extend TiVo's features into real life.

"I am so used to being able to ask, 'What did they say?' or 'What was that?' and immediately rewind the TV to get clarification. I expect to be able to do this with everything now."

"For example, I often find myself trying to 'rewind' the radio in the same way I have this capability with the TV. However, what I end up doing is simply turning down the volume—and then I remember that my radio doesn't have this functionality."

"My husband watched our cat do something crazy and funny—a spectacular jump off the railing—and immediately picked up the remote to try to TiVo the cat! Wouldn't it be useful to rewind moments in our lives to replay them?"

Chapter 8

Setting Up a Pioneer Combination TiVo and DVD Recorder

How to...

- Decide Whether the Combo TiVo-DVD Recorder Is Right for You
- Set Up Your Pioneer Combination TiVo-DVD Recorder
- Use the Audio and Video Cables and Accessories
- Work Around Common Phone Cable Problems
- Connect Your TiVo-DVD Recorder to a Cable or Satellite Box
- Use Both a Satellite Box and an RF Program Source
- Use a Cable or Satellite Box, an A/V Receiver, and a Game Console

One of the most exciting ways that TiVo technology can be integrated into a home theater system is now available: a TiVo device that can also play and record DVDs. Pioneer Electronics has combined a TiVo box with an integrated DVR recorder, allowing you to pause TV programs and schedule recording onto a hard drive as well as record shows onto a blank DVD disc.

Is the TiVo-DVD Recorder the Right Device for You?

Pioneer's combination TiVo-DVD recorder is among the first from major manufacturers to offer both DVR- and DVD-recording capabilities, and it is certain to increase the popularity of digital video recording. As of this writing, Pioneer has two models, offering either the basic 80GB drive, or a more spacious 120GB drive. As with other TiVo models, you can change the way you watch live television by pausing it, reversing it, and playing it in slow motion. In addition, both Pioneer models offer you several other ways you can enjoy your TiVo service at no additional cost:

- You can watch a recorded program from the beginning, while the recorder simultaneously finishes recording that very program.
- You can record content to the hard drive while copying different content from the hard drive to a DVD.
- Both models of the DVD recorder come equipped with a 181-channel cable TV tuner, for instant one-touch recording to the hard drive.

Making Sense of the DVD Maze

Some of the terms used in this chapter look and sound like each other. Here is a quick review of some potentially confusing terms:

- **DVR (digital video recorder)** A device that is similar to a VCR but records television data in digital format as opposed to the VCR's analog format. VCRs record on analog tapes, but DVRs encode video data in MPEG-1 or MPEG-2 format and store the data on a hard drive.
- **DVD (digital video disc)** A type of optical disk technology similar to a CD-ROM. A DVD holds a minimum of 4.7GB of data, which is enough for a full-length movie.
- **DVD-R (DVD-recordable)** A DVD-R can only record data once, and the data is permanently on the disc. The disc can not be erased or recorded onto a second time.
- **DVD-RW (DVD-rewritable)** A re-recordable DVD. The data on a DVD-RW disc can be erased, and the disc can be recorded over several times without damaging the medium.
- **DVD-RAM** A DVD format that can be recorded and erased repeatedly. These discs are only compatible with devices manufactured by the companies that support the DVD-RAM format. Avoid this format, as TiVo does not support it.

You can transfer content from VCR tapes to DVD-R or DVD-RW discs by connecting a VCR to analog inputs on the recorder.

When you find yourself with a recording you want to watch again, you can keep it on the hard drive or simply burn it to a DVD-R or DVD-RW. Both models of the Pioneer feature up to 18x recording speed. This means that you'll be able to burn a one-hour program onto DVD in Medium quality in just over three minutes.

DVDs last longer and take up less space than the older, clunky VHS tapes. You can record, or "burn," the programs you like to DVD and keep them on that medium. This also frees up your TiVo hard drive, so you won't have to delete a program you haven't watched yet to record another one.

The Pioneer models are also equipped to let you transfer old video content to DVD-R or DVD-RW discs to preserve the content for future generations. Once you've transferred the content from videotape to the hard drive, you can edit it before burning it to DVD. The newly created DVD-R can be played back on most other home and portable DVD players, as well as on DVD-ROM computer drives.

TIP *It's best to use DVD-R discs for archiving because they are write-once discs and cannot be accidentally erased.*

The Pioneer TiVo-DVD recorder models also come with a new feature, called PureCinema 3:2 progressive scan. When a DVD is created from celluloid film, the process of putting the original performance into digital format can introduce distortion. According to DVD industry insiders, this is a common occurrence when taking a movie, which is filmed at 24 frames per second, and converting it to a DVD video, which is recorded at the equivalent of 30 frames per second. Pioneer's 3:2 progressive scan technology digitally corrects this distortion and removes redundant visual information in order to display a film-frame-accurate picture.

Progressive scanning is the other part of the new feature. This form of scanning is sometimes referred to as 480p, for the number of horizontal lines that compose the video image. Because conventional DVD pictures only use 240 lines, Pioneer is able to create a picture using twice the scan lines of a conventional DVD picture. By increasing the number of lines, you also increase the level of detail you can show in a picture, thereby giving you higher screen resolution and sharper images.

NOTE *You will need an HD-ready TV to appreciate the progressive-scanning picture, as a typical television cannot display more than 240 lines on the screen. Once an HDTV screen is used, the results are dramatic and give a much more lifelike, almost three-dimensional, look to many images.*

Is a combo TiVo-DVD recorder the right machine for you? It really depends on how you plan to use your TiVo, and whether you want to create a semi-permanent collection of programs on disc. Price may also be a factor, at least for a while. At the time of writing, Pioneer charges in the neighborhood of $700–800 for its less expensive model. However, Pioneer's rivals in the consumer electronics field are already developing competing products. Both Toshiba and Sony Electronics have licensed TiVo's service and technology and plan to roll out new products that use the DVR service. TiVo also has plans to release its own branded model of DVD-R recorder.

Voices from the Community

Pioneer—the TiVo That Does It All

When the Pioneer models first came out, I was longing after the combo TiVo-DVD-R feature for quite a while. When I finally bought my 810H (the 80GB hard drive model), it was very easy to set up, with a good install manual and quick-reference pamphlet. The picture is great, especially when recording in Extreme Fine. My personal quirk is that I like to use Medium quality for animated programs, and I use Best/Extreme Fine for everything else. The image quality on the DVD pictures are excellent. I even recorded an old videotape to the DVR hard drive and then burned it to DVD, with superb results. Got to give credit to Pioneer for allowing me to save my favorite programs—and my old videos of my college friends!

—Joel Elad, Marketing and Sales Director for Top Cow Productions, and gung-ho TiVo user

NOTE *When you purchase the Pioneer TiVo-DVD recorder, you automatically receive the basic TiVo service. This does not require a monthly fee for the DVR service, but it does give you the option of upgrading to the full TiVo service so you can access additional features, such as the Season Pass recordings, WishList searches, and a 14-day program guide.*

Setting Up Your Pioneer TiVo-DVD Recorder

If you have already installed your Pioneer TiVo-DVD recorder, you can move on to the next chapter. The remainder of this chapter will provide information that overlaps portions of what is provided in the Pioneer TiVo-DVD recorder's Installation Guide or the User's Guide. We'll focus particularly on nonstandard forms of installation (for example, making the connections if you don't use a stand-alone cable box) and on troubleshooting problems that you may encounter along the way. These setup instructions will also be particularly helpful if you've acquired your machine from a friend or on eBay without the original documentation.

Cables and Accessories

As you unpack your TiVo-DVD recorder, you'll find that several cables and other items come with the unit. These components will allow you to perform most hookup procedures. If you've acquired the unit as an "out-of-the-box" purchase, you should double-check that you've received all of the basic equipment for setting up your machine.

You should have the following:

- A power cable
- The TiVo remote control
- A serial control cable
- An infrared (IR) controller
- A composite (RCA) video cable
- A coaxial (RF) cable
- A phone cable
- A phone splitter

A Word on Audio/Video Cables

Make no mistake, the job of choosing and connecting cables and components for your home theater system has become more complex with each new development. Your home theater system may now include a VCR, an A/V receiver, a home network, a wireless unit, and even a game system, such as the Sony PlayStation or Microsoft Xbox. However, the same principle applies to each of these services—you still need to get your data (pictures, sounds, text) from their source to your television screen.

Paths and Connectors

Data is sent to your television using cables, and those cables have connectors or plugs at each end that allow you to connect them to the devices. The connectors are used for specific functions and also specify whether the data will be transferred to a device (labeled "In") or from a device (labeled "Out"). When you connect your TiVo-DVD recorder to your TV or other home theatre equipment, you are creating a data path that allows information to flow from the wall socket though to your unit.

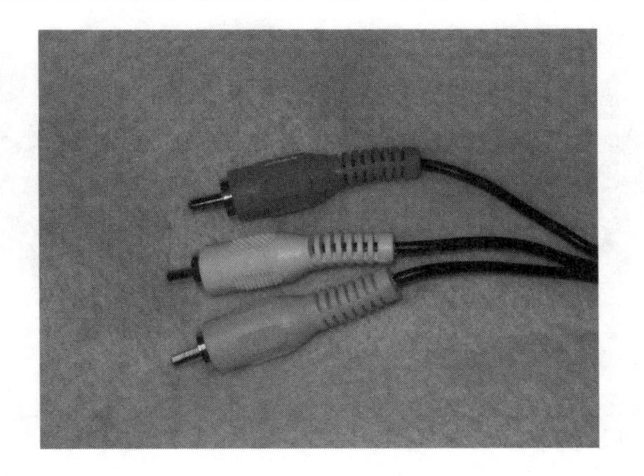

FIGURE 8-1 The composite A/V cable carries both audio and video signals.

Choosing the Right Cables

Your TiVo-DVD recorder comes with a standard composite video cable. Composite A/V cables have three plugs at each end, otherwise known as *jacks,* colored differently for easy identification. The white and red jacks are for the left and right audio signals, and the yellow jack is for the video signal (see Figure 8-1). Composite video cables provide a good quality signal, so you can connect the TiVo-DVD recorder to a television or even an A/V receiver with this cable alone.

Your TiVo-DVD recorder also includes connectors the premium S-Video input and output cables, and for the component video output. S-Video cable, shown in Figure 8-2, provides a better quality signal than the standard composite cable. Even better, component video cable, shown in Figure 8-3, is considered to offer one of the highest quality video signals available for home theaters.

FIGURE 8-2 The S-Video cable and jacks

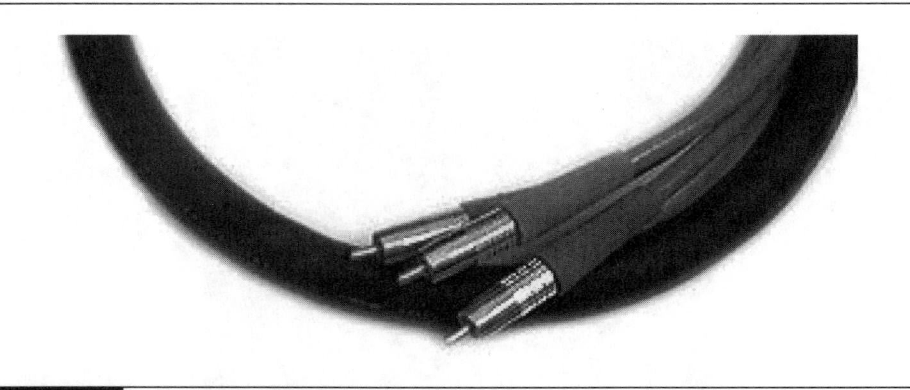

FIGURE 8-3 Component video cable and jacks

Your TiVo-DVD recorder's premium connectors for audio cables on its back panel allow you to use optical digital cable for maximum sound clarity. This cable, shown in Figure 8-4, is a single-jack connector. Using this cable won't make a huge difference on most standard systems. However, it will provide a big jump in performance if you're

Did you know?

Why Optical Digital Cable Boosts Performance the Most for Your TiVo

An optical digital cable is a lot like a thin glass tube that's coated in a protective plastic coating. The mirror-like quality of the glass creates what is called a total internal reflection. Light that travels through this glass fiber bounces at shallow angles and stays completely within it, with little or no distortion.

Light at one end of a cable switches on and off to send each bit of audio information. Modern fiber systems with a single laser can transmit billions of bits per second, as a laser can turn on and off several billions of times per second. Thus, with an optical cable, both digital signals and analog voice signals (translated into digital signals) can send information to your speakers with extremely low distortion.

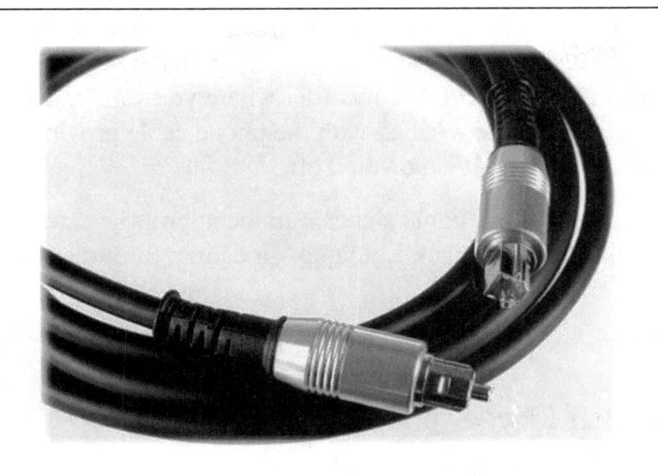

FIGURE 8-4 Optical digital cable and jack

watching DVDs or programs recorded with TiVo, you're using a Digital Theater System (DTS) for audio. It will also provide excellent sound if you're watching programs or DVDs recorded or broadcast with Dolby Digital audio.

It's worth mentioning that none of these premium cables will interfere with each other. For example, you could use a component video cable together with an optical digital audio cable to get maximum visual detail together with enough high-powered audio to guarantee that you can keep the neighbors up at night.

NOTE *These premium cables (the S-Video cable, component video cable, and optical digital audio cable) don't come standard with a Pioneer TiVo-DVD recorder. They must be purchased separately at the Pioneer Electronics Web site, other online retailers, or at a store that carries audio/video components. Because the optical digital audio cable is designed specifically for high-end home theater systems that use Dolby Digital sound, you may find this last component slightly more difficult to find outside of stores that specialize in consumer electronics, such as Circuit City, Radio Shack, or Best Buy.*

Connecting Your TiVo-DVD Recorder to the Phone Line

If you don't have a home network that's based on a shared broadband Internet connection, your TiVo-DVD recorder can use a standard phone line to connect to the TiVo service, just like any other type of TiVo box.

To connect your TiVo-DVD recorder to the phone line, complete the following steps:

1. Place the Pioneer TiVo-DVD recorder where you want it. Make sure that the phone jack can be reached with the phone cable provided with the unit, and check that the unit is powered off.
2. Plug one of the phone cable's ends into the appropriate socket on the back of the unit. The phone jack is unique—it cannot fit into any other socket on the recorder.
3. Plug the other end of the phone cable into the phone jack in the wall.

What Happens If I Need to Make a Phone Call While My TiVo Is Using the Line?

It's highly unlikely that you'll be making a phone call at the same time that your Pioneer TiVo-DVD recorder calls in to TiVo, for two reasons. First, the connection to the TiVo service typically lasts only about 10 minutes. Second, your box is preset to only call during the late evening hours when the typical household uses the phone line less often.

However, if you do need to use the phone while a connection is in progress, simply pick up the phone. The Pioneer TiVo-DVD recorder always gives priority to outgoing calls. When you pick up the phone, you will hear a sound like that of a modem dialing. Hang up and wait about 45 seconds. When you pick up the phone again, the line should be free. If it is not, just hang up, wait a moment, and try again. The DVR will sense that its connection has been broken, and it will attempt another connection later. Of course, if anyone tries to call your line during your Pioneer's connection to the TiVo service, they will get a busy signal.

Splitting the Phone Line

If you use the same phone line for TiVo as for a computer's modem or a household phone, you may need to connect more than one device to the same phone wall jack. To connect both the TiVo-DVD recorder and these other items, plug in the phone line splitter that is provided with your recorder as follows:

1. Place the Pioneer TiVo-DVD recorder where you want it, and make sure the unit is powered off.
2. Plug one of the phone cable's ends into the appropriate socket on the back of the recorder.

3. Plug the other end of the phone cable into one of the two jacks on the phone splitter.
4. Take the phone cable from whatever other device needs to be attached to the phone line (such as a household phone) and plug it into the remaining open jack in the phone splitter.
5. Plug the phone splitter into the phone jack in the wall.

Troubleshooting Common Phone-Line Problems

Your TiVo-DVD recorder needs to be connected to a phone line in order to make its regular check-in with the TiVo service. This check-in happens every 24 to 48 hours, and that's when TiVo downloads the information about when programs are scheduled to air and on what channel. This is also when your TiVo box receives additional information from the TiVo service, such as regular or special TiVo service updates.

Here are some potential phone-line problems and their solutions.

Problem: The Cord That Comes with Your TiVo Can't Reach the Phone Jack The Pioneer TiVo-DVD recorder comes with a 25-foot phone cord, which is long enough for most people's household needs, but phone cords longer than 25 feet are readily available at most hardware or electronics stores. If the phone jack in the wall isn't close enough to your TiVo-DVD recorder for the cable to reach, we strongly recommend that you purchase a single new cord in the length that you've determined you need.

It is also possible to link two or more shorter cords together with "bridges." A bridge is a small, rectangular plastic piece that has "female" connectors at both ends so that you can plug a phone cord into both ends to connect them together. You can purchase bridges at hardware or electronics stores where you can buy phone cords.

In general, we recommend that you buy a single longer cord instead of linking two or more shorter cords with bridges—the link between the two cords can come undone, and a serious knock or someone stepping on the bridge can damage it and can lead to serious degradation of the signal.

TIP

If you need to replace your phone cord, always overestimate the length of cord that you need. It's not likely that the cord will be going in a straight line across the room to your TiVo box—it will have to snake under rugs and around furniture. If you overestimate the length of the cord, you can reduce the amount of cord clutter behind your TiVo box by looping the cord and tying the loops together with a plastic twist-tie.

Problem: Your TiVo Box Is Having Problems Getting a Dial Tone If your TiVo is having trouble dialing or reaching a dial tone when you begin the Guided Setup, there are a couple of options you can try.

Restarting your machine will generally fix modems that have become "stuck" in some way. Perform the following steps if you're having problems dialing:

1. Turn the TiVo off.
2. Unplug it from the power cord.
3. Unplug it from the phone line.
4. Wait two or three minutes, and then plug everything back in.
5. Turn the power to the unit back on. This process allows the modem to be reset.

You can also change the Phone Available and Dialtone Detection options to Off. It's possible that for some reason your TiVo is having problems detecting your phone's dial tone signal. Follow these steps:

1. Using your TiVo remote, go to Messages & Setup.
2. Select Recorder & Phone Setup, as shown in Figure 8-5.
3. Select Phone Connection, as shown in Figure 8-6.

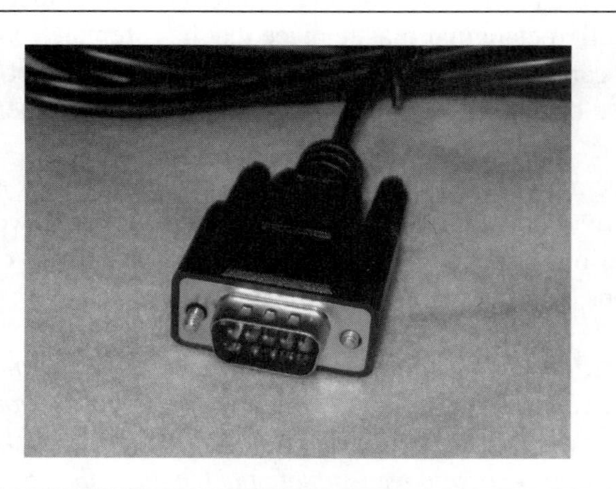

FIGURE 8-5 Selecting the Recorder & Phone Setup screen

FIGURE 8-6 Selecting the Phone Connection

4. Select Change Dialing Options, as shown in Figure 8-7.
5. Set the Dialtone Detection options to Off

Problem: Your TiVo-DVD Recorder Can't Connect to TiVo During the Guided Setup Process

If you have trouble connecting to TiVo during the Guided Setup process, try these steps:

1. Connect a household phone to the phone plug at the wall to check if there is a dial tone on the line.
2. If you have a dial tone, then repeat the test but connect the phone to the TiVo-DVD recorder's end of the long phone cable. Often, a strong dial tone at 3 feet from the wall socket will turn into a very weak, static-laden signal at a distance of 25 feet, especially if the cable runs by many operating electronic devices.

FIGURE 8-7 Selecting the Change Dialing Options

3. If your dial tone is substantially weaker and more clogged with static at the TiVo-DVD recorder's end of the phone cable, you have a few different options:

   - Move your TiVo-DVD recorder closer to the phone outlet, thus shortening the length of the phone cable. Phone cabling lacks effective electromagnetic shielding, which means that as the length of the phone cable increases, the amount of interference on the line increases. The end result is that as the signal travels over longer and longer cables, the signal gets weaker.
   - Double-check to make sure that there are no kinks in the phone cable. Kinks in the line can cause breaks in the cable over time, and it greatly increases electromagnetic interference.
   - Disconnect any other devices connected to your phone line. It is possible that they are adding to the interference by increasing the length of the cord or by putting out competing phone signals if they are attempting to dial out. In particular, try disconnecting your household phone and your home computer modems, if you are still using dial-up to access the Internet.
   - Buy phone cable with better shielding—the standard-issue phone cable with most units is the lightest, least expensive kind.
   - Leave the cabling alone. A weak signal won't prevent you from connecting to TiVo and completing the setup, but it may require you to make multiple attempts to complete the Guided Setup and run the system.
   - Buy a phone jack extension system, which allows you to bridge phone jacks using the electrical wiring in the house. You plug the master phone jack into the converter, and the converter into a power outlet. Then you plug a converter into the power outlet in the room with the TiVo, and a phone cable from the converter to the TiVo. The phone signal travels through the electrical cabling, which is present in several locations in every room in the house, unlike the phone line.

TIP

One more possibility to consider, especially if you live in a house that is more than 40 years old, is that the internal wiring simply isn't up to the task. To test this, you'll need to plug the phone line into that little box the phone company has outside your house. You can physically move your TiVo and a monitor out to the box, though that's probably impractical. A better choice is to disconnect the house phone and run a phone cable from your TiVo out to the phone jack. If it works out there, then your problem is the house's internal wiring.

Problem: The Pioneer Got Through the Guided Setup Process, But Occasionally Has Problems Connecting to the TiVo Service

If you continually have problems accessing TiVo and you have the call waiting phone service, you might want to try turning off call waiting. Call waiting interrupts the TiVo service unless you've set your box to dial the single-time call-waiting disable code. To solve this problem, simply add the Call Waiting Prefix to the number your TiVo calls to turn it off.

CAUTION *If you don't have call waiting and you enter your Call Waiting Prefix, you will get a "no dialtone" message.*

Problem: After a Nearby Thunderstorm, Your Pioneer Can't Connect to TiVo Anymore!

Lightning and other power surges can easily damage your TiVo-DVD recorder. Surprisingly, these power surges can even happen through the phone line. We recommend that you add a high-quality surge protector to your Pioneer. If possible, purchase one that also protects modem phone lines.

If you live in an area that is prone to thunder or lightning storms, you might also consider upgrading to a home battery backup system. An uninterruptible power supply (UPS) is a little more expensive than a quality surge protector, but they come with an added bonus: If you plug your cable or satellite receiver into the UPS, your Pioneer can even record programs during a power failure.

Problem: You Have One or More of the Previous Problems and None of the Solutions Seem to Work

While Pioneer makes a quality product, the odds are that at least a few devices need replacing. Start by calling Pioneer's product help line, and if necessary get your device repaired or even replaced if you have a warranty on your item.

A final option before you send your Pioneer out to be repaired is to switch to using an external modem. Typically, these replacement modems are sold as kits that include the external modem, which has been fully configured for a TiVo box, a modem power supply, an adapter to connect your TiVo serial cable to the modem, and instructions on how to set the TiVo box to use the new external device. Since these devices are now fully "plug and play," this solution doesn't even require you to open your Pioneer to install it.

There are a couple advantages to setting up your TiVo with an external modem before sending it out for repair.

- You don't have to ship your TiVo anywhere.
- If you send your TiVo to the manufacturer for a repair, they'll likely swap your unit out, meaning you will lose the recordings and settings on your drive.
- You won't be stuck waiting without your TiVo.

TIP

If you've switched to an external TiVo modem, do you still need a surge protector? Yes. External modems are more durable than the factory TiVo modems, since they are protected inside their own case as opposed to being attached with solder to the TiVo's motherboard. However, even the best computer component can be damaged by a lightning storm or an electrical surge.

Connecting Your TiVo-DVD Recorder to a Cable or Satellite Box

If your home theatre system uses a cable or satellite box, your Pioneer TiVo-DVD recorder needs to use these devices to change channels in order to record programs or display live TV. You can connect the recorder to the cable or satellite box with either a serial control cable or an IR (infrared) cable, both normally included with your TiVo-DVD recorder. Whether you'll be using the serial control cable or an IR cable depends on how the remote control for your box is set up. Refer to the documentation that came with your box to determine which is the one you should use. For example, on the Motorola/General Instruments DCT2000 series cable box the cable provider has set the box for serial control.

NOTE

If you receive cable without a cable box, then your TiVo-DVD recorder won't need a control cable to change channels and you can bypass this section.

Connecting a Serial Control Cable

There are three types of serial control cables that can be used with the Pioneer TiVo-DVD recorder. Of these, the most common is the serial control cable that comes with the recorder. The other types of cables must be purchased separately, but they all use the same jack on the recorder.

The 9-Pin Data Connector You can simply use the standard serial control cable that comes with the TiVo-DVD recorder if you have a Motorola/General Instruments DCT2000 series cable box, where the cable provider has set the box for serial control. This form of DirecTV satellite receiver uses a 9-pin data connector (named for the number of metal prongs, or "pins," in one of the connectors.

The 15-Pin Data Connector Some DirecTV satellite receivers have a connecting cable that looks almost like the 9-pin connector but has 15 "pins." Otherwise the 15-pin connector is indistinguishable from the standard 9-pin cable.

In order to use this connection, you must either purchase a 15-pin cable, or an adapter to connect the 9-pin cable to the 15-pin connector on the satellite receiver. The adapter simply screws on to the end of the 9-pin cable, converting it into a 15-pin cable.

CAUTION *Never try to force a 9-pin jack into a socket designed for 15 pins, or vice-versa. This is a more common error than you might think—always check the number of pins on the cord you're using and the number the socket will accept. The jack heads look and feel identical if you're fumbling around the back of a machine. Bending even one of the metal pins can ruin the cable; breaking off a pin in a device socket could permanently damage your TiVo-DVD recorder!*

The Home Control Connector The third variation on some DirecTV satellite receivers is the Home Control connector. Home Control is one of the fastest and most reliable forms of cable control. You'll need a special Home Control cable to complete this form of connection. You can find this connector at most audio/video stores.

It's important to remember that Home Control connectors look very different from the other forms of serial cable jacks. Instead of pins, the connection looks more like a phone jack, only slightly thinner.

To use this connector, simply plug one end of the cable into the Home Control jack on your DirecTV satellite receiver and the other end into the Channel Change/Serial jack on the Pioneer TiVo-DVD recorder.

NOTE *This connector is sometimes labeled "Low Speed Data." Don't try to connect the Home Control cable to the satellite receiver's phone jack socket, or you won't get a connection of any kind and the jack will likely fall out of the socket.*

Connecting an IR Control Cable

The IR (infrared) control cable that comes with your Pioneer TiVo-DVD recorder has a long, thin purple plug, which should be plugged into your recorder. At the other end of the cable are two IR emitters that will send infrared light signals to the IR sensor on the cable or satellite box. It's worth spending a little time to accurately place the sensors so they will operate effectively.

To set up your IR control cable, follow these steps:

1. Locate the IR sensor on your cable box or satellite box. The sensor typically looks like a tiny round light bulb covered with a plastic bubble or semi-transparent window. The IR sensor window will be on the front of the cable or satellite box. (The electronic industry is devoted to extremely dark component cases, so if you can't find the IR sensor window easily, direct stronger room light or a flashlight on the unit to locate it.)
2. Position the IR emitters so that they're centered on the IR sensor and stick out approximately an inch and a half. The IR control system works best if the IR sensor window itself is shaded from other infrared signals, such as strong sunlight.

NOTE *Don't rely on looking at the small "bulbs" on the IR emitters to determine whether they are working or not. The bulbs emit infrared light, which is beyond the light spectrum that is visible to the human eye. If you're having problems with IR control, check the connections and check whether the units are centered on the IR sensor window.*

Example Setups

The three most common TiVo setup configurations are covered in Chapter 1:

- Analog connections
- Cable box connections
- TiVo with both cable and satellite TV sources

The setup procedure is identical for both a typical TiVo system and the Pioneer TiVo-DVD recorders. However, there are two additional setups that you might encounter with the TiVo-DVD recorder:

- Connecting to a satellite box and an RF program source
- Connecting to a cable or satellite box, an A/V receiver, and a game console

This section will look at these latter two setups.

Connecting to a Satellite Box and an RF Program Source

Connecting to a cable or satellite box requires you to upgrade to TiVo Plus service. In order to upgrade to TiVo Plus, visit the TiVo web site (http://www.tivo.com/upgrade) to register and pay online. You can also upgrade by calling TiVo by phone, but be prepared for a wait time of up to 30 minutes during peak hours.

Before starting the process to upgrade to TiVo Plus, double-check to ensure that the satellite cable coming out of the wall is properly connected to the satellite box. Also, make sure that all of the equipment is turned off and that the TiVo-DVD recorder is unplugged. Changing around cable components while the recorder is plugged in could damage its hard drive, even if the recorder is not actually powered on at the time. Finally, determine whether you will be using a serial cable connection (and if so, what type) or an infrared control cable connection—this will become important in step 5. For more info about these different cables, see the "Connecting Your TiVo-DVD Recorder to a Cable or Satellite Box" section earlier in the chapter.

Follow these steps to make the connections:

1. Connect a composite A/V cable from the A/V Output connectors on the satellite box to the A/V Input 1 connectors on the back of the TiVo-DVD recorder.

2. Connect the RF coaxial cable from the antenna or cable source to the RF IN connector on the back of the TiVo-DVD recorder.

3. Connect the composite A/V cable from the A/V Output 1 connectors on the TiVo-DVD recorder to the A/V Input connectors on your television set.

4. Connect your phone line to the TiVo-DVD recorder as discussed earlier in this chapter.

5. Connect the TiVo-DVD recorder with a serial cable, a Home Control cable, or an infrared control cable:

   - **Serial control cable** If the connector for your satellite box has a 9-pin data connector, plug one end of the cable into the connector on the satellite box, and the other end into the Channel Change/Serial connector on the back of the TiVo-DVD recorder. If you have a 15-pin connection, you'll have to purchase a 15-pin adapter to be able to complete the connection.

- **Infrared control cable** If you are using the infrared (IR) control cable, plug the purple end of the control cable into the Channel Change/IR connector on the back of the TiVo-DVD recorder. This will be the uppermost, miniature jack. Then locate the IR sensor and mount the IR emitters as described in the "Connecting an IR Control Cable" section earlier in this chapter.

6. Plug in all power cords and turn on the television. If you don't see the TiVo Welcome screen, you still need to find the correct TV data input setting. Press the Input, Source, or TV/Video button on your remote control, as appropriate, until you see the Welcome screen.

7. Once you see the Welcome screen, proceed through the Guided Setup. You must complete this setup in order to play and record DVDs on your recorder. (For a detailed explanation on completing the Guided Setup, please see Chapter 2.)

NOTE *For the best video signal, you may want to use component video cables or an S-Video cable instead. If you want the best quality audio signal, consider using optical digital cables. These cables are discussed in detail in the "A Word on Audio/Video Cables" section earlier in this chapter.*

Connecting to a Cable or Satellite Box, an A/V Receiver, and a Game Console

Before starting to connect the cable or satellite box, the A/V receiver, and the game console to the TiVo-DVD recorder, double-check that all of the equipment is turned off and that the recorder is unplugged. Changing around cable components while the recorder is plugged in could damage its hard drive, even if the recorder is not actually powered on at the time. Finally, determine whether you will be using a serial cable connection (and if so, what type) or an infrared control cable connection—this will become important in step 5. For more info about these different cables, see the "Connecting Your TiVo-DVD Recorder to a Cable or Satellite Box" section earlier in the chapter.

When you're ready to start, follow these steps:

1. Connect a composite video cable and left and right audio cables from the A/V Out connectors on the cable or satellite box to the A/V Input 1 connectors on the back of the TiVo-DVD Recorder. If you don't have a cable or a satellite box, connect the RF coaxial cable from the antenna or cable source to the RF IN connector on the back of the TiVo-DVD recorder.

2. Connect the composite A/V cable from the A/V Output 1 connectors on the TiVo-DVD recorder to the A/V Input connectors on the A/V receiver. If the A/V receiver has a Digital Audio Input plug, you can connect a digital audio cable from the A/V receiver to the Digital Audio Out connector on the back of the TiVo-DVD recorder.
3. Connect the composite A/V cable from the A/V Output connectors on the receiver to the A/V Input connectors on the television.
4. Connect the composite A/V cables from the A/V Output connectors on the game console to the A/V Input connectors on the receiver.
5. Connect your phone line to the TiVo-DVD recorder as discussed earlier in this chapter.
6. Connect the TiVo-DVD recorder with a serial cable, a Home Control cable, or an infrared control cable:
    - **Serial control cable** If the connector for your satellite box has a 9-pin data connector, plug one end of the cable into the connector on the satellite box, and the other end into the Channel Change/Serial connector on the back of the TiVo-DVD recorder. If you have a 15-pin connection, you'll have to purchase a 15-pin adapter to be able to complete the connection.
    - **Infrared control cable** If you are using the infrared (IR) control cable, plug the purple end of the control cable into the Channel Change/IR connector on the back of the TiVo-DVD recorder. This will be the uppermost, miniature jack. Then locate the IR sensor and mount the IR emitters as described in the "Connecting an IR Control Cable" section earlier in this chapter.
7. Plug in all power cords and turn on the television. If you don't see the TiVo Welcome screen, you still need to find the correct TV data input setting. Press the Input, Source, or TV/Video button on your remote control, as appropriate, until you see the Welcome screen.
8. Once you see the Welcome screen, proceed through the Guided Setup. You must complete this setup in order to play and record DVDs on your recorder. (For a detailed explanation on completing the Guided Setup, please see Chapter 2.)

NOTE *For the best video signal, you may want to use component video cables or an S-Video cable instead. If you want the best quality audio signal, consider using optical digital cables. These cables are discussed in detail in the "A Word on Audio/Video Cables" section earlier in this chapter.*

If you run into any hardware problems with your machine, be sure to either visit Pioneer Electronics' web site (http://www.pioneerelectronics.com) for help, or call their tech support line at 1-800-421-1404. Remember that Pioneer will only be able to assist you on technical matters related to their TiVo-DVD recorder, such as the data inputs you need to use, or problems with the DVD-burning mechanism. If you have problems connecting to TiVo or upgrading your service contract, it's best to contact TiVo directly over the phone or through the web site (http://www.tivo.com).

Chapter 9

Playing and Recording Programs on a Pioneer TiVo-DVD Recorder

How to...

- Determine DVD or CD Compatibility
- Understand Copy Protection
- Play a DVD or CD
- Record a DVD from TiVo
- Record a DVD from a Videotape
- Record a Home Movie to DVD
- Care for Your Discs

In many ways, Pioneer's TiVo-DVD recorders are a videophile's dream: TiVo service, a large hard drive, and the ability to keep the content you record for as long as you want without recording over it when the hard drive is full. While other companies are planning to enter the market with similar devices, Pioneer holds the distinction of being the first to bring this machine to market—and does a excellent job of it. As with most TiVo devices, you can record on the built-in hard drive. Where Pioneer's machines differ from the run-of-the-mill TiVo devices are in their "next generation" video technology. This new technology gives the Pioneer TiVo-DVD recorders two advantages: enhanced picture resolution, and the ability to burn DVDs from multiple image sources.

Voices from the Community

No-Fuss Freedom of Copying from TiVo

I'm on my third TiVo, and finally it has a DVD burner built in. It is awesome. Instead of copying a show from the TiVo to a VHS in *slow* real time, I just hit the remote and it burns the show onto a DVD. My mother-in-law is constantly asking me to tape this or that for her. Now, I just TiVo it and *burn* it. No fuss, no muss. How did we ever live without TiVo?

–Andrew Malis, President, MGH Advertising, Inc., and hopeless TiVo addict

Reducing Noise and Filling Jagged Edges

Pioneer has incorporated an electronic noise-reduction circuit on all of its DVR models, which improves the sound clarity from your speakers. In addition, Pioneer has introduced a new technology on its DVRs called DCDi, which enhances the picture on your TiVo.

DCDi stands for Directional Correlation Deinterlacing, and it is an algorithm designed to smooth the ragged edges along diagonal lines in video images. Much like oversampling on high-quality CD decks, DCDi smoothes these images by creating new information to fill in image gaps in the monitor's edge transitions.

DCDi technology was introduced in the late '90s, although the $50,000 cost meant it was useful only to broadcasters who wanted to convert NTSC video signals to HDTV (High Definition Television) signals. DCDi's latest incarnation is in the consumer market as the FLI2200 chip. Pioneer's engineers have customized the chip by building it into the latest generation of DVD players. The end result is that video content has improved overall picture quality.

Go For the Burn: CD or DVD Compatibility

The Pioneer TiVo-DVD recorders allow you to watch and record DVDs, listen to CDs and MP3-CDs, and even record your home movies direct from the camera onto a DVD. Pioneer's recorders allow you to burn any video formats onto a DVD-R disc (which is permanent and cannot be erased) or a DVD–RW disc (which can be erased and rewritten many times).

Did you know?

About the Terms *Copy* and *Burn*

In case you're wondering, the terms *copy* and *burn* really mean the same thing. The "burning" term refers to the fact that the information is "written" on the disc using a laser, which *burns* the information onto the disc's surface. The terms "copy" and "burn" are used interchangeably in the CD and DVD worlds, though we tend to use the term "copy" in this book for clarity.

Recording Speed

Pioneer's model DVR-810H features a record speed of up to 18X on its DVD disc drive. The other model available from Pioneer as of this writing is the DVR-57H, which also features the same speed DVD drive. Currently, the DVR-57H differs from the DVR-810H in that it is equipped with a 120 GB drive and a heavy-duty, double-layered chassis that reduces vibrations, thereby improving signal readout accuracy.

This 18X recording speed means that you'll be able to copy a 1-hour program to DVD in basic quality mode in approximately three and a half minutes. The Pioneer TiVo-DVD recorders' advanced circuitry can even record content directly to the hard drive while dubbing separate content from the hard drive to a DVD. And as an added benefit, the content being transferred from one system to the other remains in a digital state throughout, so both the audio and video quality stay high.

Disc Compatibility for Recording

Disc Size

The TiVo-DVD recorder can copy video programs onto discs 12 centimeters in diameter, which is the standard.

Audio Format

The TiVo-DVD recorder *cannot* record onto audio CDs and MP3 music CDs. Therefore, it *cannot* use CD-R discs (which are permanent and cannot be erased) or CD-RW discs (which can be erased and rewritten many times).

Video Format

The TiVo-DVD recorder can copy onto DVD-R and DVD-RW discs, so long as the discs are DVD-RW Version 1.1 or 1.1/2x, or DVD-R Version 2.0 or 2.0/4x. The DVD disc version you are using is marked on the package, and the versions mentioned here are standard today. You should only be concerned about these numbers if you plan to use recording media that is more than a few years old.

Disc Compatibility for Playback

It's important to note that disc compatibility for playback is significantly different than that for recording. This is due in part to DVD "Region Codes." DVDs and DVD players are manufactured for different areas of the world and therefore must have matching codes. Discs marked with the region code "All" will work in any DVD player

Did you know?

DVD-R Versus DVD-RW

Which is better, DVD-R or DVD-RW? There is some contention over this in the videophile community, although the consensus is that DVD-R provides slightly better quality over a DVD-RW if the DVD-RW has been rewritten several times. DVD-Rs are often significantly less expensive, so it's generally best to buy based on whether you will really need to rewrite the same disc over and over again.

worldwide. DVD Players are most commonly marked with the region code on the back panel. Pioneer and other manufacturers for the North American market are marked with a "1" for the code. Any disc that is not marked with a "1" or an "All" will not play on the DVD player. However, it will be exceedingly rare that you will find a disc that matches this criteria.

According to Pioneer Electronics, some discs that were burned on a different DVD recorder, a CD burner, or a personal computer's DVD burner may not be playable in the TiVo-DVD recorder. This can happen even if the material was recorded with software that is compatible with the recorder.

Disc Size

The TiVo-DVD recorder can play discs that are either the standard 12 centimeters in diameter, or the mini-discs that are only 8 centimeters in diameter.

Audio Format

The TiVo-DVD recorder will play audio CD-R and CD-RW discs recorded in CD Audio format or recorded as CD-ROMs containing MP3 files.

Video Format

The TiVo-DVD recorder can play DVD-R and DVD-RW discs, so long as the discs use DVD-Video format (Video Mode).The recorder call also play DVD-RW discs that use the Video Recording (VR) format.

PC-Created DVDs

The TiVo-DVD recorder will play most DVD-R and DVD-RW discs, though there is a small chance that the software or software settings used to create the disc may be incompatible. If the disc does not play, contact the publisher of the DVD creation software for more information on whether an upgrade or more compatible version of the software is available.

Also, according to Pioneer, sometimes the TiVo-DVD recorder will have problems playing discs that have complex directory structures—in other words, discs that have been burned with large numbers of directories and subdirectories listed on the disc.

Disc Compatibility with Other DVD Players

All of Pioneer's DVD players record in Video Mode, which is the standard use of the DVD-Video format when recording on DVD-R or DVD-RW discs. This means that DVD discs copied using the Pioneer TiVo-DVD recorder can be played back on any brand of DVD player, including DVD drives installed on computers that run software compatible with the DVD-Video format.

However, it's important to note that support for playing DVD-R and DVD-RW discs (as opposed to regular DVDs) has always been optional for manufacturers of DVD player units, and some may have chosen not to implement a system that plays DVD-R or DVD-RW discs. In some cases, a hardware-based or software-based upgrade may be available. Contact the manufacturer of the player or recorder you intend to use to determine this.

MP3 Format Issues

Pioneer's TiVo-DVD recorder can play discs containing music or other audio files, such as the MP3 format. The recorder can handle the .mp3 filename extension, regardless of whether it is written in uppercase or lowercase letters.

NOTE *A file with the .mp3 filename extension isn't necessarily a music file in MP3 format. Usually they are, but anyone can give any file any name they want. A file that is not in MP3 format can have the .mp3 extension, and some valid MP3 files can be misnamed and not have the .mp3 extension. In either of these cases, the file either will not play, or it will play with substantially reduced sound quality.*

Bit Rate

The TiVo-DVD recorder is able to play MP3 files encoded at any rate. Of course, only music recorded at 128Kbps (kilobytes per second) will come close to CD audio quality. The lower the bit rate, the "patchier" and less dynamic the music will sound.

Troubleshooting: Physical Problems that can Occur to your Pioneer TiVo-DVD Recorder

Damage, dirt, and condensation are physical problems which may require you to replace or repair the disc, the drive, or the entire unit. The most common damage suffered to the TiVo-DVD Recorder is to the hard drive, from either electrical surges or from excessive weight placed on the recorder itself. To prevent these problems, install a surge/lightning protector and never place additional video or audio components on top of the recorder. Additional issues and their most common solution may include:

- Dust or dirt crusted on the disc or in the TiVo-DVD Recorder's disc drive. Solution: Follow instructions on disc case or TiVo-DVD recorder's User Guide to either physically remove dust or dirt, or to use a disc lens cleaner.
- Condensation formation in the TiVo-DVD Recorder's disc drive due to resident moisture in the room or in the air. Solution: Move the recorder to another room to allow it to dry; install a dehumidifier in the original location before returning the recorder there.
- Damage suffered to the TiVo-DVD Recorder's disc drive or a scratch on the DVD disc's data face. Solution: Refer to your warranty or return policy for a repair or replacement unit.

A Word on Copy Protection Issues

Before you set out to record and burn DVDs to your heart's content, remember that the original producers of the material you are copying retain certain legal rights. Unless you own the copyright (which is the case if you created the material, such as home videos), or have obtained permission to make copies from the owner of the copyright, or have a legal right to reproduce or copy, you may be breaking the law by copying a program to DVD. Doing so makes you subject to payment of damages to the copyright owner or other remedies.

In other words, the DVD recordings you make on the TiVo-DVD recorder *are for your personal use only*. You may not mass-copy the disc, or sell or rent your copies to others. The law is less clear about creating a single copy and displaying it to an audience, but no one individual has been prosecuted or even taken to court in such a case, so you shouldn't have to worry too much about that happening.

Sometimes copyright protection is built in. For example, when portions of a program are copy protected, they cannot be copied to DVD. The TiVo-DVD recorder will only be able to copy the potions of the program that are not protected.

How to Play a DVD or CD

Playing a DVD or CD on your TiVo-DVD Recorder is generally a very simple matter, but the Pioneer machines also have some advanced playback features that you can use.

NOTE *You must complete the TiVo Guided Setup before you can play CDs and play or record DVDs. See Chapter 1 for details on setting up your TiVo.*

Inserting and Playing a Disc

To insert a disc into the TiVo-DVD Recorder, perform the following steps:

1. Press the Open/Close button on the front panel of the recorder, to the right of the disc tray.
2. Place the disc in the tray with the label side facing up, using the disc tray guide to align it.

NOTE *For small mini-discs (8 centimeters in diameter), notice that there is a small disc tray guide that is indented into the larger, standard-size guide. Make sure when you insert a mini-disc that the disc sits properly in this smaller tray.*

3. Press the Open/Close button on the front panel again.

CAUTION *Although it may be tempting, try to avoid simply pushing or tapping on the disc tray to close it. While this method works, it increases the likelihood that the disc tray will break or become misaligned.*

4. After you insert the disc, highlight the DVD option on TiVo Central using your TiVo remote control, and simply press Play to play the disc, as shown in Figure 9-1.

More playback options are available on the TiVo-DVD Recorder's DVD screen. To get to this screen, highlight the DVD option on the TiVo Central screen and press Select.

NOTE *You can also press the DVD button on your remote to get to the DVD screen.*

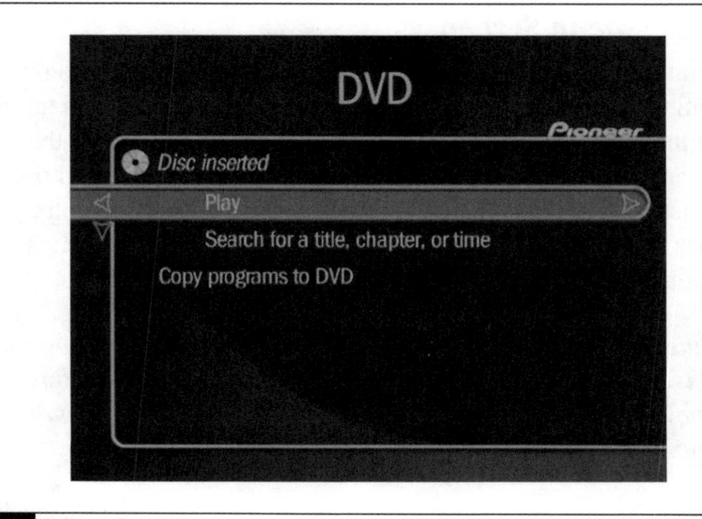

FIGURE 9-1 Pressing Play

The DVD screen may display options such as Resume Playing or Search Disc, depending on whether you've already begun playing the disc. Notice also that when you begin playing the disc, the TiVo-DVD Recorder's front-panel LCD screen will display the following information about the disc, which you can see at any time, whether the disc is playing or not:

- DVD title
- DVD chapter
- Time elapsed in the current title

NOTE *While playing MP3 files that have been burned onto a CD, or while playing an audio CD, the front panel will display the number and the elapsed time of the current audio track.*

DVD Information and Playback Settings

If the DVD you've inserted includes a menu screen, this screen will appear when you first insert the disc and press Play. However, you can open or return to this DVD menu screen by pressing either Menu or Top Menu on your remote when the DVD is playing.

Navigating the Menu Screen

When the Menu Screen is displayed, use the Up and Down arrows on your remote to move around the options displayed on the screen. Press the Select button to actually select the option. If you try to move beyond the listed options, the TiVo-DVD Recorder may emit a soft bong sound to let you know that there are no more options to be listed onscreen. Also, certain DVDs will also allow you to use the number buttons on the remote to select onscreen options. This depends on the DVD and not on the recorder's own internal software settings.

NOTE *One internal setting on the TiVo-DVD recorder that can affect how the controls work is the Parental Control setting. If you have turned on this option, you may have to enter a password to watch certain DVDs, depending on their content.*

Using the DVD Banner

If you press the Info button while a DVD is playing, the screen will show the DVD banner, as in Figure 9-2. The DVD banner shows the current time, the title and chapter number that is currently playing, and how far you are into the chapter that is being played.

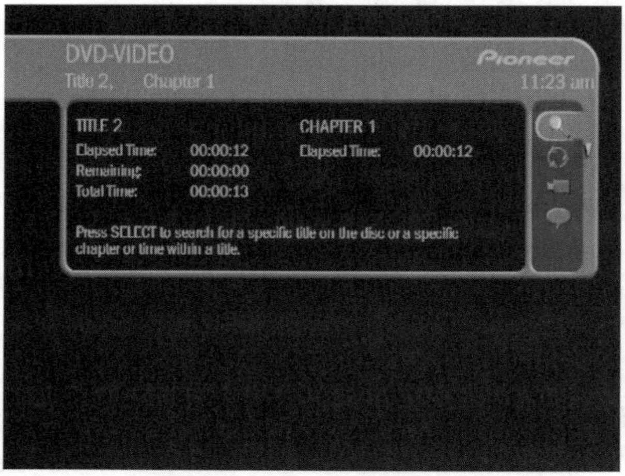

FIGURE 9-2 A sample DVD banner screen

On the right side of the banner are selectable icons that allow you to change DVD settings. Use the arrow buttons to navigate between the available icons, and press the Select button to pick one of the settings:

- **Title search** Allows you to search for a title, chapter, or time on the DVD.
- **Repeat setting** Allows you to choose a different repeat setting for the DVD—you can choose to repeat the whole disc, or only a certain chapter.
- **Camera angle** Allows you to change the angle from which you view the action on the disc. This is only available if you're watching a DVD that provides for multiple camera angles.
- **Audio track and subtitle** Allows you to change the audio track or the subtitle language to whichever languages have been added to the DVD.
- **Progressive output** Allows you to change the Progressive Output setting. This option will only be available if the DVD's Progressive Output setting is Progressive.

You can hide the banner at any time by pressing the Clear button or the left arrow button.

NOTE *When watching DVDs and other recorded programs, you can use the Slow, Frame Forward, and Frame Back controls, which allow you to move forward or backward a single frame at a time in a program.*

MP3/Audio Information and Playback Settings

Pioneer's TiVo-DVD Recorder not only plays CDs, it also plays MP3 files. This is because it is able to read music playlists that have been saved in the standard MP3 format, M3U. It can also read the rare file that has been recorded with VBR (Variable Bit Rate), but if the file is large, the navigation buttons may remain inactive.

Using the Navigation Buttons

Luckily for the technically challenged, the navigation buttons used by TiVo are the standard ones used with all CD or DVD players today. All of these controls work the same way for playing MP3s, DVDs, CDs, and recorded programs:

- **Play** Plays a disc at a normal speed.
- **Pause** Pauses the selection at a given point. Pressing Play allows you to continue.

- **Fwd** Moves forward in a selection.
- **Back** Moves back in a selection.
- **Advance** Advances to the end of a program.
- **Instant Replay** Rewinds the program approximately eight seconds.

You should keep in mind that when playing DVDs and CDs, you can also use the Skip Fwd or Skip Back buttons, which allow you to skip forward to the beginning of the next chapter or track, or to skip back to the beginning of the current chapter or track on the disc.

NOTE *When watching DVDs and other recorded programs, you can also use the Slow, Frame Forward, and Frame Back controls, which allow you to move forward or backward a single frame at a time in a program.*

Displaying Audio Information

While MP3 music is playing, the TiVo-DVD Recorder will show a blue song-information banner onscreen. If the MP3 files are tagged with the song information, it will be displayed: song title, artist, album, and total number of tracks. The software that was used to create the disc will determine what information is displayed, not the TiVo-DVD Recorder. Note that the song-information banner will automatically change positions on the screen every 20 or 30 seconds. This is the normal setting for the recorder. You can press the Clear button on your TiVo-DVD remote to hide the song-information banner at any time.

While you're playing an audio disc, press the Info button on your remote. You'll see the music banner, which is much like the DVD banner described in the "Using the DVD Banner" section earlier in this chapter. The music banner displays the track number, track title (if available), and the length of the track. The screen will disappear on its own after about six or seven seconds, or you can press the Clear button on the remote to hide the banner.

Music Play Options

While the music banner is displayed (as described in the previous section), you can press the Enter button to see the Music Play Options screen, shown in Figure 9-3.

Use the arrow buttons to navigate the menu, and press the Select button to choose the option you want. There are four to choose from:

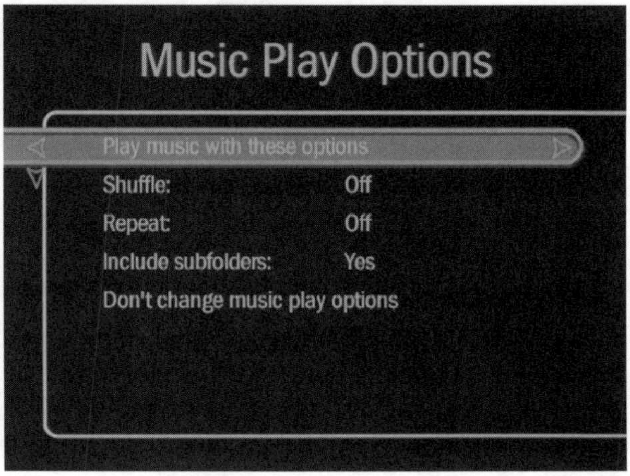

FIGURE 9-3 Setting options for playing music

- **Shuffle** Plays music tracks in a random order, as opposed to the order they are recorded on the disc.
- **Repeat** Sets the Repeat mode to either One or All. If One is selected, the current track will repeat again and again. If All is selected, the entire disc will be replayed.
- **Include Subfolders** Allows you to include on the playlist any MP3 files that are listed under separate subfolders on the disc.
- **Don't Change Music Play Options** Returns you to the default options settings.

CAUTION *Programs transferred to the TiVo-DVD Recorder from a different kind of TiVo DVR, such as one using Home Media or TiVo Plus cannot be copied from the TiVo-DVD Recorder's hard drive onto a DVD disc.*

To copy programs from the TiVo-DVD Recorder's hard drive to DVD, perform the following steps:

1. Open the disc tray, insert a DVD-RW or DVD-R disc, and close it.
2. From TiVo Central, select DVD. The screen will look like Figure 9-4.
3. Select the Copy Programs to DVD option.

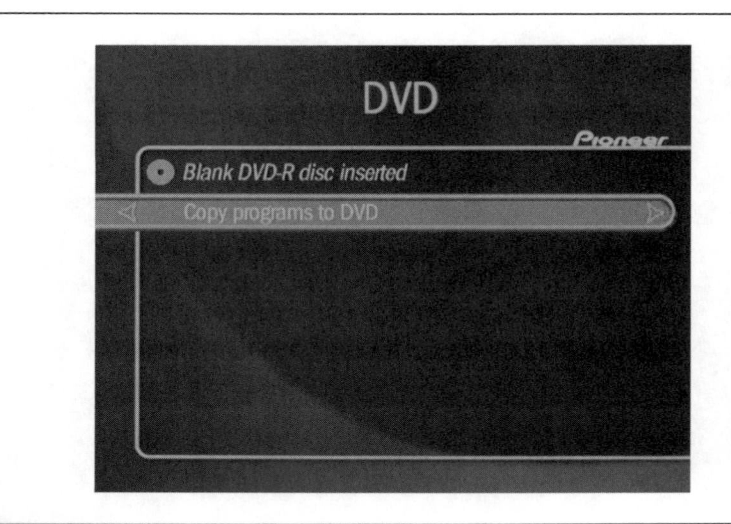

FIGURE 9-4 Copy Programs to DVD screen

4. Choose the programs you want to copy:
   - Programs listed in the Now Playing screen that cannot be copied are marked with either an X or a square with a slash through them. Available programs are marked with an empty square.
   - Programs will be recorded in the order you select them, and they will be marked with a number (1, 2, 3, and so on) as they are selected. To change the order of the selections, deselect the programs you want changed by using the Select button on the remote, and then reselect them in the order you want to record them.

How to ... Record a DVD from TiVo

Once you have completed the TiVo Guided Setup (described in Chapter 1), you can copy programs from your TiVo to DVD. Copying to DVD will, of course, use the disc tray, so you cannot watch one DVD while copying another. However, copying to DVD will not interfere with recording TV programs to the TiVo hard drive or watching live television.

- You can select up to 24 programs to copy onto a DVD. The TiVo-DVD Recorder provides a gauge as to how much space you have left on a DVD by showing the percentage on a DVD icon in the upper-right corner of the screen. Also, the recorder will mark (with an X) the selections that will not fit on the disc.

5. When you are done making your selections, select Done Choosing Programs, or press the right arrow on your remote.
6. You'll be prompted to name the DVD being created. Enter the name of your recording by using the arrow buttons on the remote and select each letter with the Select button. Select Done Entering Name when you are finished.
7. The Confirmation screen (shown in Figure 9-5) will be displayed so you can confirm your selections before starting to copy to the DVD. Select Begin Copying to CD to start copying, or press the left arrow button to return to the selection screen and make changes.
8. While the DVD is being copied (or burned), the Copying to DVD screen will automatically be displayed. You can switch to watch live television or perform other commands. Pressing the DVD button on your remote will return you to this status screen, as shown in Figure 9-6.

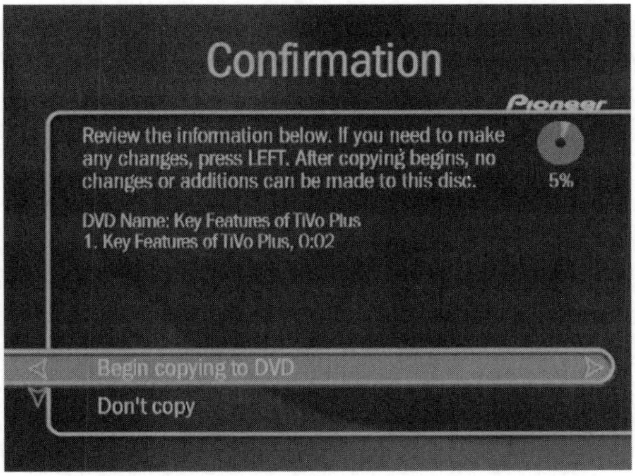

FIGURE 9-5 Confirmation screen

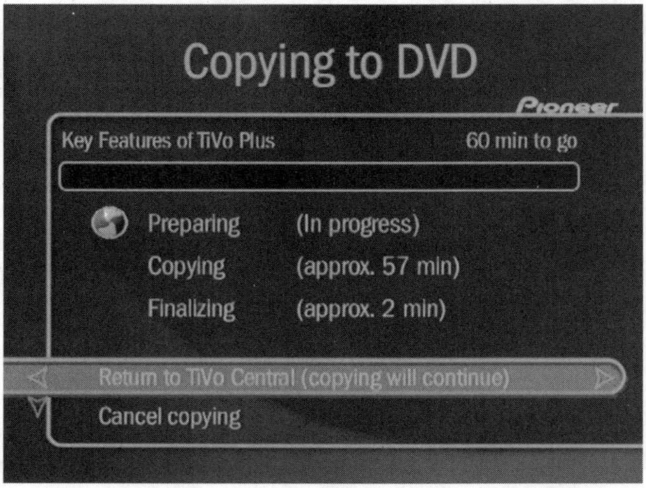

FIGURE 9-6 Copying to DVD screen

9. The TiVo-DVD Recorder "finalizes" the DVD when the disc is completed. You do not have to enter any commands for this step. Once the DVD is finalized, it appears on the onscreen menus as any other DVD does.

CAUTION *If you're concerned about adult content on a DVD, be aware that you cannot set parental-control ratings on DVDs that you create. Therefore, the TiVo-DVD Recorder (or any other DVD player) will not be able to block the material on the DVD unless the parental-control ratings on the player are set to Block All DVDs.*

One question that sometimes comes up when copying material onto DVDs is whether the medium or style of the program results in an increase or decrease in the size of the files copied. It's certainly possible that a two-hour movie filmed in black and white could consist of less data than a two-hour movie filmed in color, but in practice there's little or no difference. Where you will see large differences in the size of files is between images (movies have lots of visual data to record) and audio files (audio tracks contain less data). A picture really can be worth a thousand words, or a thousand notes at least.

How to ... Record a DVD from a Videotape

You can perform a two-step process to record a DVD from a videotape (as long as you have completed the TiVo Guided Setup described in Chapter 1). In the first step, you copy the content of the videotape to the TiVo hard drive, and then you copy the content from the TiVo hard drive onto a blank DVD-R or DVD-RW disc.

To copy a videotape to the TiVo hard drive, follow these steps:

1. Go to TiVo Central by pressing the TiVo button on your remote.
2. Select the Pick Programs to Record option.
3. Select the Record From a Video Camera or VCR option.
4. You'll be prompted to confirm this choice, as shown here. Select the Yes, Continue option.

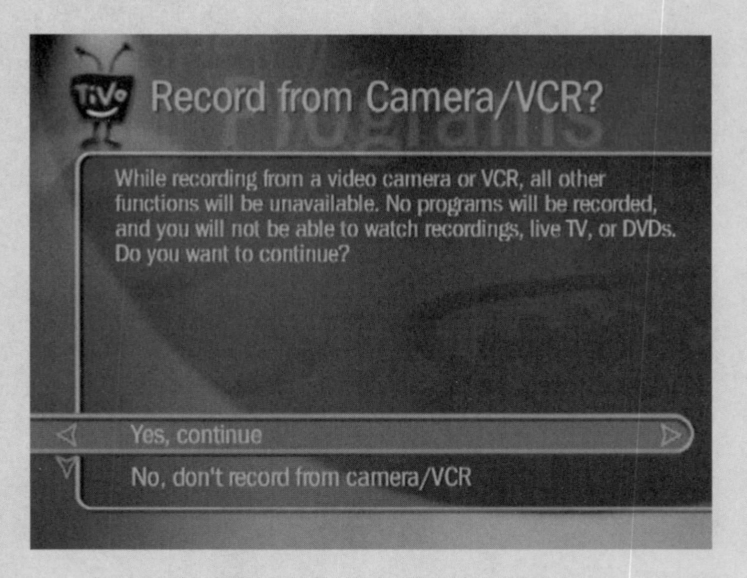

5. Connect your VCR to the TiVo-DVD recorder's Input 2 jack with a composite cable.
6. Turn on the VCR and insert your videotape.
7. On the TiVo-DVD recorder, select the Done Connecting and Turning on Camera/VCR option to continue.

8. On the Name Recording screen, enter the name you want to give this recording by using the arrow buttons to select letters with the TiVo onscreen keyboard. Select the Done Entering Name option when you are finished.

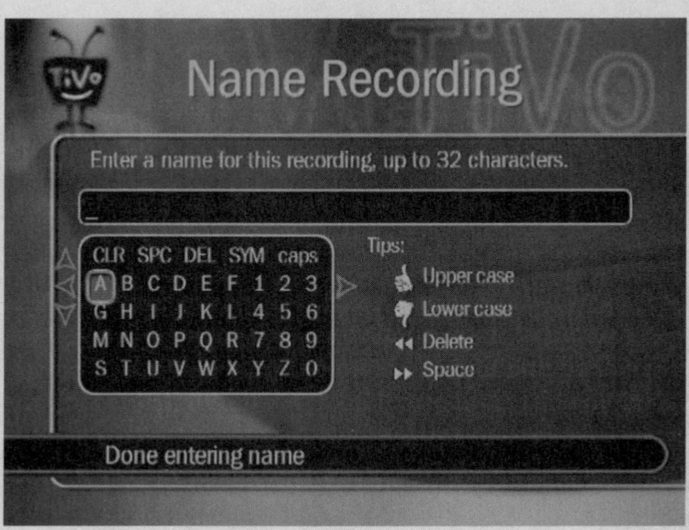

9. Specify the recording quality you want to use and the length of time the recording should be kept. Note that the TiVo-DVD recorder will indicate when it will be finished recording at the bottom of the screen.

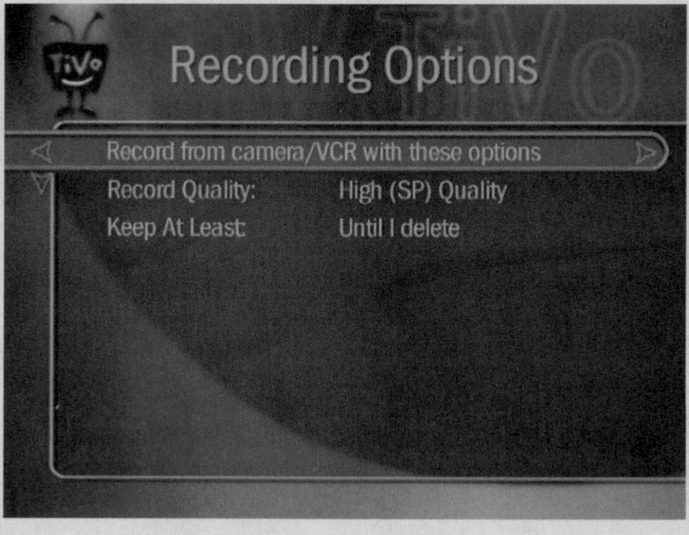

10. When the recording is complete, it will be listed in the Now Playing screen. You can now play this recording like any other regularly recorded TiVo program.

To copy your VCR program, which is now stored on the TiVo hard drive, to a blank DVD, simply follow the same steps you would for recording any program from the TiVo to a DVD. See the "How To Record a DVD from TiVo" section earlier in the chapter for the details.

How to ... Record a DVD from a Video Camera

You can perform a two-step process to record a DVD from a video camera (as long as you have completed the TiVo Guided Setup described in Chapter 1). In the first step, you can copy the content from your camera to the TiVo hard drive, and then you can copy the content from the TiVo hard drive onto a blank DVD-R or DVD-RW disc.

To copy a recording from a video camera to TiVo, follow these steps:

1. Go to TiVo Central by pressing the TiVo button on your remote.
2. Select the Pick Programs to Record option.
3. Select the Record from a Video Camera or VCR option.
4. You'll be prompted to confirm this choice. Select the Yes, Continue option.
5. Connect your video camera to the TiVo-DVD recorder's Input 2 jack with a composite cable.
6. Turn on the video camera.
7. On the TiVo-DVD player, select the Done Connecting and Turning on Camera/VCR option to continue.

8. On the Name Recording screen, enter the name you want to give this recording by using the arrow buttons to select letters with the TiVo onscreen keyboard. Select the Done Entering Name option when you are finished.
9. Specify the recording quality you want to use and the length of time the recording should be kept. The TiVo-DVD Recorder will indicate when it will be finished recording at the bottom of the screen.
10. When the recording is complete, it will be listed in the Now Playing screen. You can now play this recording like any other regularly recorded TiVo program.

To copy your video camera recording, which is now stored on the TiVo hard drive, to a blank DVD, simply follow the same steps you would for recording any program from the TiVo to a DVD. See the "How To Record a DVD from TiVo" section earlier in the chapter for the details.

Part III

Don't Say I Told You So: Hacking TiVo Systems

Chapter 10 Easy Hacks

How to...

- Enable Backdoor Mode on Your TiVo (Series1 TiVos Only)
- Hack Your TiVo Using Your Remote Control
- Add an External Modem if Your Internal One Fails

Ever since the first TiVos came out, powered by the open source (and, therefore, rather accessible) Linux operating system, devoted TiVo owners have spent hours and hours hacking their TiVos to enable them do more. Some of these hacks are very useful, and others are silly or almost useless—discovered and documented merely for the thrill of the chase.

All of the hacks in this chapter are *software* hacks—you will not need a screwdriver to try them out for yourself. However, if you are running version 3.1 or later of the TiVo operating system, you may need a screwdriver to prepare your TiVo to accept some of the hacks.

CAUTION *Just because these are software hacks does not mean they come without risk. It is possible that, during the execution of some of these hacks, you could potentially render your TiVo unbootable unless you have a backup (these hacks are noted). Some, or even all, of these hacks may invalidate your warranty. Exercise your own judgment and take precautions, especially with the hacks marked as potentially dangerous.*

You will need to know what model TiVo you're working on and what version of software it's running. Also, you'll need to enable a special mode on your TiVo called *backdoor mode* to use many of these hacks. When you enable backdoor mode, your TiVo knows to pay more attention to the key combinations you press. Without backdoor mode enabled, most of these remote control hacks won't work, because the TiVo won't be "listening."

Backdoor mode is fairly easy to enable on Series1 TiVos but much more difficult on Series2 TiVos. Enabling backdoor mode on Series2 TiVos is so risky that it is beyond the scope of this book to explain it.

Identifying Your TiVo

Unless you know it already, you'll need to find out the model number of your TiVo. This information is located on the TiVo's rear panel near the power cord. Series1 TiVos include Philips HDR models, Philips DSR 6000, Sony SVR 2000, Sony SAT

T-60, and the Hughes GXCEBOT models. A complete chart of current TiVo models can be found on the Internet at http://tivo.upgrade-instructions.com/step1.php.

Any TiVo model may be running any of a variety of operating system versions, and identifying which version you're running is the next step. Go to Messages & Setup and select System Information. The software version should appear on the first page of information.

As mentioned before, many of these easy hacks require enabling backdoor mode on your TiVo. If your TiVo is a Series2 and you're determined to use these tweaks, you will need to do your own research about enabling backdoor mode and be very, very cautious. (Don't forget to back up your TiVo, as described in Chapter 12!) It is possible that easier solutions for enabling backdoor mode in Series2 TiVos may crop up in the future, so it's a good idea to subscribe to TiVo bulletin boards or join user groups to keep up to date.

Unlocking the TiVo Backdoor Mode

Here's one of those times where knowing the version of your TiVo's software comes in handy. There are different ways to enable backdoor mode, depending on whether your TiVo operating system is version 3.0 and older or version 3.1 and later.

TiVo OS Version 3.0 and Earlier (Series1 TiVos Only)

If you're a lucky soul with TiVo operating system version 3.0 or earlier, enabling backdoor mode is criminally simple: all you have to do is type in a code and go. Follow these steps:

1. Go to TiVo Central (or DirecTV Central).
2. Select Pick Programs to Record.
3. Select Search by Title.
4. Select All Programs.
5. Refer to Table 10-1 and find your operating system (OS) version. Enter the corresponding backdoor code into the TiVo's Title field exactly as it appears in the table, spaces included. To enter a space, choose SP from the letter menu.
6. Press the Thumbs-Up button; you'll hear three tones, and a confirmation message will appear.

TiVo OS Version	Backdoor Code
1.3 (US)/1.5 to 1.5.1 (UK)	0V1T
1.5.2 (UK)	10J0M
2.0	2 0 TCD
2.5 (US)/2.5.5 (UK)	B D 2 5
2.5.2 (DirecTiVo)	B M U S 1
3.0	3 0 BC

TABLE 10-1 Backdoor Codes for Series 1 TiVos running OS 3.0 and Earlier

Keep in mind that if you reboot your TiVo or it loses power, you will need to re-enable backdoor mode. The flip side of this is that if you want to disable backdoor mode, restarting your TiVo is the easiest way to do so.

TiVo Remote Control Codes

Below you'll find a number of hacks you can use to customize your TiVo using only your remote control. Although simple to program and occasionally useful, you will not find these methods mentioned anywhere in the owner's manual that came with your TiVo. So enjoy...

NOTE *Not all of the codes described in this section work on every version of the TiVo operating system. If you're sure you entered the codes properly, and they still don't work, you may be out of luck. Where the limits are known, they have been noted.*

Select-Play-Select (SPS) Codes

Select-Play-Select (SPS) codes are the simplest but most useful TiVo codes to try, and they do not require backdoor mode to be enabled. To enter an SPS code, start a recorded show playing and then press the buttons quickly: Select, Play, Select, the number of the hack, and then Select once more. If you entered the codes successfully, you will hear three Thumbs-Up tones. The changes that result from the code may not take effect immediately; you may need to back out of what you're doing, head back to TiVo Central, and then return to what you were doing to see the changes.

To disable these hacks, just re-enter the same code and wait for the tone (and again, they may not be disabled immediately; don't panic, back out of the menus to TiVo Central and then check again).

TIP *It's easier to enter these codes while playing a recorded show, but you can enter them during live TV too.*

30-Second Skip Button

One of the presumed advantages of ReplayTV over TiVo is the 30-second skip feature—pressing a single button lets you skip ahead in 30-second increments. While TiVo has three fast-forward speeds, it's still not the same as being able to skip a whole commercial in a single bound. But TiVo *does* have a 30-second skip—you just have to look for it.

To change your Jump to End button to a 30-Second Skip button, do the following:

1. Select Watch Live TV or go to your Now Playing list and begin playing back a previously recorded program.
2. During playback, press the following button sequence on your remote: Select, Play, Select, 3, 0, Select.

Once you've entered the full sequence, you should hear three repeated confirmation tones. If you heard the tones, go ahead and press the Jump to End button to make sure that it worked. If you didn't hear the tones, chances are you'll have to try it again.

NOTE *This hack will also work on Series2 units with the exception of Toshiba models.*

Voices from the Community

Touchdowns for TiVo!

The 30-second skip is useful in many situations other than burning through commercials. For example, when I watch my alma mater play football on TV, this is a fantastic feature. The Cal Bears seem to use/need every second on the clock between plays, so the time between a tackle and the next snap is nearly always 30 seconds. And if for some reason they decide to rush things, I can use the instant replay button to quickly jump back a few seconds.

—*Jeff Shapiro, weaKnees.com*

TiVo Clock

TiVo truly is a multipurpose device. For example, you can turn on a clock in the bottom right corner of your TV screen so you don't have to go to the Live Guide or crane your neck to look at the clock on your wall to know what time it is.

To enable the clock, enter this code: Select, Play, Select, 9, Select.

This is one of the hacks that may require backing out of what you're doing. Return to the main menu if the clock does not appear. Your clock may have a black background; some of the other remote control hacks have the side effect of removing this background. In some versions of the TiVo OS, this hack enables an elapsed-time indicator.

NOTE *Is your clock tilted or partially hanging off your screen? The "geometry" of your TV needs adjustment. This is not something you can usually adjust yourself, so you'll have to either take the TV to a repair shop or live with it. If you hadn't noticed until now, just turn off the clock and return to your previously blissful existence.*

TiVo Status

If you have some desire to know what your TiVo is doing at any given time when you're sitting in front of it, you can turn on a text-based status indicator at the bottom of the screen. It's not a very pretty indicator, but it is potentially useful (or not useful, but neat to look at nonetheless).

To enable this TiVo status mode, use this code: Select, Play, Select, Instant Replay, Select. To disable it, re-enter the code. You'll need to visit and return from another menu for the change to take effect.

Quicker Timeline Bar

Are you annoyed by that pesky timeline bar that appears at the bottom of the screen when you're rewinding, fast forwarding, or otherwise manipulating a live or recorded program? With this little hack, you can hasten its disappearance.

Enter the following code to remove the timeline bar: Select, Play, Select, Pause, Select.

The timeline bar will now disappear almost instantly and won't remain on the screen while programming is paused, fast forwarded, or rewound.

Triple-Thumb Codes

These codes work only in TiVo OS versions 2.5 and later. Most of these do not require the backdoor to be enabled.

Scheduled Suggestions and Hidden Recordings

To see Scheduled Suggestions in your To Do list, enter the following code: Thumbs-Down, Thumbs-Down, Thumbs-Up, Instant Replay.

To have the hidden promotional recordings show up in the Now Playing list, enter the same code in the Now Playing screen.

Clips on Disk

To get to a hidden TiVo section called Clips on Disk, enter the following code while in the Now Playing screen: Thumbs-Down, Thumbs-Up, Thumbs-Down, Instant Replay.

From the Clips on Disk screen, you can see all of the hidden promotional recordings separated into clips used in Showcases. This is not very useful, unfortunately, and it may force reboot your TiVo.

Hidden Showcases

TiVo frequently stores promotional content on your TiVo DVR. It is hidden from view until promoters have requested it be presented. To make any hidden Showcases become visible, enter this code while in the Showcases screen: Thumbs-Down, Thumbs-Up, Thumbs-Down, Record. This is not a particularly useful hack, unless you're a Showcase junkie.

Italic Fonts

To make the TiVo fonts italic, enter this code in TiVo Central: Thumbs-Down, Thumbs-Up, Thumbs-Down, Clear. Enter the same code again to disable the italics … it's not exactly easy to read that way!

Debug Messages

TiVos log error messages just like computers. If you're possessed by an urge to see what errors your TiVo may be generating, enter this code: Thumbs-Down, Thumbs-Up, Thumbs-Down, Enter. It doesn't matter where you enter it.

Clear-Enter-Clear Codes

These Clear-Enter-Clear codes require the backdoor to be enabled, although most of them are still not particularly useful. Most of the useful functions that they enable can also be done through regular TiVo menus on most TiVos.

NOTE

Any code you enter with the sequence Clear, Enter, Clear will make the confirmation tones described earlier in the chapter, but that doesn't mean it's a valid code that actually does something.

Turn On Scheduled Suggestions

Like the triple-thumb code listed previously, this hack turns on the display of Scheduled Suggestions in the To Do list. To enable it, enter this code: Clear, Enter, Clear, 2. You can then delete upcoming Suggestions that you aren't interested in.

NOTE

Remember that the recording of TiVo Suggestions is a standard menu option, and many users find their TiVo operates a little more efficiently with this option disabled.

Turn Off Scheduled Suggestions

In case you get sick of your To Do list being full of TiVo's Suggestions, you can hide Scheduled Suggestions again by entering this code: Clear, Enter, Clear, 0.

Rebuild Suggestions

Has your TiVo picked up some bad habits by studying your recording history? Is it suggesting programs you don't really want? Enter this code to reset your TiVo's learned behavior: Clear, Enter, Clear, 4.

NOTE

On later TiVo systems, you can do this through the regular menu system.

Turn Off Correction in FF/RW

Overshoot correction is that adjustment (sometimes good, sometimes annoying) that the TiVo makes when you fast forward or rewind to take into account how fast you're scanning and how quick human reflexes really are. You can turn it off by entering this code: Clear, Enter, Clear, 5. This hack is probably not a great idea unless you have superhuman reflexes.

Browse Log Files

This hack allows you to browse the log files on the TiVo instead of just seeing the most recent error messages: Clear, Enter, Clear, Thumbs-Up.

The log files will disclose such fascinating facts as when your TiVo has connected to the TiVo service. Scroll through the log info using the Channel button. The right arrow moves from log to log. When you're done, press the TiVo button to exit the logs.

Shut Down TiVo's Operating System

Enter this code to shut down the operating system on the TiVo: Clear, Enter, Clear, Thumbs-Down.

This will leave you with nothing. The only course of action left for you is to reboot—you have to actually pull the plug! This little hack is not really recommended.

Reboot TiVo

Enter this code to reboot or reset your TiVo: Clear, Enter, Clear, Fast Forward. You can also do this through the menus from the Messages & Setup screen.

View Node Navigator

Node Navigator is an interesting aspect of the TiVo's guts, but the problem is that you can render your TiVo useless by merely *looking* at the numbered nodes. (If you have a backup, you should be able to restore your TiVo from the backup and return it to functionality, but even that may be a gamble.) We only mention this code as a warning—*don't* enter the Node Navigator. The code to avoid is: Clear, Enter, Clear, 6.

Other Software Hacks

There's more to tweaking TiVo than the codes you've seen so far. Most of the following hacks are still remote-based, but they require a little more explanation than the very simple ones described so far in this chapter.

AutoTest Mode

AutoTest mode is a feature the TiVo folks added (and hid) for testing a TiVo that's having problems. It emulates normal use by generating random button-pressing. The actual usefulness of this, as of many other the non-backdoor hacks, is debatable, but it can be somewhat amusing to watch if there's *really* nothing on TV.

Enter the following code into the Now Playing screen to enable AutoTest mode: 1, 2, 3, Channel Down. You'll get a confirmation message. Then press 4 to start the test, 5 to change tests, and 7 or 8 to adjust the delay between (simulated) key presses.

When you're sick of watching your TiVo act possessed, press 4 once more to stop the test. This may reboot your TiVo.

Sort Now Playing in OS 3.0

In OS versions 4.0 and later, you can easily sort your Now Playing list any way you want by simply pressing Enter. But there's good news for OS version 3 users: if you enable the backdoor and then execute this little hack, you, too, can sort your programs at will. This is the code to enter: Slow, 0, Record, Thumbs-Up.

NOTE *You may notice that the code spells out SORT: S(low), 0, R(ecord), T(humbs up). Someone had way too much fun thinking of that one.*

Once you apply this hack, there will be a note at the bottom of the Now Playing screen that tells you to press Enter to see the sorting options You'll have the choice of sorting by recording date, by expiration date, or alphabetically by show title. You can also just use the number buttons 1, 2, and 3 to quickly choose a new sort method.

Enable Advanced WishLists

WishLists are immensely powerful, but there's an even more powerful WishList option that the TiVo programmers chose to hide. Advanced WishLists are hidden, and they're just about perfect, because you can specify multiple criteria, such as a certain actor *and* a certain director.

In OS versions 2.5 and up, just press 0 when you are on the WishList screen to make an Advanced WishList.

If you're running an older version of the operating system, you'll have to do a little bit more work. You need to use the Node Navigator hack (the one described as so potentially dangerous earlier in this chapter), so back up your TiVo first and exercise *extreme caution.*

CAUTION *As mentioned earlier in this chapter, Node Navigator is dangerous and could disable your TiVo. You should be able to restore your TiVo's functions from a backup, but there are no guarantees.*

Enter this code to enter Node Navigator: Clear, Enter, Clear, 6.

Once you've entered Node Navigator, resist the urge to look around. Merely looking at some of the nodes can leave your TiVo unusable. Carefully enter Node 30 and select the expert interface for WishLists. Next, select Pick Programs to Record, then Search Using WishLists and finally Create New WishList. The Advanced WishList option will appear at the bottom of the WishList varieties. Once you select it, you'll be able to enter your advanced search parameters. When you're finished, you'll be out of Node Navigator mode. Reboot the TiVo and the Advanced WishList feature will remain enabled.

Adjust Fast Forward and Rewind Speeds

Not only can you disable or tweak overshoot correction, you can adjust the speed of rewind and fast forward. By combining the two different hacks, you can really take control of your TiVo. Once you enter one of the speed codes, you can adjust the speed in multiples of 100 (with 100 being the same as 1x speed, or regular playing speed). As you know, your TiVo will scan forward or backward in three speed increments, with each press of the Fast Forward or Rewind buttons increasing the speed by one increment. The defaults for the three speeds vary on different TiVo operating systems, but they are always around 300 (3x) for Speed 1, 1,800 (18x) for Speed 2, and 6,000 (60x) for Speed 3. You can't see these defaults—you can only enter a new number.

The following codes must be entered in the Search by Title screen. Once you've entered a code successfully, the recording indicator light will turn on (your TiVo will not actually begin recording, nor will it stop recording if it's currently recording a show). You can then type in a new speed value in multiples of 100 using the number keys on your TiVo remote. Once you've entered the new value, re-enter the code you used to get the prompt, and the recording light will go off again and the value will be saved. Note that if you restart your TiVo, you'll need to re-enter these values.

To set the rate for Speed 1 (the speed for the first press of Rewind or Fast Forward), enter this code: Enter, Enter, 1.

To set the speed for Speed 2 (the speed for the second press of Rewind or Fast Forward), enter this code: Enter, Enter, 2.

To set the speed for Speed 3 (the speed for the fastest Rewind or Fast Forward, achieved by pressing the button three times), enter this code: Enter, Enter, 3.

NOTE *Want to play a little trick on someone? Enter a value of 1 for Speed 1 on their TiVo. That will make their rewinding or fast forwarding a crawling 0.01x speed! To reset the value, either use the hack again or restart the TiVo.*

Remote Shortcuts

You already know that you can press the TiVo button on your remote twice to jump to the Now Playing screen, but there are other shortcuts as well. These shortcuts let you jump all over the TiVo menu systems without all that time-consuming menu navigation. Table 10-2 lists several button sequences that start with the TiVo button and perform different menu options.

Button Sequence	Action
TiVo, 0	Plays the TiVo's start-up movie, usually seen only when plugging in the TiVo. Different models have different "boot movies," but most of them feature the animated TiVo character.
TiVo, 1	Displays the Now Playing screen (same as pressing TiVo, TiVo)
TiVo, 2	Displays the To Do list
TiVo, 3	Displays the WishLists
TiVo, 4	Displays the Search by Name screen
TiVo, 5	Displays the Browse by Channel screen
TiVo, 6	Displays the Browse by Time screen
TiVo, 7	Displays the Record Time/Channel screen (manual record)
TiVo, 8	Displays the Suggestions screen
TiVo, 9	Displays the Network Showcases screen
TiVo, Slow	Displays the Messages & Setup screen

TABLE 10-2 Shortcuts to Use with the TiVo Button

Using an External Modem

Standalone TiVo models need to dial in to TiVo Central to stay current on programming information. Series2 DirecTV TiVos get their programming information via the satellite, but unless you order pay-per-view programming over the Internet, DirecTV users will need the use of a modem as well.

If your TiVo's modem fails, don't despair. You can buy an external modem for next to nothing, and it can be configured to make all of your dial-in calls. An external modem will work with any TiVo running software version 3.0 or later.

Alternatively, if you don't want to forage for parts and configure your own modem, you can purchase a complete kit on the Internet. Do an Internet search for "TiVo" and "modem."

Materials

To hook up an external modem to your TiVo, you'll need a few things:

- **Modem** You'll need to pick up a modem, of course. The tried and true product is a US Robotics Sportster modem, usually with a baud rate in the 19,200–33,600 Kbps range. If you want to use a different modem, you'll need to do some research to determine the proper dip-switch settings.

- **Computer and software** Windows users can use a terminal emulation program such as HyperTerminal to configure their modem. Mac users can use ZTerm.
- **Computer cable** An RS-232 cable (DB9 female to DB9 male) will connect a Windows computer to the modem using an adapter (see below). Mac users with older models can use the same cable, but late model users will also need a USB to serial adapter in addition to the modem adapter (again, see below).
- **Stereo to 9-pin serial cable** Once the modem has been configured on your computer, you'll need a different cable to connect the modem to the TiVo. This cable was supplied with Series1 stand-alone TiVos, but if you've lost it or need one, a replacement isn't expensive. One end has a mini-headphone jack and the other has a DB9 male connector.
- **Modem adapter** You'll need a 9-pin female to 25-pin male adapter to connect your serial cable to the modem.

Configuring the Modem

You'll need to connect your modem to a PC and enter a command string to configure it for use with your TiVo. This will configure the modem for all TiVo models.

Follow these steps to set up a US Robotics modem:

1. Set the modem's dip switches as follows:

   1-down, 2-up, 3-down, 4-up, 5-down, 6-down, 7-up, 8-down

2. Connect the modem to your computer with the appropriate cable.
3. Plug in the modem's power cable and turn the modem on.
4. Start up the terminal emulation program on your computer.
5. Type the following string to reset the modem to factory settings:

   AT&F

6. Press ENTER. After you receive an OK response, enter the following string to configure the modem for use with your TiVo:

   AT&R1&D0&I0&H0M0S0=0&W&W0&W1

7. Press ENTER and wait for the OK response. Your modem is then configured and can be disconnected from the computer.

NOTE *If you are using a different type of modem, you will need to research the correct settings and commands for use with the TiVo.*

Connecting the Modem and Configuring the TiVo

Once the modem has been configured, you need to connect it to the TiVo and change the TiVo's dialing preferences. Follow these steps:

1. Connect the modem to the TiVo with the stereo-to-serial cable and the modem adapter.
2. Remove the telephone wire from the TiVo and connect it to the line-in port on the modem.
3. Navigate to the Dialing Preferences screen on the TiVo and enter the following dialing prefix:

   comma-pound-3-1-9

   For the *comma* you'll have to press the Pause button, and for the *pound* press Enter. If you have any other dialing prefixes you usually use, such as one for disabling call waiting, you should enter them after this new prefix.
4. Your modem and TiVo are now both configured. Test the connection to make sure it works—listen in on the phone line while the TiVo makes its call.

TIP *Sometimes a very long phone line will cause problems. And often you'll need to slow down the modem by using an alternative prefix to the one listed above. In this case you might try using comma-pound-3-9-6.*

Chapter 11 Upgrading Your TiVo's Hard Drive

How to...

- Diagnose a Hard Drive Problem in your TiVo
- Determine How Much Drive Space your TiVo will Hold
- Open your TiVo and Replace the Hard Drive
- Back-up the TiVo Image from Your TiVo's Hard Drive
- Format New Drives to Add to, or Replace, Your Existing Hard Drive
- Unlock a "Software-Locked" Hard Drive

Of the million and a half TiVos out there recording and pausing live television every day, it is estimated that some 10 percent have been upgraded in some way. How do we know? Because just as your TiVo will report back to TiVo corporate that you replayed Janet Jackson at the Super Bowl five hundred times in slo-mo, it will also report how many recording hours you have. Surprisingly, the majority of these upgrades have been performed by the owners themselves, and that statistic stands as a testament to how easy the procedure actually is.

Numerous pioneers have led the way in discovering and perfecting the upgrade, and today the marketplace has ample choices for TiVo owners wishing to follow suit. Whether your interest is in spending a few hours gathering parts and running the Linux software on your own, or if, instead, you don't want to lift so much as a screwdriver, every contingency has been accounted for. Most upgraders have opted for the "kit," a prepackaged bundle containing formatted drives, parts, and instructions, which takes all of the guesswork out of the process. Even if you have no electronics experience whatsoever, the kit should take you well under an hour to install, should you decide to go that route.

This chapter will provide you with a variety of options for performing this advanced hack, and it has something of a "choose your own adventure" quality to it. This is mostly because there are a large number of variables that apply to any particular upgrade. If you are new to the process or are just interested in understanding what it's all about, we recommend the short read: start at the beginning and read through to the "Surgery" section. If you decide you're up for the task, you'll then want to locate the architecture model that matches your unit (there are roughly six) and read through the short instructions. At the halfway point in those instructions, you will be directed to other sections in the chapter where you are free to choose from a selection of options and find one that best suits your needs. (We'll be giving you advice on these choices long before you need to make them.) After that, all that remains is to join the thousands of others who've discovered the spacious world of the TiVo upgrade.

Reasons for Upgrading

You don't really need a reason, but we thought we'd highlight a few of the common motivating factors behind the decision to upgrade drive capacity on a TiVo. Despite some slight nuances, the reality is that people decide to undertake this hack for one of two main reasons; either they *want* to or they *have* to.

Reason 1—More Space

Call it wiggle room. Call it manifest destiny. However you spin it, the primary reason people upgrade their TiVo is to allow for more recording capacity. Whether you live alone and simply want a limitless selection of programs on your Now Playing list or whether you have a houseful of people with completely different tastes, the end result is the same—you need more space. Not everyone wants to watch a show and erase it. And there are as many 40-year-olds as there are 4-year-olds who like to replay their favorite episodes.

Your TiVo will tell you how many hours of recording time you have at various quality levels (see Figure 11-1). When it comes down to it, 14, 20, or even 35 hours may simply not be enough. And that's often at Basic quality. DirecTV customers may find that their anticipated number of hours diminishes when they record lots of prime time or sports programming, which tend to use up more hard drive space. There certainly are larger capacity units coming on the market, but as it turns out, you will get a lot more space for the dollar by buying a lower-end TiVo and beefing it up. And more space means less deleting. So in the end, increasing your drive capacity will keep you, and everyone under your roof, happy. (Dare we say sedated?)

Reason 2—Dead Drive

When it comes to TiVos, the dead can rise and walk again. It happens all the time. No, your TiVo is not coming back as a zombie, but you can make it bigger and better than it was before. And it's completely legal. Many people upgrade their TiVo simply to get it working again. Your TiVo is essentially a computer with a hard drive, so it stands to reason that if the hard drive dies, you can simply replace it and move on with your life. Drive problems aren't the only reasons TiVos die prematurely, but they do account for a huge percentage of problems.

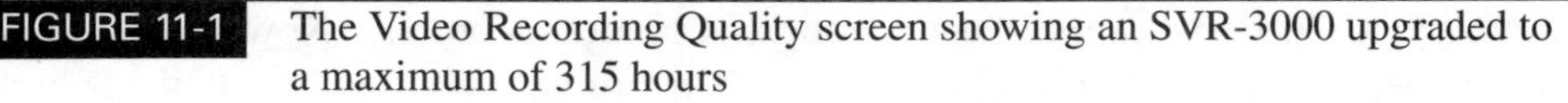
FIGURE 11-1 The Video Recording Quality screen showing an SVR-3000 upgraded to a maximum of 315 hours

NOTE *If you are replacing a dead or corrupted drive, you'll likely want to opt for a preformatted kit. See the "Useful Web Resources" section at the end of the chapter for some vender suggestions.*

Unlike the hard drive in a normal computer, your TiVo's hard drive works around the clock. It spins 24 hours a day, recording your Season Passes and buffering live television. Even when you put it in Standby mode, it's actually still working. The picture feed to the television may have stopped, but the TiVo and its hard drive have not. After a couple of years, those spinning magnetic platters have given a lot of service, and sooner or later the drive's going to call it quits.

Whether your drive is dead or dying, you'll likely find that replacing it yourself is not only more cost effective but also far less painful than sending it in for service, which—gulp—could take weeks! Don't forget, your Lifetime Service Agreement is attached to the TiVo, *not* the hard drive, so replacing the drive will not affect your service.

How Do I Know if My Hard Drive Is the Problem?

The longer a TiVo stays in active duty, the greater the risk it will develop drive problems. It's best to recognize the symptoms of drive failure early, so that you can act ahead of time and avoid the panic of losing your programming or of being without your DVR for a single must-see TV night.

TIP *If you're in doubt about whether your hard drive is at fault, you might consider purchasing a self-install kit. Because most problems are drive-related, and because most reliable vendors allow for returns, trying a drive kit is generally more cost-effective (and painless) than sending your TiVo in for repair. At worst, you might be out a restocking fee; at best, you'll have your TiVo fixed with only a half hour of downtime.*

Symptoms of Drive Failure

There are many potential signs that a TiVo hard drive is failing. By learning to recognize the signs early, you'll be in a better position to do something about it before your TiVo dies and your programming is lost. If you are experiencing one or more of these problems, it may be time to replace your drive.

- **Stuck on the "Welcome. Powering Up..." screen** Your TiVo will display this screen whether it has a drive in it or not. This image isn't stored on the drive; it is an indication that the TiVo is looking for a hard drive to read. If it stays on this screen for longer than a few minutes, the drive may have some serious problems.
- **Frequent rebooting** This is when you're watching TV and the TiVo just restarts on its own. It may be an anomaly. It may be a power problem. But if it starts happening a lot, it's usually a drive problem.
- **Picture problems** This is when the image sticks, freezes, or skips ahead without any instruction from the remote.
- **Pixellation** Pixellation is characterized by blocky, garbled chunks of color that look much worse than the image you're accustomed to at the quality setting you're watching.
- **Green Screen** Insiders refer to this as The Green Screen of Death (GSOD). It is a green screen that reads "A severe error has occurred," and it pleads with you to wait while the TiVo sorts itself out (see Figure 11-2). If you see this screen, you may have drive corruption or other drive-related problems. Wait as long as the screen instructs, but if all is not well beyond that, you probably need a new hard drive.
- **Noise inside the TiVo** This could be an early sign of drive malfunction. There are only two moving parts inside your TiVo: the fan and the drive. Early units have drives with ball bearings, as opposed to the fluid bearings used in their successors. Ball-bearing drives tend to get louder (and more annoying!) as time goes by.

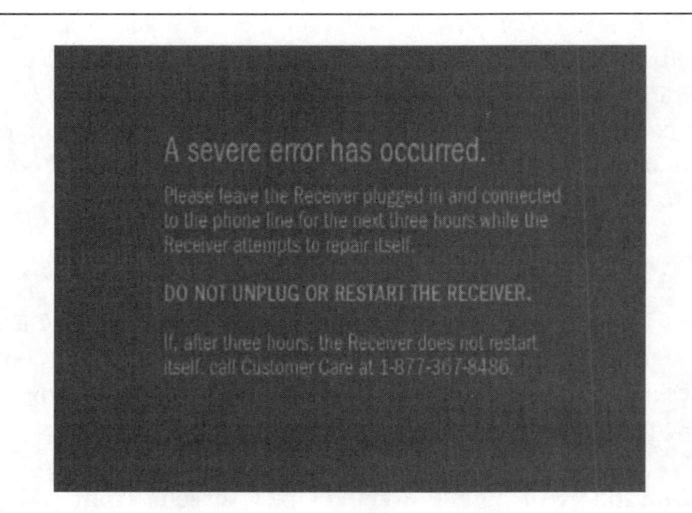

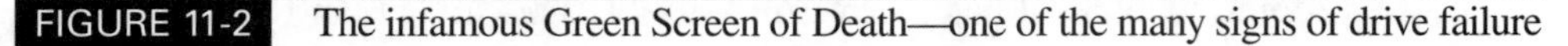
FIGURE 11-2 The infamous Green Screen of Death—one of the many signs of drive failure

TIP *You didn't hear it from us, but there is a simple way to test the fan noise on some units. If your TiVo happens to have a rear-mounted fan with vents on the back panel, you might try unplugging the unit and wedging a toothpick through the vents and into the fan blades to keep them from spinning. Plug the unit back in. If the noise you were hearing continues, it's likely that the drive is singing its death rattle. (Remember to remove the toothpick after you've tried this little experiment.)*

All of these symptoms indicate a dead or dying hard drive *or* that there are problems with the information *on* the drive. The TiVo does have a built-in correction protocol for minor corruption of the TiVo software—the behind-the-scenes activity going on while you're stuck staring at the Green Screen of Death. But this self-repair is not always effective, to say the least. When it fails, all bets are off as to whether the problem is related to software or hardware, unless you can test the drive on a PC. You can attempt to fix the problem yourself by choosing either "Repeat Guided Setup" or "Clear and Delete Everything" from the System Reset window. But if these built-in repair mechanisms don't work, your choice may be clear: Replace that drive! Remember, the earlier you're able to recognize the symptoms, the more successfully you may be able to do something about it.

NOTE *Software corruption doesn't necessarily indicate that the drive needs to be replaced, but you will still need to either send the drive somewhere to be reformatted, or you will need to acquire a clean version of the TiVo software to restore onto your drive. Unfortunately, unless you really know what you're doing, the cost of simply replacing the drive often outweighs the more complicated option of repairing the existing one. On the up-side, though, you're bound to end up with a newer, larger drive in the end.*

Symptoms That Do Not Indicate Drive Failure

The following problems are likely unrelated to the hard drive:

- **No lights anywhere** If your TiVo has lights on the front panel, lights-out may indicate a power supply problem.
- **Black screen** If no image is being sent to the television whatsoever, or if all you see is fuzz, you may have power problems, motherboard problems, or simple cable problems.
- **The DVR won't successfully complete its daily call** This problem could indicate an external wiring problem or a modem failure. If the modem is the problem, go on the Internet and search for "TiVo" and "modem." You should be able to find an external modem that you can use with your TiVo to bypass the broken internal one.
- **Sound and picture synchronization problems** If your TV image is frozen but the sound continues, this may be indicative of motherboard problems. Although it is similar, this is a distinct problem from the "picture problems" discussed in the list of drive-failure symptoms.
- **No Input on Sat 1 or 2** If you have a DirecTV TiVo, and you lose the ability to connect to one of your satellite feeds, it's probably not a drive issue. Check your cabling first, and then look into having your unit repaired or replaced.

TIP *After a power outage, if your TV shows a blue screen and you're running the TiVo thru a VCR, the problem might be as simple as turning on the VCR! When in doubt, simplify the way your TiVo is connected to the TV to rule out wiring problems.*

Can My TiVo Even Be Upgraded?

Almost without exception, the answer to this question is "yes." It makes more sense to publish a list of TiVo models that *can't* be upgraded than ones that can. And to date, the non-upgradeable list only includes one entry—the Toshiba SDH400. By the time you read this, even that one may be upgradeable, so check the Internet.

Planning for a New Hard Drive

If you've diagnosed a drive problem or just want to add more capacity to your TiVo DVR, the next step is to formulate a game plan. It's time to decide matters such as how much capacity to add and what method you'll use to go about the hack.

How Much Hard Drive Space Do I Need?

Well, *need* may not be the right word. *Want* is probably more accurate, but why argue over semantics? When deciding how much capacity to add to your TiVo, there are several factors to consider. Start with the number of people who will be storing programming on this particular DVR. Is it just you, or are there others: a spouse, some kids… your mother, perhaps? Young children tend to like watching the same things over and over, so you might want to block out a certain amount of space for their favorites. If they're a little older, you've potentially got their shows plus a few movies. For yourself, you need to think about all the shows you normally watch in a week, from movies to news to sitcoms to reality TV. And if you're using a stand-alone TiVo, you also want to think about what quality setting you normally record at. The advantage of adding lots of capacity is that you never need to record at a lesser image quality, and doing so can be something of a nuisance if you have a large-screen TV.

If all this calculating is making your head spin, think of it another way. Take the current capacity of your TiVo and consider how restricted you feel. Are you hurting badly for space, or would you just like a little more room? A simple 80GB bump might very well suffice. For DirecTV models, figure you'll get about 0.875 hours of recording time for every gigabyte you have. If you have a stand-alone TiVo, you'll get about 1.1 hours per gigabyte at Basic quality, or about 0.33 hours at Best quality.

If you really want to max it out, here are the limits. Keep in mind that, except where noted, the recording hour limits are based not on drive capacities but on predetermined restrictions built into the TiVo.

Model	Max. Number of Drives	Max. Recording Hours	Model	Max. Number of Drives	Max. Recording Hours
Hughes GXCEBOT	2	260	Samsung SIR-S4040R	2*	240
Hughes HDVR2; HDVR3	2*	240	Samsung SIR-S4120R	2*	240
Hughes HR10-250	2*	77 (HiDef) or 515 (Standard)***	Sony SAT-T60	2	260
Hughes SD-DVR40/120	2*	240	Sony SVR-2000	2	344**
Philips DSR6000 and 7000 series	2*	260	Sony SVR-3000	2*	320**
Philips DSR704/708	2*	240	TiVo TCD130040	2	344**
Philips HDR series and PTV series	2*	344**	TiVo TCD140060	2	344**
Pioneer 810H/57H	1	315**, ***	TiVo TCD230040 or TCD24 series	2*	320**
RCA DVR39/40	2*	240	TiVo TCD54 series	2*	697**

*Installing a second drive requires an additional or third-party bracket.

**Recording time is at Basic quality; highest quality allows about 1/3 as many recording hours.

***Maximum capacity presumes 300GB drive limits; the Pioneers, the Hughes HR10-250, and the TCD54 series have no drive-size restrictions, so as drive capacities increase, so does your TiVo's recording capacity.

You may decide that you want to replace your old drive with a single drive now and then add more capacity down the road. Adding another drive later is fairly simple if you are doing the formatting yourself. You'll basically be following the instructions in this chapter for adding another hard drive. For those going the kit route, however, it's not quite as simple. Kit users can add a preformatted drive to their *factory* drive easily. The same goes for *replacing* the factory drive. But if you want to add capacity after replacing your factory drive, you'll need to use a PC yourself or send the drive you wish to continue using back to the kit provider for formatting. You won't lose your programming as long as the drive is in working order.

Lastly, don't ignore the option of simply buying an additional TiVo. Prices are coming down on DVRs and TiVo service all the time, so your best solution may simply be to invest in an additional unit.

TIP *DirecTV currently honors a lifetime service agreement by applying it to all units in the house. Additional units can be had at low cost because they are predicated on your agreement to remain a DirecTV customer for a reasonable service term.*

Add Another Drive or Replace the Old One?

Once you've determined how much space you'd like to have, it's time to start thinking about how to get there. You know you need at least one new drive, but should you add it to the one you have or, instead, replace it?

Adding a drive has a number of advantages. Most obviously, you will be adding capacity to the amount you already have. Adding a 120GB drive to your 40-hour TiVo will bump cable users up to 185 hours of recording time at Basic quality. Not too shabby for spending well under $200 (at the time of writing, 120GB add-kits cost approximately $160). In addition to the added room, there's another big advantage—your settings don't change. The process we'll cover for adding a drive increases your recording capacity while also allowing you to keep all your settings and recorded shows. Your preferences, your local access information, your Season Passes, and your Now Playing list will all be unaffected. Just plug in your TiVo and you're good to go.

But… (There's always a "but," isn't there?) In the case of adding a drive, there are a few "buts." First, you can only *add* a drive to your TiVo if there's only one drive in it to begin with. (A good resource for determining how many drives your TiVo has can be found at http://www.weaknees.com/faq.php#howdoi, but if you don't find the information there, you might need to open your TiVo to figure this out.)

Second, be aware that if you add a drive to the one that's already in there, they become a "married" pair. The factory drive will always look for the added drive upon startup, and it won't work until it finds it. Similarly, the added drive needs the information on the factory drive in order to function. On its own, configured without the TiVo software, the added drive is useless. (Along these lines, you can't simply replace one master or slave drive with another. It must be a clone of the original and of the same capacity or greater.)

Third, should either of the drives fail, you will need to reformat the remaining one (either by itself or with a new drive) in order to get it working again. You'll certainly be losing your programming under those circumstances, and depending on your capabilities, you may need to send them both somewhere to determine which one's still good and to have it reformatted. To that extent, you need to consider the condition of your factory drive before you decide whether you should add to or replace your drive.

Remember that your TiVo's drive has been spinning nonstop since the moment you plugged it in. It's been writing and erasing, trying to stay cool, and possibly dealing with power fluctuations. It's had a hard life. If your drive is more than a year old, it's not unreasonable to think about retiring it. You can put it out to pasture, or perhaps even give it a nice desk job. If it's significantly older than a year—say two or three—definitely bid it farewell. There's no sense in taking a brand new, expensive hard drive and pairing it with one that may give out at any moment. That's not to say the new drive may not die during infancy, but we're playing the odds here, and the odds are stacked against an older drive. Besides, a 40GB drive isn't really contributing much in the scheme of things when you're about to add three or four times that amount. Don't get greedy.

NOTE *There's an excellent discussion on this subject the subject of adding a drive versus replacing the old one at http://www.weaknees.com/add_replace.php.*

Our advice: Don't hem and haw about this one too long. Just be objective and realistic about it. And keep in mind that if you decide to replace your factory drive while it's still in decent shape, you can always keep it aside for a rainy day. If your new drive goes south, you'll have a backup of the TiVo software and a formatted drive to use while you decide on your next move.

Preformatted Kits vs. Doing It Yourself

This starts getting us into really nerdy territory. Get ready to be either the envy of your classmates or the butt of their jokes. For those of you who don't feel like using a PC or spending some significant time gathering all the parts and software you'll need, the best option is to buy a kit from any of a number of online dealers. Your level of satisfaction will be nearly as great, if not greater. And you'll likely have some security in knowing that your upgrade is guaranteed to work. For anyone out there wondering how to avoid doing anything at all, there's a solution for you, too. See the "Hands-Off Options" section that follows shortly.

Do-It-Yourself: What to Expect

If you decide to go ahead with buying drives for your TiVo, formatting them on your PC, and installing them in your TiVo, we're very impressed. Aside from the formatting issues, you'll be dealing with gathering your own parts, troubleshooting the installation, and dealing with drive warranties. But the path has been well traveled, so you should be in good company.

Everything you need to know is included in this chapter, from descriptions and pictures of parts to locations where you can secure the proper software. In addition, the "Useful Web Resources" section at the end of the chapter lists some excellent resources for you to refer to in a pinch or where you can shop for tried-and-true upgrade components.

Before you make a decision, though, look over the following list of requirements for the do-it-yourself procedures described in this chapter:

- You have a working TiVo. You will need a functioning drive with uncorrupted software in order to restore information to your new drive or drives. (If your drive is corrupted, malfunctioning or dead, you will probably need to order a third party kit.) The other option is that you might be able to acquire a software image on the Internet, in which case you can follow along in the chapter and leap in at the appropriate moment.
- You are working with a TiVo that has never been upgraded.
- Your TiVo is running version 2.5 software or later.
- You will either be adding one drive to your factory drive or replacing your factory drive with either one or two drives.
- You will be using a new, or at least tested, drive whose capacity is equal to or greater in size than the total combined capacity of your existing drive(s).

The Upgrade Kit Option: What to Expect

Over the past few years, the TiVo upgrade kit has been conceived, modified, and perfected. It is now an easy-to-find, user-friendly option for thousands of TiVo customers across the country. The current crop of upgrade kits is designed with simplicity and reliability in mind. As there are several companies out there with unique versions, we'll start by discussing what you should reasonably expect from any of them.

Drives You'll want to start your research with the hard drive, which is the most expensive and crucial part of the upgrade kit. Although any of the half dozen or so drive manufacturers out there have a product that will technically work, there are some subtleties to consider.

A standard desktop computer drive will fit any model of TiVo, but the function of the drive may not be suited to its intended purpose. Remember that the drive will be working 24 hours a day, 365 days a year. To that extent, you'll want a drive that can operate quietly in close quarters without overheating. TiVo specs only require 5400 rpm, and early models of TiVo DVRs had drives running even slower than that, which, by all measures, performed just fine. Any faster than that is both unnecessary and could result in higher temperatures. That's not to say there aren't plenty of TiVos out there running fine at, say, 7200 rpm, but it isn't necessary. Also, some drives read and transmit information differently. In the case of streaming audio and video, try to find a drive that is optimized for this purpose.

TiVo DVRs, TiVo the company, and other DVR manufacturers (Hughes, Philips, and so on), primarily use a line of drives from Maxtor (see Figure 11-3). In fact, these drives are in the vast majority of TiVos out there, and according to Maxtor statistics, they are in about 80 percent of all DVRs in the world. This DVR-friendly brand of drives, called QuickVIEW, is designed specifically for DVR use and has come to be recognized for its performance and reliability. Other DVRs ship with Western Digital drives or Seagate drives—both lines specially tuned for this purpose—which have also been known to perform well.

When searching for a kit vendor, know that prices are often quite similar from vendor to vendor, but the quality of the drives can vary markedly. Whichever drive your kit provider uses, be sure to ask about its track record and do your research.

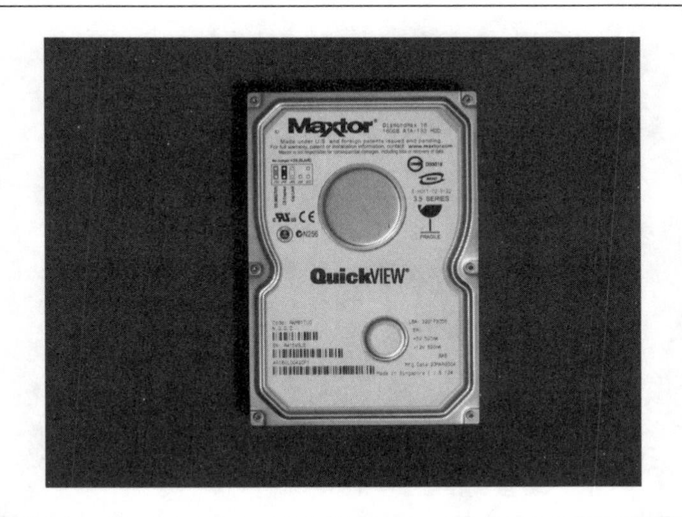

FIGURE 11-3 Maxtor's QuickVIEW drive is optimized for DVRs and is found in 80 percent of all TiVos.

A good place to start is the highly regarded third-party TiVo forum, located at http://www.tivocommunity.com.

Instructions Any kit should come complete with printed instructions to walk you through the installation, step by step. TiVos are a little busy under the hood, so find out whether there are pictures and diagrams to guide you. Be sure to read through the instructions before you start. Chances are they'll only be a few pages long, but it really helps to know what you're in for.

Also, the instructions will include a list of do's and don'ts, some of which will keep you from frying yourself or your TiVo. In most cases, the instructions are designed with the layperson in mind. And as it turns out, the people who tend to mess up their installations most often are the ones who claim to have "computer experience" and assume they already know what they are doing.

Parts Any reasonable kit will include all of the parts and odd tools required to complete your specific install (see Figure 11-4). Be wary of any kit seller that won't tell you in advance exactly what you're buying, along with the make and model of your *new* drive.

When the parts arrive, inspect them for damage. Did your drive come packed in bubble wrap or peanuts? If so, there's a decent chance it was damaged during shipping. You might consider asking the vendor in advance whether they use a manufacturer-approved drive box. Reputable vendors will; more careless operators may not.

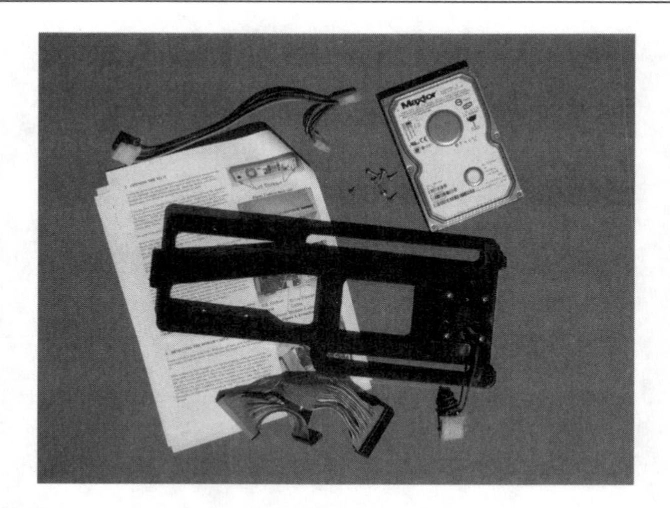

FIGURE 11-4 Parts required to complete an install

Compare the parts you received with the parts listed in the instructions. Look for cracks, nicks, exposed or cut wires, and loose connections. Many of these parts are produced overseas in enormous batches. Flaws might be subtle, so pay attention. Do the yellow, red, and black wires all start and end in the same order? You'd better check before you plug it in.

It is worth mentioning that eBay sellers are notorious for not including correct parts, so our advice is to stay away from eBay for upgrade kits.

Warranty Just because you're voiding the warranty of your TiVo doesn't mean the kit you're purchasing shouldn't come with its own guarantee. In all likelihood, the drive that it comes with is warranted by its manufacturer for anywhere from one to three years.

That said, the kit provider should stand behind its product for a certain period of time, as well. Certainly they should guarantee that it won't come dead on arrival, but find out for certain what the warranty period is. You'll want to make sure that if the drive starts clicking two months down the road, you won't be expected to shell out for a new one.

Customer Satisfaction Due to the nature of Internet business, many online retailers have to go the extra mile to assure potential customers of their legitimacy. TiVo upgrade kit providers are no exception. Read through their customer comments, which should be current and plentiful. For further proof, you'll find less-carefully screened opinions on the various user forums. Again, check out http://www.tivocommunity.com.

Hands-Off Options

If you have little time or interest in performing your own upgrade, or if you lack the confidence that you'll be able to pull it off, there's an easy solution. Several upgrade providers will allow you to send in your TiVo and have the upgrade performed at their location. For what should be a reasonable fee, you can prepurchase your kit, send in your TiVo, and have the unit shipped back once the surgery has been performed. Your TiVo will arrive back looking fresh and ready to work.

If this option appeals to you, make sure of a few things. Find out whether there are any additional charges you could be facing if everything doesn't go according to plan. Will your previous content be saved, or is that extra? Is return shipping included? Will return shipping be insured and trackable? What's the turnaround time? And what parts, if any, do you need to send with your unit?

When sending out your unit to be upgraded, you will definitely want to take a few precautions. Remember that your TiVo is essentially a computer. Keep this in mind when packing it up, and be aware that it's going to get knocked around in transit. You should assume that the delivery driver is going to turn it upside down, drop it, throw it, and put other packages on top of it. Knowing this, you'll want to provide plenty of impact protection on all sides of your TiVo and pay for outbound insurance, just in case.

But don't worry too much. Every piece of electronic equipment is designed with the understanding that it will eventually be transported. So long as you make its trip comfortable, there should be little cause for concern.

Precautions and Warnings

Whether you're upgrading from a kit or doing the upgrade from scratch, there are a few things to be aware of ahead of time.

Safety Measures

Before you get started with your upgrade, take a moment to review these precautionary measures. They will save you time, money, and might even save your life.

- **Don't die** First and foremost: *Disconnect the power cord!* Okay, it seems obvious, but you'd be surprised how often the obvious gets overlooked. At any rate, be aware that death or serious injury can occur if you come in contact with the live, internal power supply. Also, give the unit a few minutes to discharge before you open it up. Do not plug the TiVo back in until the upgrade is complete and the lid has been replaced.
- **Work neatly** You'll want to move your TiVo onto a flat, uncluttered surface. Some of the parts you're dealing with are small and will try to roll away and disappear. The only things you need nearby are your TiVo, your installation parts, instructions, and tools. No Big Gulps or Grande Mochaccinos, please.
- **Get grounded** As an added safety measure, more for your TiVo than for yourself, you may want to invest in a simple static-guard device (see Figure 11-5). These are inexpensive items that ground you and prevent you from delivering a potentially harmful electrostatic charge to the hard drive. You could do without it, but after spending all that money on your TiVo and hard drive, are you really going to sweat the extra five bucks?
- **Don't kill your TiVo** You need to be especially careful when reconnecting your TiVo in order to ensure that all cables are firmly seated in their respective connection ports. You can damage your TiVo if you power it on when there are still cables loose. In some cases, you will fry a chip or two on the motherboard by powering up with a loose cable, so check them all twice.

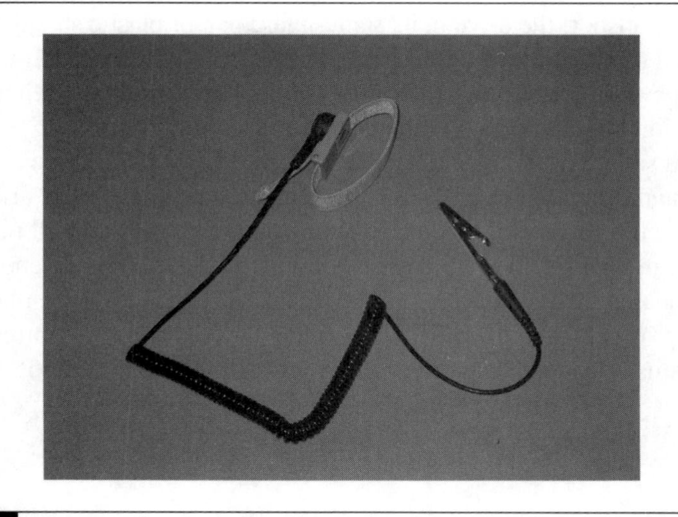

Callout missing

FIGURE 11-5 Inexpensive static guards attach to the wrist with an elastic loop and ground potentially harmful charges.

A Note on Warranties

In case you didn't notice it, there's probably a sticker on the back of your TiVo that says something to the effect of, "Warranty Void if Seal Is Broken." Guess what that means?

Whether you have a sticker or not, you need to be aware that the procedures described in this chapter will void your warranty. If you bought an extended warranty, this may mean something. If not, the parts and labor warranty on your TiVo probably only lasted about three months to begin with. After that, there's very little cause for concern. And if you go really slow, you may not even break the seal. Wink, wink.

Drive Handling

In the earlier "Safety Measures" section, we suggested getting grounded and mentioned the dangers of electrostatic charge. Now let's review a few other safe-handling practices for hard drives.

The average hard drive consists of one or more magnetic platters in a suspension mechanism contained within a heavy metal casing. Outside the casing are all sorts of busy-looking circuits and chips. Inside, there are tiny arms that are designed to move across the surface of the platters, reading and writing information as they go. When the arms are at rest, they hover in a location called the *landing zone*. Even when your drive is not plugged in, it is possible to damage these internal parts, either through an electrostatic charge or by force or impact.

Be sure to transport the drive in its static-guard bag or plastic shield at all times. And when you remove it, you should be grounded in some way. If you don't have a grounding device, you should first touch a grounded object to discharge any current you might be carrying, just in case.

When you set the drive down on a surface, *be gentle*. If the drive receives any sort of significant impact, it transmits vibrations to those read/write arms, which will bounce on the surface of the landing zone and may chip a small piece of the material from its surface. If that happens, tiny particles will be loosened and will float around near the platter surfaces. This can cause scratches on the platter, which is bad. A drop of one edge of the drive to the table from about 4 centimeters high delivers 3 grams of force to the drive, and 5 grams is enough to kill it. Get the picture?

Surgery

We are about to begin the physical act of opening the TiVo and performing the upgrade. Users who have decided to purchase a kit can certainly keep reading, as much of the information contained herein will apply to you. However, your personalized kit instructions should supersede ours, and the kit provider should be contacted in case you run into problems.

This section contains the do-it-yourself instructions for each of the six TiVo design categories. In some cases, one design covers half a dozen or more models. In others, it refers to only one or two. Find the architecture that refers to your model and read through the appropriate section in the following pages:

It is our strong recommendation that you read through all of the instructions first before you remove a single screw from your TiVo. This includes referring ahead, when instructed, to the Installation Details section where you'll find a list of needed parts as well as drive formatting procedures.

- **Architecture 1** Philips PTV and HDR models and Sony SVR2000
- **Architecture 2** TiVo TCD130040 and TCD140060, Hughes GXCEBOT, SonySAT T-60, and Philips DSR6000
- **Architecture 3** Sony SVR3000
- **Architecture 4** TiVo TCD23x, TCD24x, and TCD54x models, Hughes HDVR2/3 and SD-DVR-40/120, Philips DSR7x models, Samsung SIR-S4120 and SIR-S4040, and RCA DVR 39/40

- **Architecture 5** Hughes HR10-250
- **Architecture 6** Pioneer 810H and 57H

Each architecture description has sections called "Parts," "Disassembly," "Drive Formatting," and "Reassembly." After you've completed "Disassembly," if you are formatting your own drive, refer to the "Hard Drive Formatting" section later in the chapter. There you'll find instructions for the optional step of backing up your software and for your choice of drive formatting, whether it be single or dual, with or without saving your programming. If you've ordered a kit with a preformatted drive, you should follow the instructions that came with it; however, the procedure for disassembly and reassembly should essentially remain the same.

All TiVo's essentially contain the same basic parts (see Figure 11-6) with variations in the internal layout. Inside you can expect to find a large, green motherboard containing computer chips and circuitry, a power supply board, a cooling fan, and a hard drive (or drives), mounted on a metal bracket. Each of these components is self-contained, with various wires and cables connecting them all together.

TIP *For color pictures demonstrating the opening of each TiVo, you can preview the installation instructions on the WeaKnees Web site at http://www.weaknees.com/upgrade_instructions.php.*

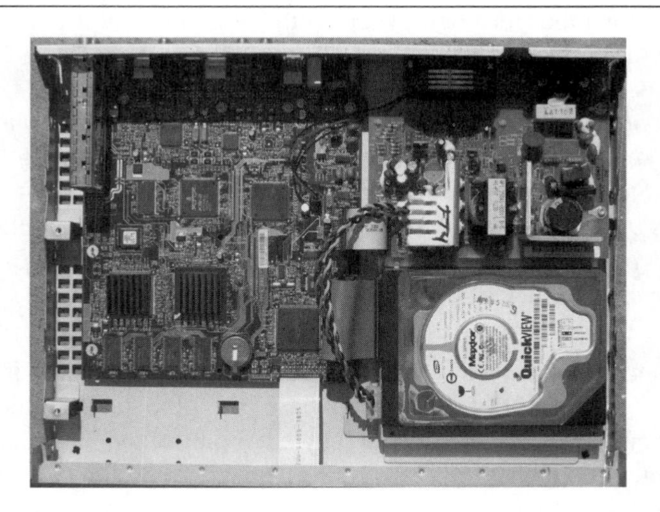

FIGURE 11-6 The inside of a TiVo consists of a motherboard, power supply, hard drive, and cooling fan.

Architecture 1

This architecture includes all Philips PTV and HDR models and the Sony SVR2000.

Parts

You will need Torx T10 and T15 screwdrivers and, if you're adding a drive, a Series 1 metal bracket with four rubber grommets, four metal casings, four drive screws, and two additional T10 screws to hold the bracket in place.

Disassembly

The HDR series and SVR2000 are the only TiVo models that present something of a physical challenge to take apart. The lid is somewhat stubborn, so get ready to grunt.

Follow these steps:

1. Unplug the unit, turn it around so you're facing the rear, and unscrew the four Torx T10 screws that hold the lid in place. There are two along the top edge and one at each of the side edges.
2. From a seated position, lift up the unit from the sides and prop the back bottom edge against your thighs, a few inches back from your knees. With the unit almost vertical (with the faceplate pointing upward), pull down firmly along the side edges, keeping the pressure on until the lid releases and snaps back an inch or so. Set the unit back on the table and take a breath.

TIP *As an alternative, you might try the WeaKnees.com "tried and true" method to remove the lid. Take out the lid screws and set the unit on the floor. Stand over the TiVo from the front and lean down on it, resting your hands on either edge of the lid, halfway between the front and back. With a few sharp, downward pressing motions, the lid should release with a snap as the metal claws unseat from the rear panel.*

3. Take a moment to look inside your TiVo and locate the circuit board, the drive, the fan, and the power supply. Be careful to avoid contact with the power supply or, for that matter, any other part of the TiVo that you aren't instructed to touch.
4. Unscrew the two Torx T10 screws that are holding the drive bracket in place. These are side by side, facing upward between the drive and the faceplate.

5. Detach the power cable and IDE cable from the drive.
6. Note that the drive bracket is held down on one side by two tiny metal tabs, which, in conjunction with the two Torx screws you just removed, hold the drive in place. Remember this for later, when you'll need to put the drive back.
7. Slide the drive and bracket toward the power supply about a quarter inch, and lift them out as one attached unit, setting it on the table beside the TiVo.

Drive Formatting

You must now proceed to the "Installation Details" section before continuing here. Start with the "MFS Tools" section to acquire the Linux software you'll need for the job.

Then continue to the "Hard Drive Formatting" section. If you're planning on backing up your drive, start with the "Making a Backup" section (this is an optional step). Otherwise skip ahead to one of the four formatting options.

When that's all done, continue on with the next section to reassemble your TiVo.

Reassembly

If you're simply replacing your old drive with one new drive, all you're doing is taking the old one off the bracket and putting the new one on. The drive is held on by four drive screws that pass through the bottom of the bracket where four rubber grommets and metal casings hold the drive securely in place and reduce vibration.

If you are putting in two drives, you'll mount the second drive on the second bracket, and secure that one on the bay beside the first one. Note that this architecture already has an additional power cable and IDE port for a second drive. The cable simply needs to be cut loose from where it is tethered to the bottom with a white cable tie. Attach the master drive on the outside bay and the slave on the inside bay. (See the description of jumpers in the "Parts Summary" section later in the chapter—it explains the master/slave drive setup.)

Once you've figured out where the drive (or drives) go, follow these steps:

1. Align the drives in the bays and screw them down with the T10 screws (two apiece). You may need to adjust their location a bit to get them aligned properly.
2. Reconnect the power cables to the drives (not the main power *cord*) along with the IDE cables.
3. Fasten the lid back in place, screw it on, and you're done.

All that remains is to hook your TiVo back up to your television and plug it in. Go through Guided Set-Up as if you've just bought a brand new TiVo. If you did it right, you should be watching TV inside of 35 minutes. (If you copied your settings or are simply adding a second drive to work with your factory drive, you won't need to go through Guided Set-Up again.)

Architecture 2

This architecture includes the TiVo TCD130040 and TCD140060, Hughes GXCEBOT, SonySAT T-60, and Philips DSR6000.

Parts

You will need Torx T10 and T15 screwdrivers and, if you're adding a drive, four drive screws. Some units require a long Y power splitter and a two-port IDE cable.

Disassembly

This TiVo architecture has a few minor variations between the different models, but the nuances are slight enough that the TiVos deserve to be grouped together. Once you open the TiVo's lid, you may discover that you already have a power splitter or that your IDE cable can accommodate two hard drives. If this is the case, congratulations. For some SAT T-60 owners, you'll have an extra screw to wrangle with, but once you remove it, you won't need to put it back. In general, this is a simple TiVo architecture to upgrade, with the exception of one, annoyingly placed screw.

Follow these steps:

1. Unplug the TiVo's power and turn the TiVo around so you're facing the rear panel.
2. With a Torx T10 screwdriver, remove the four screws that hold the lid in place. There are two along the top edge and one along each of the sides.
3. Grasp the lid by the sides, pull it back an inch or so, and then lift it off. You can set it aside until later. Sometimes it's a good idea to keep your removed screws and other pieces in the overturned lid.
4. Take a minute to familiarize yourself with the inside of the unit. Along the left side (as you look at it from the back) you'll see the power supply. Perpendicular to that, near the front, is the drive bracket. It contains either a single drive with an empty drive bay beside it, or there will be two drives,

side by side. Additionally, you'll see a silver box that conceals the TiVo's fan. Note how the drive bracket is held at the right by two T15 screws, and at the left it pokes through the power supply wall with two metal prongs resting on two rubber grommets.

5. If you are working on a SAT T-60, you need to remove that annoying screw we mentioned. (Not all SAT T-60s have this screw, but take a look just in case.) The front metal prong of the drive bracket is sometimes held in place by a screw that passes through the rubber grommet from the power supply side. Use either a long, thin Torx T15 screwdriver or an L-shaped Torx T15 key to carefully remove the screw. It's about half an inch long, and to get it out you may need to slightly spread apart some busy-looking power supply gizmos. Just be careful, and don't move anything too far. Once the screw is out, you won't need to put it back in. Estimated time… two minutes.
6. Disconnect the fan's power cable from the circuit board. It's on the near side of the drive bracket, consisting of red and black wires leading to a plastic fastener. Detach the plastic fastener by pulling it straight up.
7. Unscrew the two T15 screws holding the bracket on the right side. Try to leave the metal casings in place where they sit in the screw holes.
8. The bracket is now somewhat free to move. Tilt it up from the right side and pull the metal prongs out from the other side, just until they're free. Don't let it drop! You'll still need to release some cables, so either flip the bracket up and over the front lip of the unit, or pull it toward you to make some room near the faceplate.
9. Disconnect the power supply and IDE cables from the drive, and remove the bracket and drive(s) from the TiVo completely.

Drive Formatting

You must now proceed to the "Installation Details" section before continuing here. Start with the "MFS Tools" section to acquire the Linux software you'll need for the job.

Then continue to the "Hard Drive Formatting" section. If you're planning on backing up your drive, start with the "Making a Backup" section (this is an optional step). Otherwise skip ahead to one of the four formatting options.

When that's all done, continue on with the next section to reassemble your TiVo.

Reassembly

If you're simply replacing your old drive with one new drive, all you're doing is taking the old one off the bracket and putting the new one on. The drive is held on by four drive screws that pass through the bottom of the bracket, directly into the base of the drive.

If you're putting in two drives, put the slave where the old factory drive was (near the power supply), and put the master on the right near where the two screws came out. (See the description of jumpers in the "Parts Summary" section later in the chapter—it explains the master/slave drive setup.)

The first step in reassembling your TiVo is to reconnect the power supply and IDE cable. For a one-drive upgrade, this is relatively simple, but for a dual-drive upgrade, there is some preliminary work to be done. Follow these steps if you are installing two drives:

1. Set the drive bracket aside for the moment, and look at the IDE cable. If it happens to have two drive connectors, you can leave it connected and continue with step 3. Otherwise remove the old IDE cable from its port on the front of the circuit board.
2. In place of the old IDE cable, attach the blue end of your new, two-port IDE cable to the circuit board. (If the IDE connectors aren't in ideal locations, you may need to connect the opposite end to the circuit board, but be sure that the drives have their jumper pins set correctly.)
3. Check the power cable for the drive. If your unit has a two-drive power cable, your cables are ready to go. Otherwise, attach your long Y power splitter to the power cable to create two new power supply ports.

Regardless of the number of drives in your TiVo, complete the reassembly with these steps:

1. Pick up the drive bracket and flip it upside down. Viewed from the rear of the unit, the fan box should be at the left, near the power supply. Carefully rest the drive bracket on its back on the front edge of the TiVo, holding the bracket carefully with one hand while you connect the power supplies and the IDE cables to the drives. Make sure you've got the bracket aligned correctly with the fan and prongs toward the power supply side of the TiVo.
2. Carefully flip the drive bracket back over, taking care not to pull out the cables. Stuff the excess cabling between the bracket and the faceplate, and work the bracket back into place.

3. Insert the two prongs through the blue grommets, and then lower the screw holes over the metal screw receptacles on the right.
4. Put the two T15 screws back in place, and reconnect the fan's power cable (see Figure 11-7).
5. Replace the TiVo's lid, screw it on, and you're done.

All that's left now is to hook your TiVo back up to your television and plug it in. Go through Guided Set-Up as if you just bought a brand new TiVo. If you did it right, you should be watching TV inside of 35 minutes. (If you copied your settings or are simply adding a drive to work with your factory drive, you won't need to go through Guided Set-Up again.)

Architecture 3

This architecture includes the Sony SVR3000.

Parts

You will need a Phillips-head screwdriver and, if you're adding a second drive, four drive screws, a long Y power splitter, and a two-port IDE cable.

FIGURE 11-7 Don't forget to reconnect your cooling fan—it attaches to two tiny prongs sticking up from the motherboard.

Disassembly

This is a rather easy installation to perform, but there will be lots of Phillips-head screws to keep track of, so keep it tidy!

Follow these steps:

1. Turn the unit around so the rear panel is facing you and remove the six screws securing the lid. Two are located at the back top edge where the lid folds down over the rear panel, and the rest are located on the sides of the unit. Note that the side screws are slightly larger.
2. Grasp the lid by its sides and lift it gently, tilting it up from the rear, and then back from under the ridge of the faceplate. Set it aside for now—you might even want to flip it upside down and keep all of the loose screws in it.
3. Take a moment to familiarize yourself with the inside of the TiVo. Note the green circuit board, the long, brown power supply running down the left side (as viewed from the rear), and, most prominently, the drive bracket. Note that the drive bracket contains a drive at the center with an open bay beside it for a second drive. It also contains a silver box concealing the TiVo's fan.
4. The fan's power cable runs from the fan box to the circuit board where red and black wires terminate with a white plastic mount. Pull the mount straight up to disengage it from the circuit board.
5. Remove the four Phillips-head screws at the corners of the drive bracket. They are seated in four brass collars, which you'll want to leave in place.
6. Once the screws are out, gently lift the drive bracket up and toward you a few inches to stretch out the cables running to the hard drive.
7. Disconnect the power supply and the IDE cable from the drive, and then remove the bracket completely from the unit.

Drive Formatting

You must now proceed to the "Installation Details" section before continuing here. Start with the "MFS Tools" section to acquire the Linux software you'll need for the job.

Then continue to the "Hard Drive Formatting" section. If you're planning on backing up your drive, start with the "Making a Backup" section (this is an optional step). Otherwise skip ahead to one of the four formatting options.

When that's all done, continue on with the next section to reassemble your TiVo.

Reassembly

If you're simply replacing your old drive with one new drive, all you're doing is taking the old one off the bracket and putting the new one on. The drive is held in the bracket by four screws that pass through the bottom of the bracket, directly into the base of the drive.

If you're putting in two drives, put the slave where the old factory drive was, and put the master on the right; the side opposite to where the power supply board is located.

The first step in reassembling your TiVo is to reconnect the power supply and IDE cable. For a one-drive upgrade, this is relatively simple, but for a dual-drive upgrade, there is some preliminary work to be done. Follow these steps if you are installing two drives:

1. Set the drive bracket off to one side.
2. Take your Y power splitter, and fasten the single end to the factory power cable, creating two new positions for attaching drives.
3. Gently pull the IDE cable out from the circuit board, and replace it with your new two-port IDE cable, attaching the blue end to the circuit board.

For either the single or double drive configurations, reattach the power and IDE cables and complete the reassembly as follows:

1. Align the drive bracket correctly, with the fan on the left near the power supply, and then flip it over on its back and rest it carefully on the front lip of the TiVo.
2. Keeping hold of the bracket, connect the power and IDE cables to the drives. If you have a dual-drive configuration, connect the end of the IDE cable to the right, master drive.
3. Flip the drive bracket back over, carefully stowing the excess cabling between the bracket and the faceplate, then lower the bracket into position over the brass screw holes. This will take some polite shoving, so be careful and don't get rude.
4. Reattach the four bracket screws and reconnect the fan's power cable to the circuit board. The fan is very important!

5. Take the lid and slide it into place over the TiVo. Start by tilting the front end forward so that it engages under the lip of the faceplate. Then lower the back end into place, taking special care not to scratch the sides.
6. Replace the two tiny rear screws, then the four larger side screws. And that's it!

You may now plug the power cable back in and reconnect the TiVo to your television. Run Guided Set-Up, which should take about 35 minutes. (If you copied your settings or are adding a drive to work with your factory drive, you won't need to go through Guided Set-Up again.)

Architecture 4

This architecture includes the TiVo TCD23x, TCD24x, and TCD54x models, Hughes HDVR2/3 and SD-DVR-40/120, Philips DSR7x models, Samsung SIR-S4120 and SIR-S4040, and RCA DVR 39/40.

Parts

You will need Torx T10 and T15 screwdrivers. If you're planning on adding a second drive, you'll need a third-party bracket to accommodate it, along with a Y power splitter, a two-port IDE cable, and any hardware required to attach the drive and bracket. (TCD23x and TCD24x owners should also consider a PowerTrip—see the "Parts Summary" section later in the chapter for a description of this device.)

Because this unit wasn't designed with a second drive in mind, you should also consider adding additional cooling. Refer to the "Useful Web Resources" section at the end of the chapter for suggestions on where to get this hardware.

Disassembly

This architecture has one of the easiest lids to take off, and removing the drive is rather simple, as well. Follow these steps to disassemble the unit:

1. Turn the TiVo around so you're facing the back panel.
2. Remove the four or five T10 screws that hold the lid in place. They are located in the back around the perimeter where the lid folds over the back panel.
3. Grasp the lid by the sides, and wedge your thumbs against the back of the unit. Pull the lid back backwards until it loosens (about half an inch), and then lift it free of the base. Set it aside for now.

4. Inside the unit, locate the large green circuit board, the square power supply, the fan, and the drive bracket.

5. To detach the drive bracket, locate and remove the two tiny T10 screws between the drive and the faceplate that hold the drive bracket in place. Carefully pull the drive and bracket toward the power supply to disengage the bracket from the metal tabs that hold it in place, and then lift it up. Note how the bracket unseats so you can correctly replace it later for single drive configurations. Be careful not to yank it because the cables are still attached.

6. Disconnect the power and IDE cables which are both plugged in at one end of the drive (see Figure 11-8).

You can release the drive bracket and remove the cables as described above, or you can disconnect the drive's cables first. Both methods work.

Drive Formatting

You must now proceed to the "Installation Details" section before continuing here. Start with the "MFS Tools" section to acquire the Linux software you'll need for the job.

FIGURE 11-8 The hard drive attaches to the motherboard with a wide 40-pin IDE cable and to the power supply with a 4-wire Molex connector.

Then continue to the "Hard Drive Formatting" section. If you're planning on backing up your drive, start with the "Making a Backup" section (this is an optional step). Otherwise skip ahead to one of the four formatting options.

When that's all done, continue on with the next section to reassemble your TiVo.

Reassembly

If you're simply replacing your old drive with one new drive, all you're doing is taking the old one off the bracket and putting the new one on. Follow these steps:

1. Secure the drive to the bracket with four T15 screws—two on either side.
2. Reseat the bracket by first engaging the tabs near the power supply, and then reattach the two T10 screws on the faceplate side.
3. Reconnect the power and IDE cables.

If you are doing a dual-drive install, you will need to carefully follow the instructions provided by your bracket manufacturer. Note that you will be replacing the short, single-port IDE cable with a longer, two-port cable. You'll also need a Y power splitter.

Once the drives are secured, all that remains is to replace the lid:

1. Align the lid on top of the unit, starting an inch or so too far back and with the lid flat against the top of the box.
2. Slide the lid forward so that lid makes a tight seal with the faceplate.
3. Refasten the four or five lid screws.

Note for TiVo TCD23x and TCD24x Owners

If your TiVo is model TCD23x or TCD24x , you may want to consider investing in a PowerTrip. This third-party device delays the startup of the master drive because the TiVo's power supply may have difficulty producing enough energy to start both drives at once. This is explained in more detail in the "Parts Summary" section later in the chapter.

When connecting the power supply to your drives, you'll want to attach the PowerTrip between one branch of the Y power splitter and the master drive.

Note for TiVo TCD540140 and TCD540040 Owners

The TCD54x models have an IDE cable that plugs into the motherboard in a different location from other models, necessitating a longer cable to reach between the motherboard slot and the first drive connector. You may need to reverse the direction of the IDE cable, but be certain to have the end connector attached to the master drive.

You may now reconnect your TiVo to the television and run through Guided Set-Up. (If you copied your settings or are simply adding a drive to work with your factory drive, you won't need to go through Guided Set-Up again.)

Architecture 5

This architecture includes the Hughes HR10-250.

Parts

You will need Torx T10 and T15 screwdrivers. If you're adding a second drive, you will also need a Y power splitter, a two-port IDE cable, and a third-party bracket to hold the second drive inside the unit. See the "Useful Web Resources" section later in the chapter for suggestions on sources for this hardware.

Disassembly

This unit looks a lot like the SD-DVR40 on the outside, but the inside has been reconceived to accommodate the parts necessary for high-definition broadcast. Follow these steps to disassemble the unit:

1. Position the unit so that you're facing the rear panel.
2. Remove the screws that secure the lid. There are five screws holding it in place, all of which are located in the back, along the perimeter where the lid folds over the case.
3. Grasp the lid by the sides and slide it toward the back about half an inch or so, until it releases. Then lift the lid up and away from the base. Set it aside for now.
4. Take a moment to familiarize yourself with the TiVo's interior. You'll see the large green circuit board, the power supply, the fan, and the hard drive.

5. Locate and remove the two tiny T10 screws between the drive and the faceplate which hold the drive bracket in place.
6. Carefully push the drive and bracket toward the power supply to disengage the bracket from the metal tabs that hold it in place, and then lift it up. Note how the bracket unseats so you can correctly replace it later for single drive configurations. Be careful not to yank it because the cables are still attached.
7. Disconnect the power and IDE cables from the end of the hard drive.

NOTE *You can release the drive bracket from the TiVo unit and then remove the cables, as described above, or you can disconnect the cables from the drive first. Both methods work.*

Drive Formatting

You must now proceed to the "Installation Details" section before continuing here. Start with the "MFS Tools" section to acquire the Linux software you'll need for the job.

Then continue to the "Hard Drive Formatting" section. If you're planning on backing up your drive, start with the "Making a Backup" section (this is an optional step). Otherwise skip ahead to one of the four formatting options.

When that's all done, continue on with the next section to reassemble your TiVo.

Reassembly

If you're simply replacing your old drive with one new drive, all you're doing is taking the old one off the bracket and putting the new one on. Follow these steps:

1. Secure the drive to the bracket with four T15 screws—two on either side.
2. Reseat the bracket by first engaging the tabs near the power supply, and then reattach the two T10 screws on the faceplate side.
3. Reconnect the power and IDE cables.

If you are doing a dual-drive install, you will need to carefully follow the instructions provided by your bracket manufacturer. Note that you will be replacing the short, single-port IDE cable with a longer, two-port cable. You'll also need a Y power splitter. A bracket that can accommodate a second fan is strongly recommended.

Regardless of the number of drives you have installed, replace the lid on the TiVo base making sure that the lid fits snugly all around, and then reinsert the five T-10 lid screws around the perimeter of the rear panel.

You can now reconnect your TiVo to your television. You'll need to repeat Guided Set-Up, which will take about 35 minutes. (If you copied your settings or are simply adding a drive to work with your factory drive, you won't need to go through Guided Set-Up again.)

Architecture 6

This architecture includes the Pioneer 810H and 57H models.

Parts

You will need a Phillips-head screwdriver. (Yep, that's it.)

Disassembly

This is an extremely simple install, mainly because you won't need any new parts and because there's only room for a single drive in the unit. However, it's worth noting carefully how everything comes apart, particularly the lid, so that putting it back together again is all the easier.

Follow these steps:

1. Position the TiVo so that the rear panel is facing you.
2. Remove the eight screws holding the lid in place. Four are located along the rear perimeter, where the metal lid folds over the back panel, and two larger screws are located on each side of the unit.
3. Lift the back of the lid up so it just clears the rear panel and fan box. Then pull it toward you to unseat it from the groove in the faceplate. The lid should now be free, so simply lift it all the way off and set it aside. You may find it useful to keep all of the loose screws in the overturned lid.

CAUTION *When removing the lid, take special care because the metal is very pliable and prone to scratching.*

4. Take a moment to familiarize yourself with the inside of the TiVo. At the center is a silver box containing the DVD recorder. At the back left is the power supply, and the hard drive is at the front right. Avoid coming into contact with any parts you are not specifically instructed to touch.
5. Remove the drive and bracket by unscrewing the four Phillips-head screws holding the drive bracket in place around the base of the drive.

6. Note the position of the various markers stamped on the bracket, which indicate how it should be aligned in the unit.
7. Lift the drive and bracket slightly and disconnect the power and IDE cables, freeing the drive from the unit completely.

Drive Formatting

You must now proceed to the "Installation Details" section before continuing here. Start with the "MFS Tools" section to acquire the Linux software you'll need for the job.

Then continue to the "Hard Drive Formatting" section. If you're planning on backing up your drive, start with the "Making a Backup" section (this is an optional step). Otherwise skip ahead to one of the single-drive formatting options, since your Pioneer only has room for one drive (options 1, 3, and 4).

When that's all done, continue on with the next section to reassemble your TiVo.

Reassembly

Essentially all that needs to be done is to take the bracket off the old drive and attach it to the new one. Then you just reverse the disassembly steps to reconnect the drive and seal up the unit.

Follow these steps:

1. Remove the old drive from the bracket by unscrewing the four screws that hold it in place, and use the screws to attach the bracket to the new drive.
2. Set the drive bracket into place on the drive bay, keeping the cable ports pointing toward the rear of the unit and being sure the drive bracket settles into the dimples on the drive bay. There are two sets of dimples, and only one set will allow you to see the screw holes through the bracket.
3. Replace the four screws that hold the bracket in place.
4. Reconnect the power and IDE cables.
5. Replace the lid, which you'll remember fits into place very securely. Start by tilting the front down and under the groove of the front faceplate. Once in place, grasp and spread the sides slightly around the outside body of the base, being careful not to scratch it. The rear of the lid fits in between the fan box and the rear panel, so lower it into place slowly. Once it's all the way down, all the screw holes should line up pretty closely. Watch the tabs that have to slide into the faceplate—they can be a nuisance.

6. Replace the four rear screws first, and then replace the four larger screws on the sides. You can now reconnect the TiVo to your television, plug it in, and power it on. Running Guided Set-Up will take you about 35 minutes. (If you copied your old settings to the drive, you won't need to run Guided Set-Up.)

Installation Details

This section contains useful information for those who have decided to forgo the kit and, instead, will be acquiring parts and formatting their own drive(s). As we mentioned earlier, we suggest reading through this section before you open your TiVo so you'll have an idea what to expect.

First, we've included a list of all the parts you're likely to need to perform your upgrade. Pictures are included, as well, and although you may find slight variations in shape and color once you start shopping, you'll at least know what to look for. Next comes a primer on the Linux software you'll be using, MFS Tools. We'll show you where to find it and how to prep the software for use in your PC. Last comes the actual hard drive formatting. Each upgrader has a variety of options available to them, so the choice is yours. You'll have the option of making a back-up of your TiVo's factory installed software or you can skip ahead and get right to prepping your new drive(s).

Parts Summary

The following is a list of parts that you may need for your upgrade. Check the parts listing for your specific TiVo model to find out what you need, and then refer to the "Useful Web Resources" section at the end of the chapter for some suggestions on where to find them.

- **Drive bracket** Factory installed brackets come in a variety of designs (see Figure 11-9). While most current TiVo models are not designed to accommodate two drives, there is enough physical room inside, should you decide to add one. Several third-party dealers have stepped up to provide for this contingency (see Figure 11-10). The brackets are made either of sheet metal or vibration-dampening polymer, and they are frequently sold separately or in kits with preformatted drives. Be sure to find one that includes all the hardware required to secure it safely to your TiVo without risking possible damage to the motherboard or the drive. The "Useful Web Resources" section will get you started, but you can also do a Web search for "TiVo" and "bracket."

- **Drive screws** If your TiVo has a bracket that can accommodate two drives, you will need to get four extra screws to connect the second drive to the bracket. Often the drives come with these screws, but if not, you're looking for standard #6-32 (3/16 or 1/4 inch) hard drive screws. The head type doesn't matter. If you're using a bracket that requires different size screws, the bracket supplier will likely be able to provide them.
- **IDE cable** This is another fairly common part—your hard drive will connect to the motherboard with a flat ATA/66 IDE cable (see Figure 11-11). The length should be about 18 inches with connectors positioned at 0, 12, and 18 inches. The connectors attach in only one direction, with the two small ridges on one of the flat sides fitting into a notched space in the connection port. The cable is wired so that the blue connector attaches to the motherboard and the far end goes to the master drive. However, if the position of the connectors doesn't seem to fit the location of your drives, you'll need to reverse the cable, with the blue end connecting to the master drive. In this case, you must have the drives jumpered as "master" and "slave" rather than "cable select." (Read the description of jumpers that follows if you need a tutorial.) Regardless of how you wire the drives, make sure that the connectors have all 40 holes open to accommodate the drive pins. Should you end up with a blocked hole, you may be able to clear it with a heated safety pin.

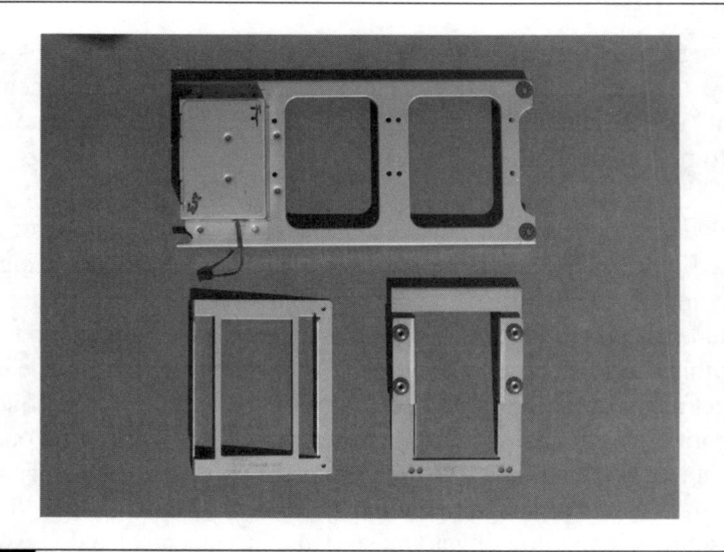

 FIGURE 11-9 Manufacturers' brackets for architecture 2 (top), architecture 4 (bottom left), and architecture 1 (bottom right)

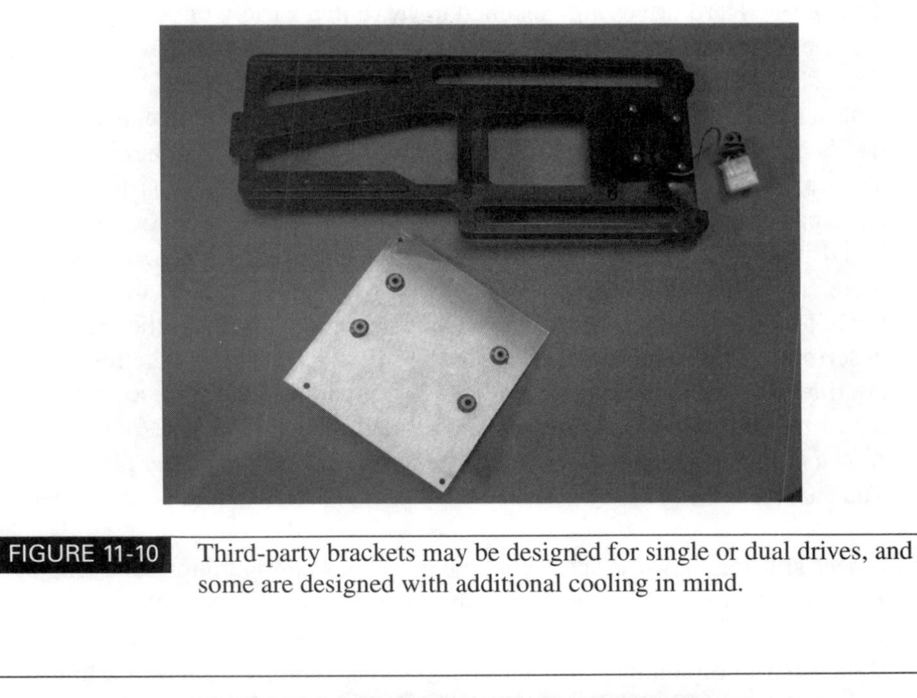

FIGURE 11-10 Third-party brackets may be designed for single or dual drives, and some are designed with additional cooling in mind.

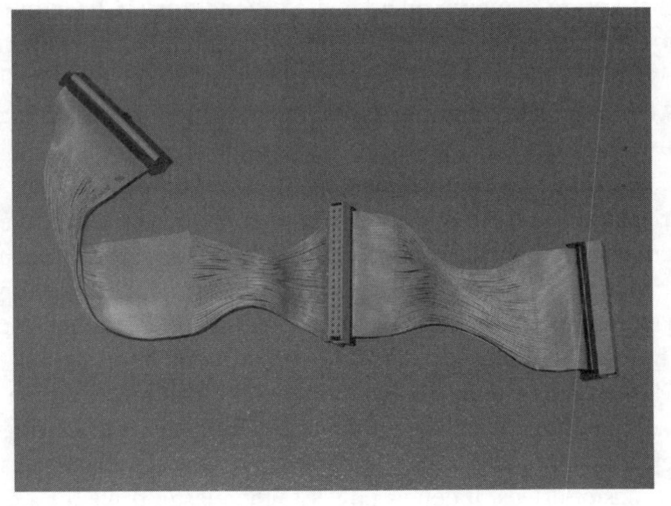

FIGURE 11-11 An IDE cable designed to connect to two hard drives

- **Jumpers** Hard drives are designed to serve in a variety of configurations, and the jumpers allow you to designate the specific role each drive will serve. Drives can be designated variously as "cable select," "master," "slave," or "master with slave present," and few other ways, as well. The master/slave designations are useful in configuring two drives for use together in one TiVo. (Note that once two drives have been "married" into a master and slave pair, you cannot separate them without reformatting them or replacing one drive with a clone of itself.) Among the hard drive's exposed pins, there are usually two separate rows of four or five pins, which are meant for holding the jumper(s), and most drives have a diagram on the face label describing the proper alignment of the jumpers. The jumpers themselves are tiny black or white plastic and metal fittings that slide into position on top of two pins at a time. They can be affixed vertically or horizontally, as shown in Figure 11-12, so always refer to the drive label for proper placement. You can use either your fingernails or tweezers to remove and position them.

- **Power splitters** These splitters come in various configurations and lengths. A power splitter connects to the main power connector inside the TiVo and splits it into two power connectors—one for each drive (see Figure 11-13). It's a fairly common part. Most models will need either a short or a long Y splitter if you're planning on adding a second drive. Check online or ask your local retailer for a standard IDE hard drive power splitter. Otherwise, the "Useful Web Resources" section at the end of the chapter lists some places that will carry them, often in lengths and dimensions customized for TiVo use.

- **PowerTrip** The "strike" energy required to spin up two drives at once is, on occasion, greater than what some TiVos are equipped to handle. Once they're spinning, the drives use less energy, though, and the regular power supply is more than sufficient. (Many electrical devices are the same way. Light bulbs, for example, strike with more energy than they need to stay lit, which is why they frequently blow right when you turn them on.) This being the case, you may be instructed to get a PowerTrip (see Figure 11-14). This simple device connects on the path of the power supply between the slave and the master drives. It delays power to the master drive for a few seconds, allowing the slave to start up first. Since the master must see the slave on startup, it's critical for the master to start up second. Refer to the "Useful Web Resources" section for a supplier.

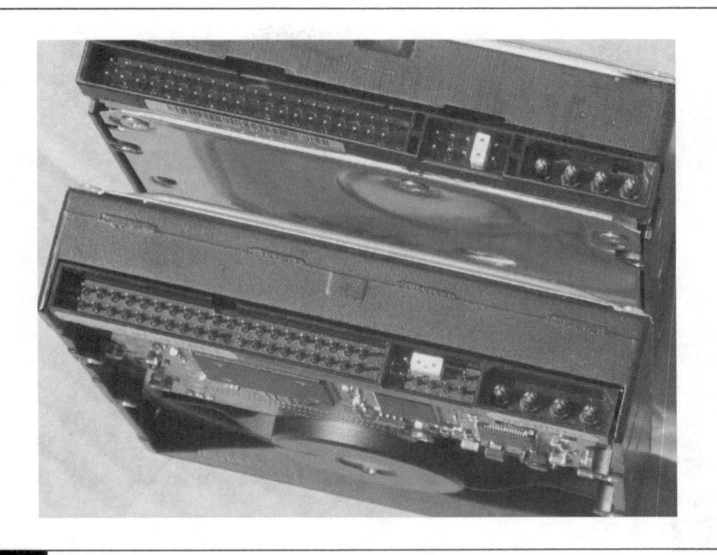

FIGURE 11-12 The jumper (white) may be aligned in a variety of ways.

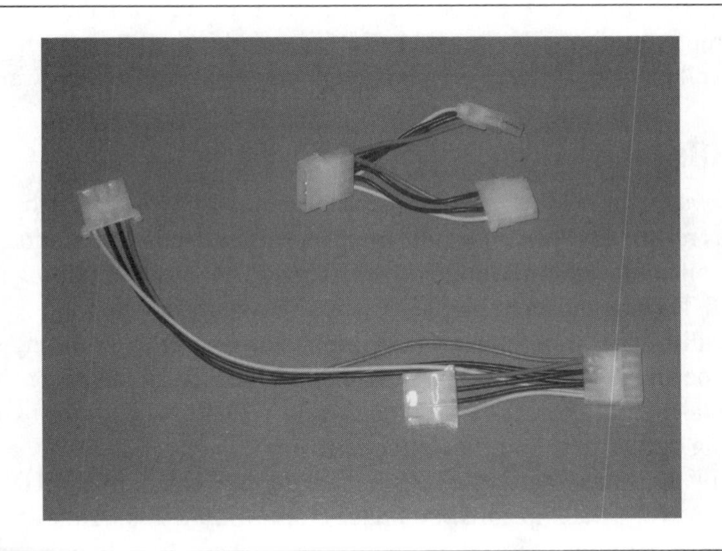

FIGURE 11-13 Long and short power splitters

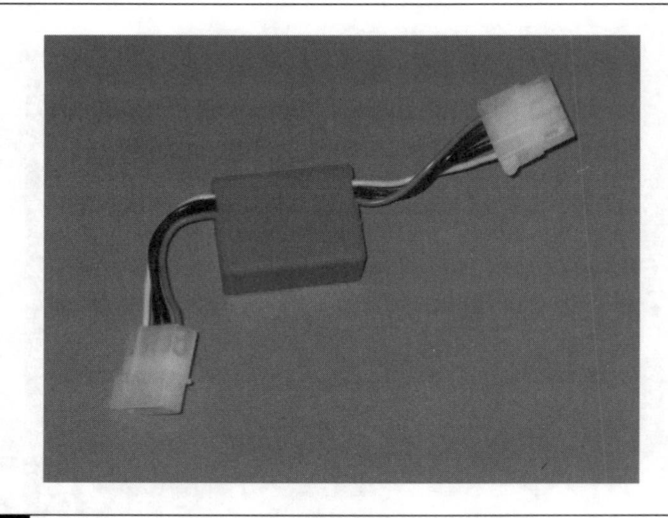

FIGURE 11-14 A PowerTrip designed to delay the startup of the master drive

- **Torx Tools** Most TiVo are assembled with a star-shaped screw variety called a Torx screw (see Figure 11-15). They heads come in a range of sizes, but without exception, the T-10 and T-15 sizes are the only ones currently used in TiVo assembly. You should be able to find Torx drill bits as well as manual Torx screwdrivers in any well-stocked hardware store.

MFS Tools

MFS Tools was written by a TiVo-user screen-named Tiger, and version 2.0 of the program was released in June 2002. It is an indispensable Linux application, providing a powerful, all-in-one solution for backing up, restoring, and upgrading capacity on nearly all TiVo models to date.

Specifically, MFS Tools will back up and restore the TiVo operating system, it will create manageable backups that fit onto CDs for storage, it will enable TiVo owners to make use of larger hard drives, and it will allow TiVo users to copy programming and settings to new, larger drives.

The software is a utility, and it requires a working version of the TiVo software on a hard drive (or a backup software image) for it to function. Although it won't work on a Mac, the Linux boot cd will boot on almost any PC, On older computers, you may need to go into the BIOS settings and set drive detection to "None." In the following sections, you will find a description of precisely what you will need in order to take advantage of this program.

FIGURE 11-15 Torx screws are frequently used in TiVo assembly. Torx screwdrivers can be commonly found in your hardware store.

Basic Assumptions

We need to make a few assumptions in order to reduce the large number of permutations possible when upgrading a TiVo, and these assumptions include the manner in which your PC is configured. Any variation in your computer configuration from the standard will require you to make changes in the command strings for running MFS Tools. For example, in this command, the `hdc` presupposes you're backing up a drive connected in the C position on your PC.

```
mfsrestore -s 127 -zxpi /mnt backup.bak /dev/hdc
```

If the drive is in a different position, you will need to change the drive letter accordingly. For convenience, the variable characters in the command strings have been underlined.

The following list identifies some other assumptions we have made about the PC's setup and the status of the TiVo:

- The computer has a drive that can read CD's (CD-ROM, CD-RW, DVD-ROM, etc.) and open IDE ports for connecting drives. The IDE ports are typically accessed by opening the PC's case.
- The computer's boot drive is connected in the primary master position (known as the A position).

- The CD drive is connected in the primary slave position (known as the B position).
- The TiVo drive(s) will be connected to the first open IDE connectors, which the computer recognizes as the C and/or D positions.

If you open your TiVo and discover you have two drives inside, you *will* be able to make a backup of the software, but the procedure for saving your programming *will not* be discussed here. See the "Useful Web Resources" section for help with this.

Acquiring the Software (MFS Tools 2.0)

To run MFS Tools, you need to download the software and burn the image file of the software onto a CD. Follow this link to download a copy of the free software: http://hellcat.tyger.org/MFS/2.0/mfstools2noJ.iso.

NOTE *If your TiVo is a Pioneer model, a Hughes HR10-250, or a TiVo brand TCD54x model, you will need to make a boot CD from a different image containing LBA48 support. To find a current copy of the program, do an Internet search for "LBA48" and "TiVo." The rest of the formatting instructions are still applicable to your TiVo model.*

Once you've downloaded the software, burn the image file (the .iso file) as an ISO image to a CD. Any standard CD burning software should do the trick, but the software needs to recognize the ISO as an image as opposed to merely a file. (If you have difficulties, use data mode 1, block size 2048.) This will be your boot CD for all subsequent operations.

Hard Drive Formatting

The time has come to format your hard drive. In this section, you are given a number of options to choose from. With your old and dew drives connected to the PC and with an MFS Tools 2.0 boot disk in the CD drive, you'll be able to copy either the TiVo software itself or everything, on the old TiVo drive, including your settings and programming. You can copy to either to a single drive or, for maximum capacity, to a pair of drives (unless your Architecture instructions stated otherwise).

Preceding the four options are instructions for making a back-up. This step is not mandatory, but may be useful if you want to safe a copy of your TiVo software for use down the road. Unless you take this step, you will have to skip Option 1.

- **Option 1** Restoring a backup to one or two drives
- **Option 2** Adding a drive to your TiVo's factory drive

- **Option 3** Copying from your TiVo drive to a single or to dual drives (saving your programming and settings)
- **Option 4** Copying from your TiVo drive to a single or to dual drives (without preserving programming or settings)

Making a Backup (Optional)

You can back up your TiVo's software and, later, write it to another hard drive or just keep it on hand in case of later disk corruption. This backup saves all your settings, but not the programming. If you restore from this backup, your old programming will appear in the Now Playing list, but it won't actually play.

NOTE *If your PC is running Windows, its hard drive needs to have a FAT partition in order to write the backup to its disk. On many PCs you can right-click on the hard drive to see its properties and determine this. If you don't have a FAT partition, consider getting a program such as Partition Magic to set one up. To be safe, you will need at least 1GB of space on a properly formatted drive. If you don't feel like dealing with this, you may want to skip the backup process and simply copy the necessary files directly to the new drive.*

To make a backup, follow these steps:

1. Make sure your computer is set to boot up from a CD when there's a boot disc in the CD drive. This setting can be changed in the BIOS window on startup. Once it is set, you may want to test and verify that the PC boots from the CD before attaching your TiVo drive (or drives). Bad things will happen if TiVo drives boot into Windows.
2. Boot from your MFS Tools CD with your TiVo's drive(s) connected to the PC. At the command prompt, press ENTER. Once the Linux operating system is loaded from the CD, you will see a # prompt. Scroll up using SHIFT-PAGEUP and check to make sure the PC recognized your drive(s). As you scroll up, you should see text similar to the following:

```
hda: XXX, ATA DISK drive
hdb: IDE/ATAPI CD-ROM 48X, ATAPI CD/DVD-ROM drive
hdc: QUANTUM FIREBALL CX13.6A, ATA DISK drive
     ide0 at 0x1f0-ox1f7, 0x3f6 on irq 14
hda: _______ sectors (YYY MB) w/ZZZ KiB Cache, CHS=1111/222/
     33, UDMA(33)
hdb: ATAPI 17X CD-ROM drive, 128kB Cache, UDMA(33)
hdc: 26760384 sectors (13701 mb) w/418KiB Cache, CHS=1665/
     255/63 UDMA(33)
```

NOTE *XXX, ___, YYY, ZZZ, 1111, 222, 33 are variables which denote the number of sectors on your drive and the drive capacity, all of which will vary, as will the size of your drives at hda, hdb, hdc, and hdd.*

3. If you discover that one of your drives has been recognized at a far lower capacity than it ought to be, that hard drive may be locked. You'll need to power down with CTRL-ALT-DELETE and unlock the locked drive. See the "Unlocking TiVo Hard Drives" section later in this chapter for help with this.
4. Mount your computer's hard drive into the Linux file system with this command:

```
mount /dev/hda1 /mnt
```

5. Mount and copy the software from the TiVo drive to your computer with this command:

```
mfsbackup -f 9999 -1so /mnt/backup.bak /dev/hdc
```

Or, if you're copying from two drives, use this command:

```
mfsbackup -f 9999 -1so /mnt/backup.bak /dev/hdc dev/hdd
```

This will take a few minutes to back up. When it's finished, you should shut the computer down with the CTRL-ALT-DELETE command. If your PC automatically reboots, interrupt it by turning it off with the power button. You may now disconnect your drives.

Option 1—Restoring a Backup to One or Two Drives

If you've created a backup of your TiVo's software, you can restore it to either one or two drives with the following procedure:

1. Attach your new TiVo drive to the C position of the computer (or two TiVo drives to the C and D positions). One drive should be designated "master" and one "slave" by setting the jumper pins according to the instructions on the label of your specific drive. (A discussion of jumpers and master/slave drive settings can be found in the jumpers description in the "Parts Summary" section earlier in the chapter.)
2. Boot from the MFS Tools CD. You will see the following once your PC is started up.

```
<enter> default setup: Kernel with DMA enabled, but no
byteswapping.
Nodma same as above, but with DMA disabled.
swap DMA disabled, byte swapping enabled.
Dmaswap DMA enabled, byte swapping enabled. (Dangerous)
boot:
```

3. At the `boot:` prompt, press ENTER. A huge bunch of text will scroll by, eventually leaving you with a # prompt.

4. Check to see that your drive (or drives) mounted properly. Use the SHIFT-PAGEUP keys to scroll up and confirm that your drives have been recognized and at their correct sizes. (Keep in mind that drives are often recognized as having +/– 10 percent variation from their known capacity.) As you scroll up, you should see text similar to the following:

```
hda: XXX, ATA DISK drive
hdb: IDE/ATAPI CD-ROM 48X, ATAPI CD/DVD-ROM drive
hdc: QUANTUM FIREBALL CX13.6A, ATA DISK drive
     ide0 at 0x1f0-ox1f7, 0x3f6 on irq 14
hda: _______ sectors (YYY MB) w/ZZZ KiB Cache, CHS=1111/222/
     33, UDMA(33)
hdb: ATAPI 17X CD-ROM drive, 128kB Cache, UDMA(33)
hdc: 26760384 sectors (81964 mb) w/418KiB Cache, CHS=1665/
     255/63 UDMA(33)
```

11

NOTE *XXX, ___, YYY, ZZZ, 1111, 222, 33 are variables which denote the number of sectors on your drive and the drive capacity, all of which will vary, as will the size of your drives at hda, hdb, hdc, and hdd.*

5. If you discover that one of your drives has been recognized at a far lower capacity than it ought to be, that hard drive may be locked. You'll need to power down with CTRL-ALT-DELETE and unlock the drive. See the "Unlocking TiVo Hard Drives" section later in this chapter for some help with this.

6. Mount your computer's hard drive into the Linux file system with this command:

```
mount /dev/hda1 /mnt
```

7. Enter this command for restoring to one drive:

```
mfsrestore -s 127 -zxpi /mnt/backup.bak /dev/hdc
```

Or enter this command for restoring to two drives:

```
mfsrestore -s 127 -zxpi /mnt/backup.bak /dev/hdc /dev/hdd
```

This will take a few minutes. When it's complete, as indicated by the # prompt, shut down the computer with the CTRL-ALT-DELETE command. If your PC automatically reboots, interrupt it by turning it off with the power button.

You may now disconnect your new drive(s) for reinstallation in your TiVo. Return to the reassembly instructions for your specific model. You'll want to retain the file backup at least until you have verified that the backup actually works, and possibly keep it for future use.

Option 2—Adding a Drive to Your TiVo's Factory Drive

This procedure allows you to add capacity to your TiVo by "slaving" a drive to your TiVo's existing drive. Note that for most models of TiVo, you can only add capacity up to a maximum of 137GB. Therefore, there's little use in adding anything larger than a 160GB drive. The exceptions to this rule are the Hughes HR10-250 and the TiVo TCD54x models, which will allow drives of any size.

NOTE *Technically, the Pioneer 810H and 57H have no size limits either, but there is only room to mount a single drive inside these units*

To add a drive, follow these steps:

1. Change the jumper position on your TiVo's drive to "master," according to the instructions on its label. On some drives, the label may instead provide directions for setting the jumper to "Master with Slave present." If you have a slave present, go ahead and set the jumper for the master drive to "master with slave present." (For a discussion of jumpers and master/slave settings, refer to the jumpers description in the "Parts Summary" section earlier in the chapter.)

2. Set the jumper position on the drive you will be adding to your TiVo to the "slave" position, following the instructions on its label.

3. Connect the two TiVo drives to the C and D positions on the computer.

4. Boot from the MFS Tools CD. You will see the following output once your PC is started up:

```
<enter> default setup: Kernel with DMA enabled, but no
byteswapping.
Nodma same as above, but with DMA disabled.
swap DMA disabled, byte swapping enabled.
Dmaswap DMA enabled, byte swapping enabled. (Dangerous)
boot:
```

5. At the `boot:` prompt, press ENTER.

6. A huge bunch of text will scroll by, eventually leaving you with a # prompt. At this point, you should check to see that your drive (or drives) mounted properly. Use the SHIFT-PAGEUP keys to scroll up and confirm that your drives have been recognized at their correct sizes. (Keep in mind that drives are often recognized as having +/– 10 percent variation from their known capacity.) As you scroll up, you should see text similar to the following:

```
hda: XXX, ATA DISK drive
hdb: IDE/ATAPI CD-ROM 48X, ATAPI CD/DVD-ROM drive
hdc: QUANTUM FIREBALL CX13.6A, ATA DISK drive
     ide0 at 0x1f0-ox1f7, 0x3f6 on irq 14
hda: _______ sectors (YYY MB) w/ZZZ KiB Cache, CHS=1111/222/
     33, UDMA(33)
hdb: ATAPI 17X CD-ROM drive, 128kB Cache, UDMA(33)
hdc: 26760384 sectors (81964 mb) w/418KiB Cache, CHS=1665/
     255/63 UDMA(33)
```


NOTE

XXX, ___, YYY, ZZZ, 1111, 222, 33 are variables which denote the number of sectors on your drive and the drive capacity, all of which will vary, as will the size of your drive at hda, hdb, hdc and hdd.

7. If you discover that one of your drives has been recognized at a far lower capacity than it ought to be, that hard drive may be locked. You'll need to power down with CTRL-ALT-DELETE and unlock the drive. See the "Unlocking TiVo Hard Drives" section later in this chapter for some help with this.

8. Enter this command:

```
mfsadd -x /dev/hdc /dev/hdd
```

This command will run rather quickly, and when the computer is finished, you should be at a # prompt. Shut down the PC with the CTRL-ALT-DELETE command. If your PC automatically reboots, interrupt it by turning it off with the power button.

You may now disconnect your new drive(s) for reinstallation in your TiVo. Return to the reassembly instructions for your specific model.

Option 3—Copying From Your TiVo Drive to a Single Drive or Dual Drives (Saving Your Programming and Settings)

This option allows you to save your programming and format your new drive or drives in a single step. You should consider retaining your old drive in case something eventually goes wrong with your new ones. This method *does not* utilize the backup that you may or may not have already made.

If you're copying to a dual-drive configuration, one drive in the pair should be designated "master" and one "slave" by setting the jumper pins according to the instructions on the labels of your drives. (For a primer on jumpers and master/slave settings, refer to the discussion of jumpers in the "Parts Summary" section earlier in the chapter.)

Once that's done, follow these steps:

1. Attach the old TiVo drive to the A position on the computer (where the PC's drive normally connects).
2. Attach the new drive to the C position, or if there are two drives, attach them to the C and D positions.
3. Boot from the MFS Tools CD. You will see the following output when your PC is started up:

```
<enter> default setup: Kernel with DMA enabled, but no
byteswapping.
Nodma same as above, but with DMA disabled.
swap DMA disabled, byte swapping enabled.
Dmaswap DMA enabled, byte swapping enabled. (Dangerous)
boot:
```

4. At the `boot:` prompt, press ENTER.
5. A huge bunch of text will scroll by, eventually leaving you with a # prompt. At this point, you should check to see that your drives mounted properly. Use the SHIFT-PAGEUP keys to scroll up and confirm that your drives have been recognized at their correct sizes. (Keep in mind that drives are often recognized as having +/– 10 percent variation from their known capacity.) As you scroll up, you should see text similar to the following.

```
hda: XXX, ATA DISK drive
hdb: IDE/ATAPI CD-ROM 48X, ATAPI CD/DVD-ROM drive
hdc: QUANTUM FIREBALL CX13.6A, ATA DISK drive
     ide0 at 0x1f0-ox1f7, 0x3f6 on irq 14
hda: _______ sectors (YYY MB) w/ZZZ KiB Cache, CHS=1111/222/
     33, UDMA(33)
hdb: ATAPI 17X CD-ROM drive, 128kB Cache, UDMA(33)
hdc: 26760384 sectors (81964 mb) w/418KiB Cache, CHS=1665/
     255/63 UDMA(33)
```

NOTE *XXX, ___, YYY, ZZZ, 1111, 222, 33 are variables which denote the number of sectors on your drive and the drive capacity, all of which will vary, as will the size of your drive at hda, hdb, hdc and hdd.*

6. If you discover that one of your drives has been recognized at a far lower capacity than it ought to be, that hard drive may be locked. You'll need to power down with CTRL-ALT-DELETE and unlock the drive. See the "Unlocking TiVo Hard Drives" section later in this chapter for some help with this.

7. If it works, you'll see another # prompt. Enter this command for copying to one new drive:

   ```
   mfsbackup -Tao - /dev/hda | mfsrestore -s 127 -zxpi - /dev/
   hdc
   ```

   Or enter this command for copying to two new drives:

   ```
   mfsbackup -Tao - /dev/hda | mfsrestore -s 127 -zxpi - /dev/
   hdc /dev/hdd
   ```

This may take up to several hours, depending on the amount of material on your old drive and the condition it's in. Ultimately, you'll be able to scroll up and see the text indicating the number of hours on the new drives. When the process is complete, as indicated by the # prompt, you should shut down the computer with the CTRL-ALT-DELETE command. If your PC automatically reboots, interrupt it by turning it off with the power button.

You may now disconnect your new drive(s) for reinstallation in your TiVo. Return to the reassembly instructions for your specific model.

Option 4—Copying From Your TiVo Drive to a Single Drive or Dual Drives (Without Preserving Programming or Settings)

This procedure allows you to copy the TiVo software to your new drive (or drives) without dealing with a backup and without preserving your programming or settings.

If you're copying to a dual-drive configuration, one drive in the pair should be designated "master" and one "slave" by setting the jumper pins according to the instructions on the label of your drives. (For a discussion of jumpers and master/slave settings, refer to the jumpers description in the "Parts Summary" section earlier in the chapter.)

Once that's done, follow these steps:

1. Attach the old TiVo drive to the A position on the computer (where the PC drive normally connects).
2. Attach the new drive to the C position, or if there are two drives, connect them to the C and D positions.
3. Boot from the MFS Tools CD. You will see the following output when your PC is started up:

```
<enter> default setup: Kernel with DMA enabled, but no
byteswapping.
Nodma same as above, but with DMA disabled.
swap DMA disabled, byte swapping enabled.
Dmaswap DMA enabled, byte swapping enabled. (Dangerous)
boot:
```

4. At the `boot:` prompt, press ENTER.
5. A huge bunch of text will scroll by, eventually leaving you with a # prompt. At this point, you should check to see that your drive (or drives) mounted properly. Use the SHIFT-PAGEUP keys to scroll up and confirm that your drives have been recognized at their correct sizes. (Keep in mind that drives are often recognized as having +/– 10 percent variation from their known capacity.) As you scroll up, you should see text similar to the following:

```
hda: XXX, ATA DISK drive
hdb: IDE/ATAPI CD-ROM 48X, ATAPI CD/DVD-ROM drive
hdc: QUANTUM FIREBALL CX13.6A, ATA DISK drive
     ide0 at 0x1f0-ox1f7, 0x3f6 on irq 14
hda: _______ sectors (YYY MB) w/ZZZ KiB Cache, CHS=1111/222/
     33, UDMA(33)
hdb: ATAPI 17X CD-ROM drive, 128kB Cache, UDMA(33)
hdc: 26760384 sectors (81964 mb) w/418KiB Cache, CHS=1665/
     255/63 UDMA(33)
```

NOTE *XXX, ___, YYY, ZZZ, 1111, 222, 33 are variables which denote the number of sectors on your drive and the drive capacity, all of which will vary, as will the size of your drive at hda, hdb, hdc and hdd.*

6. If you discover that one of your drives has been recognized at a far lower capacity than it ought to be, that hard drive may be locked. You'll need to power down with CTRL-ALT-DELETE and unlock the drive. See the "Unlocking TiVo Hard Drives" section later in this chapter for some help with this.

7. Enter this command for copying to one new drive:

```
mfsbackup -o - /dev/hda | mfsrestore -s 127 -zxpi - /dev/hdc
```

Or enter this command for copying to two new drives:

```
mfsbackup -o - /dev/hda | mfsrestore -s 127 -zxpi - /dev/hdc/
dev/hdd
```

This will take a few minutes. When it's complete, you'll be at another # prompt. Shut down the computer with the CTRL-ALT-DELETE command. If your PC automatically reboots, interrupt it by turning it off with the power button.

You may now disconnect your new drive(s) for reinstallation in your TiVo. Return to the reassembly instructions for your specific model.

Unlocking TiVo Hard Drives

If your TiVo drive mounts on your PC (using MFS Tools) at a far lower capacity than you know it can actually hold, it may be locked. That is to say, the computer that the drive was in may have put a software lock on the drive to prevent new data from being written to it. An example would be a 40GB drive showing up as 9GB. In this case, you will need to power down the computer and take steps to unlock the drive.

One method of unlocking, which involves creating a boot disk of a utility called DiskUtil, is detailed here:

1. Go to http://www.bootdisk.com and download a file for creating, you guessed it, a boot disk. Once at the site, you should locate the "Bootdisks" link and follow it. We suggest using the DOS 6.22 bootdisk image. Download the image file. Click it and the computer will prompt you to insert a floppy disk. The PC will write the image to the disk you've inserted, thus creating a boot disk.

NOTE *Afterwards, you will likely need to delete any Read Me files on the disk in order to leave enough room for the DiskUtil software.*

2. Go to http://www.upgrade-instructions.com and take a look at the list of files hosted on the site. There you will be able to download DiskUtil. This should be copied onto your newly created boot disk.
3. Once you've made your DiskUtil boot disk, insert it into your PC, and restart the computer which should be configured to boot from a floppy disk when it sees one in the disk drive.
4. Connect your locked hard drive in the secondary master IDE position (referred to as the "C" or "2" position) and boot the PC.
5. At the A:/> prompt, enter this command:

```
diskutil /PermUnlock 2
```

Be attentive to capitalization. Also, the `2` refers to the IDE position on the PC, so if you had connected your drive to the primary master position, you would enter `0` instead.

You should then receive a message that your drive has been successfully unlocked, and you will need to reboot the computer for it to take effect. Shut down the computer, eject the disk, and continue with whatever process you had going on before.

Useful Web Resources

As might be expected, the Internet provides an invaluable resource to anyone attempting to hack their TiVo. Many people have collaborated via the Web to help one other problem solve and tech support this difficult hack. In the spirit of the Internet, most of the information you will need is provided for free. However, where the opportunity has arisen for products and services to be sold, entrepreneurs have naturally responded to the call. Below we've provided you with many of the most popular sources for information, products and services; both free and otherwise.

Information Sites

Included below are links to some of the most reliable sites for finding shareware as well as exhaustive information on the subject of TiVo and TiVo hacking.

- **http://www.tivocommunity.com** This is the home page for thousands of TiVo addicts looking get the latest gossip and to trade information on all things TiVo.

- **http://www.upgrade-instructions.com** Michael Adberg of WeaKnees has created an comprehensive online resource for do-it-yourself upgraders. This site will march you step by step from model number to instructions based on highly specific choices.
- **http://www.newreleasesvideo.com/hinsdale-how-to/** This link leads you to Bill Rugnery's upgrade instructions, including boot CD and floppy instructions.
- **http://www.bootdisk.com/utility.htm** This is a great site for finding drive utilities and, obviously, boot disk images.
- **http://www.weaknees.com/repair.php** This is a thorough troubleshooting guide to provide further help diagnosing your ailing TiVo.

Kit and Part Dealers

Below is an alphabetical listing of Internet businesses providing products and services for your TiVo. Without endorsing any in particular, we recommend browsing them all thoroughly to compare before you decide to make a purchase. If you need some advice, you can start with the link we provided you above to the TiVo Community Forum.

- **http://www.digitalrecorder.com** This site sells upgrade kits in addition to entire units.
- **http://www.9thtee.com** The 9th Tee site carries an eclectic range of products, including TiVo upgrade kits, heating pads, and pool supplies.
- **http://www.ptvupgrade.com** PTVupgrade stakes a claim to being the first upgrade dealer on the market.
- **http://www.tvrevo.com** This is a straightforward upgrade dealer covering every make and model.
- **http://www.weaknees.com** WeaKnees.com is an innovative company that has a full range of TiVo parts, products, and services, many of which are not available elsewhere.

Part IV

What's Next?

Chapter 12

Finally… High-Definition TiVo

How to...

- Understand High Definition
- Understand the Delay in Bringing High Definition to the DVR Market
- Purchase a High-Definition TiVo
- Set Up a High-Definition TiVo
- Find Shows Recorded in High Definition and Dolby Digital 5.1
- Get the Most from Your Hard Drive Space

For years, high definition (HD) has been available in various formats in other countries, and in an analog format in Japan for nearly twenty years! In the mid-to-late 1990s, the United States finally entered the high-definition world, and the way you watch television will never be the same. With TiVo, our world has already changed; no longer must we sit tirelessly through an entire football game waiting for the oh-so-brief commercials to use that far-away restroom, the one that should have been built directly adjacent to your couch. Now we finally have the best of both worlds—a window to that high-resolution TV world, and a way to pause and record that window long enough to keep your kidneys healthy. Enter high-definition TiVo.

Did you know?

It Is the Size That Counts?

High-definition programming is recorded at a much higher data bit rate, thus requiring a much larger hard drive than your standard 480I material (480I signifies the current standard of 480 lines of vertical resolution used in the current system for over 60 years) Even current hard drives, which are now larger than imagined a few years back, are still not large enough to hold a whole collection of high definition films. As a matter of fact, you can hold about 35 hours of high-definition content on the current DirecTV HD unit, but you can hold 200 hours of standard definition programming.

The Basics of High-Definition TV

Standard definition TV, the television programming we have watched for nearly 60 years, has used the same standards since the very beginning. The National Television Standards Committee (NTSC), has set specific guidelines for image resolution, color standards, and a whole host of other details relating to the actual image that is broadcast.

Now we have High Definition TV (HDTV), and the American Television Standards Committee (ATSC) has specified a whole new host of standards and, most importantly, a dramatic increase in television resolution and overall color reproduction and clarity.

The old NTSC standard calls for a maximum of 480 lines of vertical resolution, or roughly the same resolution as a computer monitor of 640×480 lines. Ten years ago, computer monitors could only support 640×480, and now we have monitors that go well over 1920×1080! But no matter how large your older television was, even if it said it had over 1,000 lines of resolution, the maximum resolution it displayed was always 480 lines before high definition. Older 4:3 ("square") televisions had 640 lines of horizontal resolution as opposed to the 1,920 or 1,280 lines found today on most HDTVs.

The ATSC adopted nearly 18 resolutions as acceptable forms of high definition broadcasts. These are some of the most common: 480P, 525P, 540P, 576P, 720P, 768P, 960I, 960P, 1024P, 1080I, and 1080P (the current "super HDTV" available on only a few sets). What exactly does each number mean, and what does the "P" or "I" stand for?

The *I* stands for *interlaced.* A normal interlaced image is scanned (displayed on your screen) in two passes, with the first pass displaying some of the video image and the second pass creating the remaining portion of the image—your eye and brain assemble the image created by the two passes. The interlaced scan is what causes the "scrolling line" effect you see when a recorded image of a computer monitor is shown on TV. Interlaced images leave artifacts such as noise and picture break-up, and while they are superb for some material, they are not as good for faster action.

The *P* stands for *progressive.* With progressive scanning, the image is displayed in one pass, and the image is therefore processed by our eyes and brain faster, resulting in a "flicker-free" image. The progressive image suggests a much higher resolution to our mind's virtual eye, and it is much more pleasing to the eye for sports and fast-moving action.

The first and lowest HD resolution, 480P, falls into the category of extended definition television (EDTV), and while it is part of the ATSC standard, it still is not considered true HD by industry professionals. The same can be said for 525P,

540P, and 576P, as most HD televisions and projectors see these scans as 480P internally, and treat them exactly the same, which can potentially be a problem. Each different scan rate can result in different positioning of the image on your display. For example, on your PC monitor, if you switch between resolutions such as 640×480 and 800×600, the image might shift on the screen from one side to the other, leaving you to manually adjust the image horizontally or vertically with the built-in menu of your monitor, or using the software controls of your video card. The same can be said for HDTVs, as all true digital displays (Fixed Pixel Displays, or FPDs) are basically enhanced computer monitors. Each resolution may sync independently (meaning, each resolution may be fully supported, as in the case with a "multiscan" computer monitor), or your HDTV may only allow you to adjust some or none of the scans.

The resolutions of 720P and above are considered true high definition, with the two most popular resolutions being 1280×720P and 1920×1080I. Most videophiles and professionals agree that these two are the standard for HDTV, and manufacturers obviously agree since most newer HDTVs are only found in these two varieties.

Most of the newer HD sets come in different flavors, such as DLP (Digital Light Projection), LCD (Liquid Crystal Display), L-COS (Liquid Crystal on Silicon), and D-ILA (a variation on L-COS). These sets are considered FPDs (Fixed Pixel Displays), and they have a set native scan rate or resolution and do not rely on phosphors (like plasma screens) or cathode ray tubes (CRTs) to produce the image.

DLP (Digital Light Projection) is a Texas Instruments technology that relies on an array of millions of micro-mirrors and a color wheel (in most cases) to produce an image. LCD (Liquid Crystal Display) is the same technology found in laptop computer screens and new flat panel computer monitors. L-COS (Liquid Crystal on Silicon) is a variant of LCD and was developed by Hitachi, Hughes, and JVC; it has many offshoots, including JVC's D-ILA and Sony's S-XRD (Silicon Crystal Reflective Display) panels. DLP is becoming the most common, but newer technologies such as S-XRD are starting to find their way into high-end applications.

Both plasma screens and CRTs suffer from the "burn-in" effect, which, in essence, leaves any high contrast or slow-moving image "burned" on the screen temporarily. This results from the slow refresh (the speed of the scanning lines) and the nature of the elements used to produce the image. FPDs do not have this problem and therefore are preferred in most installations for ease of use and longevity of the product. Most current DLP- and LCD-based HDTVs have a native scan rate of 1280×720 and a 16:9 aspect ratio, which means the television is 1.78 times wider than it is tall. The standard 4:3 televisions of the past 60 years have all been 1.33 times wider than tall.

Most of the analog rear-projection HD sets on the market today have scan rates of 540P or 1080I. To the eye, 540P and 1080I look virtually identical. Most HD broadcasts on television today are 1080I, while only a select handful choose 720P. The irony is that 720P is the native rate of the vast majority of consumer-level HDTVs available today, and only a few HD sets use 1080I as their native rate. However, all HD sets will cross-convert, meaning that they will shrink down 1080I to 720P or blow up 720P to 1080I. It is better though, and highly recommended, to have the native rate of your television displayed without any scaling (shrinking or stretching).

To understand these resolutions, think of 640×480 (standard definition) as a small rectangle—you can fit more than four of these on one 1080I screen! There are more than 2,000,000 pixels (or dots) in a 1920×1080 display, while an NTSC 640×480 TV set displays only 307,200 pixels. HDTV has an amazing 6¾ times the number of pixels, which means you are seeing over 6¾ times more resolution!

HDTVs with real HD broadcasts create an amazing degree of realism. For that reason alone, DirecTV's HDNet broadcasts of "Bikini Destinations" is currently the number-one show being recorded on the current batch of HD TiVos.

Did you know?

HDTV Looks Better Than Film?

Most people agree that HDTV looks better than film in the theater, but this has less to do with resolution or color fidelity than with poorly-set-up movie theaters and the human eye's lack of memory. When you flip channels from a standard definition broadcast to a high definition broadcast, your eye immediately focuses on the new material, and you are amazed. When polled, most people, including DVD and movie reviewers, feel like the DVD and D-VHS transfers of film look better than the film looked in the theater, but that's only because most people do not get the chance to directly compare the two formats (most people don't have HDTVs in the movie theater, or theaters in their homes). Also, everything looks better when you are relaxing in the comfort of your own home.

In fact, film has far more resolution than any HD standard, and that's why companies like Sony have developed their new 4K projectors for movie theaters, which almost double the maximum resolution of today's best HD sets. These new 4K projectors will push cinemas into the future, and set a new standard for home theater to reach, and maybe one day surpass.

Why It Took So Long to Market HD DVRs

Since TiVo's arrival along with ReplayTV in the late 1990s, early adopters have thirsted for an HD-TiVo. Of course, it could be done, but there were massive stumbling blocks:

- There was the question of whether HD-TiVo was really needed.
- Studios feared piracy, so bulletproof copy protection had to be created.
- A new cable for HDTV was needed, because many consumers were unhappy with the analog component video cables previously used to transmit HDTV (which used three separate cables instead of one).

When HDTV rolled out in the late 1990s, there was very little high-definition material available for early adopters. Initially, only a few select channels had HD material at all, and it was mostly large sporting events, such as the Super Bowl or NBA Finals. DirecTV then rolled out HD-HBO and HDNet and eventually added HD-Showtime. The FCC, while encouraging networks to increase HD content in the 1990s, was not overly forceful because the cost of equipment required to switch from SD (standard definition or 480 lines of vertical resolution) to HD is not cheap (a good HD camera is over $100,000), and the demand by consumers was just beginning.

However, since about 1999, HDTV purchases have grown exponentially, and as a result, networks have added more HD programming, with most prime-time material now being available in HD 95 percent of the time. Some channels, such as Fox, have gone the EDTV route, broadcasting their material in a 16:9 480P format (that is changing, with the 2004–2005 NFL season on Fox being the network's first true HD broadcast). According to Twice.com, HDTVs are now in 1 of every 25 homes in 2003, as opposed to 1 of every 50 homes in 2001, and TiVos (or other DVRs) are in 1 of every 15 homes. The need for HD-TiVos has finally caught up with the demand from early adopters.

Piracy is one of the biggest concerns for the movie studios, artists, and related manufacturing companies. As is evident from the lawsuits launched by the Recording Industry Association of America (RIAA), the Industry takes piracy very seriously, and that piracy is far more prevalent than once thought. According to the RIAA's published studies, the majority of high speed Internet users have, at one time or another, downloaded at least one illegal MP3 or WMA file. In terms of dollars and cents, imagine each consumer walking into a local music store and stealing one CD single—the RIAA considers this a huge problem.

Movie studios, directors, and actors alike are in the business not only for their art but to make money, and they are extremely protective of their finished, and sometimes unfinished, products. The Motion Picture Association of America and the Directors Guild of America went to court over DVD screeners (frill-free advance copies of their nominee hopefuls on DVD for reviewers only) of their Academy Award nominees, and the big studios are still in court today against the makers of DVD copying software. With the advent of high-definition equipment, consumers could have an almost perfect digital copy of the original HD master, at near half the resolution to the eye, a proposition that makes studios very uncomfortable. Studios, nor directors, want perfect digital copies of their films in the market, and have had at least one step down in video quality to the consumer format with each new format. HD is the closest we have come to the original 1080P master. Film-to-digital transfers are done in 1080P, the highest home theater resolution so far, and the idea of pirates having 1080I versions of films makes studios and filmmakers very uneasy.

DVDs have Macrovision (the copy-protection that prevents DVDs from being copied on a VCR), as did many VHS tapes, but it is not a bulletproof copy-protection method. Studios have been scrambling, along with DirecTV, DISH Network, and dozens of smaller companies, to create one or more new copy-protection methods that cannot be broken, or at least not as easily as the one used on DVDs, which was cracked by a 15-year-old boy. DVDs still contain Macrovision and CSS (another form of digital copy protection), and although both methods have been broken and pirating has occurred, we are not going to see a switch to a new method as all current players only support these two.

JVC, a true pioneer in the digital business, came up with their HD copy protection method, called D-Theater. Originally an idea of 20th Century Fox, D-Theater was originally going to be a 480P digital tape format and go head-to-head with DVD. As a result of DVD's immediate success, D-Theater was redesigned and implemented as a high-definition format and copy-protection method, and it is available today on nearly 100 HD D-VHS titles (D-Theater D-VHS is a competing format to DVD whereas the motion picture is encoded in high definition on a digital tape instead of an optical disc). HDCP (High-Bandwidth Digital Copy Protection) can use D-Theater as one of its internal CP encryption techniques, but is not as of late.

HDCP was quickly ushered in. This copy-protection method uses a direct digital connection (a digital video cable), and it relies on the satellite box, DVD player, or DVR to house the HDCP circuit, which engages (preventing copying) only when it sees an HDCP flag on the source material. HDCP is a multipart system, and it requires the following:

- An HDCP-compliant satellite box, DVD player, or DVR
- An HDCP-compliant port on the display device
- A direct digital video cable

Devices that do not support HDCP will only display 480P in some cases, or, at least for now, force the consumer to watch HD via the three analog component video cables. Pre-2001, DVI ports were digital inputs only, and did not support the new HDCP copy protection method. DirecTV does not use HDCP on every channel, and HDCP is still in the testing phase on most channels. You will, however, need an HDCP-compliant port on your projector or HDTV, or you will get the dreaded gray static screen. HDCP must be used when using a one cable digital solution for HDTV, as DirecTV, Dish and Voom (a fledgling competitor to DirecTV and Dish offering 30+ HD channels) support HDCP for DVI and HDMI ports.

The adoption of HDCP led to the need for a digital video cable—a one-cable solution for HDTV. Since laser discs came into use, consumers have had the option of adding a Serial Digital Interface (SDI), and on some higher-end players it was a manufacturer's option. SDI bypasses all processing, including copy protection, so it could not be used for HDCP. Another digital video cable had been used for PC and Mac applications—the digital video interface (DVI) cable—and this quickly became a standard of sorts for HDCP, although many consumers do not care for the large connector on the end of the cable, and the cables are generally limited to 27 feet for resolutions above 720P.

Ingenious companies, such as Gefen Technologies in California, quickly developed the High Definition Multimedia Interface (HDMI), which is a DVI cable with a smaller connector that carries full bandwidth HDCP HDTV, as well as two-channel audio. The newer connector is roughly the size of USB and is quickly bypassing DVI as the new HDTV video interface standard. HDMI is fully compatible with current DVI ports, and consumers can easily purchase a DVI-to-HDMI cable, or vice versa, so early adopters are not penalized.

With HDCP and HDMI now in place, the studios and networks have signed off on HD recording, and DISH Network was the first to market with a satellite DVR, the troubled non-TiVo PVR 921. DirecTV, never one to be outdone, quickly issued a press release for the HR10-250, a DirecTV DVR. Stand-alone HD TiVos may or may not follow, as there are intimations in the community that the studios and networks will not allow a stand-alone HD DVR for fear of copy protection violation and piracy.

How to … Record Two HDTV Broadcasts at One Time

As with all DirecTV DVRs, the HR10-250 can record two shows at once. This unit has four tuners: two DirecTV HD tuners and two over-the-air (OTA) HD tuners. This handy feature is easy to use, and it makes missing your favorite shows nearly impossible.

Purchasing a High Definition TiVo

In April 2004, DirecTV released a handful of the new HR10-250s into the U.S. market. At that point, authorized online retailers had preorders for months and months, and with thousands of people waiting, only a tiny percentage could be accommodated. As with any extreme-demand and low-supply scenario, prices on online auction sites immediately went through the roof.

On eBay, two days after release, there were over a dozen units available at nearly $3,000 each, three times the manufacturer's list price. Two months later, in June, supply had not increased, and the price was still way over the $1,200–$1,400 mark on most online auction sites. DirecTV's largest online retailers had only received a minute percentage of their preorders, so many early adopters were left waiting.

If you are new to DirecTV and want the first HD TiVo on the market, it is strongly suggested that you get in one of the many authorized DirecTV retail preorder lines, as you will get priority on your shipment being a new activation, and will get the required 3 LNB dish as well as standard professional installation included for the retail price plus shipping. The units cannot, for the time being, be ordered from DirecTV or TiVo at all.

The HR10-250 is currently the only available model of HD TiVo, and it is a better performing unit than any of its non-TiVo competitors. Every single owner of an HDTV with DirecTV is a potential customer for the HD TiVo, as this is the only unit at present that can record HD from DirecTV without purchasing a modified older HD set top box along with a D-VHS deck. DirecTV is working on increasing the stock to help out with the huge demand, and it may have companies such as RCA and Samsung build them under the DirecTV banner.

The phone connection on the HR10-250 is, unfortunately, the only way the TiVo unit can receive pay-per-view shows and system updates. The HR10-250 is

How to ... Find the Best Deal on an HD TiVo

Always go to the manufacturer or to an authorized retailer first. Buying an HD TiVo from an unauthorized retailer or an online auction site can leave you without a warranty, which is not a good situation to be in if you need repairs on your $1,000 (or more) box. Going to DirecTV's Web site, http://www.directv.com, is always a good idea, as it is easy to locate DirecTV's authorized online and brick and mortar stores. Speaking with DirecTV's customer service will also help point you in the right direction, as they have access to some of the retailers' real-time stock quantities, thus giving you the shortest wait to choose from.

a full-fledged TiVo Series2 DVR, but without most of the bells and whistles that accompany the Series2. Just like the other DirecTV DVRs on the market today, the HR10-250 has two USB 2.0 ports, but unlike the other stand-alone units on the market, these are reserved for future use only and do not allow for networking, wired or wireless, at this time. Most industry insiders believe these will be used for external USB hard drives produced by Maxtor and Seagate (both are now making large capacity external drives specifically for DVRs), and not for the Home Media Features or networking any time soon. DirecTV has a closer alliance with the studios than TiVo does, and the studios consider the Home Media Features to be too risky (in terms of piracy) for their liking. Unfortunately, features like the new XM Radio PCR (PC Radio) streamed to your TiVo are not in the cards for the HR10-250 any time soon.

Setting Up a High-Definition TiVo

Unlike the majority of new consumer electronics and developing technologies that hit the market, the HR10-250 comes with 90 percent of the pieces you need, and it really is quite a breeze to set up.

Before You Start

Here are a few things you need to know before you start setting up your new HD TiVo:

- The native rate of your HDTV, and your TV's aspect ratio
- Is your HDTV HDCP-compliant?

- Does your HDTV have a DVI port or a HDMI port?
- Do you have a surround-sound receiver? And if yes, is it capable of Dolby Digital?
- Does your surround-sound receiver have an open optical or TOS Link jack?

NOTE *Your HDTV must be HDCP-compliant for you to receive an image from your TiVo. Check this out on the manufacturer's Web site or on one of the HT-geared Internet forums, look in your owner's manual, or ask the manufacturer directly.*

As mentioned earlier, the native rate of your HDTV is your TV's base scan rate, which is similar to the optimum scan rate of a computer monitor. Like a monitor, your HDTV is probably capable of displaying 1080I and 480P, and most new sets also do 720P. If you have a newer DLP or LCD rear-projection HDTV, your TV's native rate is most likely 1280×720P, so 720P is the recommended output for the TiVo.

If you have an older HDTV or an analog set, 1080I will be your best bet. If you purchased an EDTV or a set that is limited to 480P, such as many of the nicer 4:3 tube televisions, 480P will be your optimal resolution. This scan rate will still look nice for HDTV, but keep in mind that the picture will have been dramatically scaled down from the original 1080I or 720P source—visual artifacts will be created and resolution will be lost.

Next, you will need to determine the aspect ratio of your television. If you set this incorrectly, your TV could distort the image by stretching or squishing it, giving you a picture you will not want to watch. Since 2000, 99.99 percent of HDTVs come with a 16:9 aspect ratio (much wider than they are tall), and this is quite easy to tell. Even TVs with "square" resolutions, like 1024×1024, such as some of the 42 inch HD plasmas on the market, are still most likely 16:9.

Assuming your HDTV is HDCP-compliant (it must be to work with TiVo), it will have one of two types of HD ports: a DVI port or an HDMI port (or possibly both). The DVI port is easy to identify—it looks like a large printer port you might find on your PC. The HDMI port is significantly smaller and resembles a USB port, but it will be labeled "HDMI."

NOTE *If your DVI port says "Computer Input Only," that port is definitely not HDCP-compliant. You will be out of luck if it doesn't also have an HDMI port. Some of the newer HDTV plasma screens on the market use an external media box that has a PC DVI port and an HDMI port for video.*

If you have an HDTV, the odds are that you also have some sort of surround-sound system. If you do not have a surround-sound receiver, you are missing out on the outstanding audio that accompanies most HDTV broadcasts.

If you have a surround-sound receiver, does it handle Dolby Digital 5.1? As with all digital sound formats, Dolby Digital comes in multiple flavors:

- Dolby Digital 5.1EX
- Dolby Digital 5.1
- Dolby Digital 5.0
- Dolby Digital Surround 4.1
- Dolby Digital Surround 4.0
- Dolby Digital Surround 2.0
- Dolby Digital Stereo 2.0
- Dolby Digital Mono 2.0
- Dolby Digital Mono 1.0

More confusing numbers! Unlike the HD scan rates, the Dolby numbers are relatively simple to understand. Think of the number of channels (the 5.1 in the case of Dolby Digital 5.1) as the number of speakers in your system. If you have a 5.1 broadcast, you have left, center, right, surround right, surround left, and subwoofer channels. The subchannel is the .1 channel, and it is not considered a full channel as it usually only contains information below 80 Hz. The subchannel is also referred to as the low-frequency effects (LFE) channel. The subwoofer handles the low end or bass sound.

Each number signifies the number of channels used, or some variation on that. For instance, Dolby 4.1 uses all five of your speakers and your subwoofer, but it has mono sound information for the surround channels, so they are identical and thus make up one channel instead of two.

Dolby Digital EX, on the other hand, is a bit more complicated and involves a 7.1 speaker system! The information encoded in the surround channels on an EX track contains some information that causes your receiver to create the sixth and seventh channels, which go behind your head. Not every new release includes this information—it is only available on a small percentage of the soundtracks released today.

Most of the non-HD channels and some of the films shown in 1080I on channels like HD-HBO or Showtime occasionally use 2-channel surround, though they will still display the Dolby Digital logo on your surround receiver. Your receiver will switch into Dolby Pro Logic, Dolby Pro Logic II, or Dolby Pro Logic IIx to decode this sound. Dolby Pro Logic II converts any 2-channel source to 5.1 channels using an intelligent algorithm, while Dolby Pro Logic IIx converts any 2 or 5.1 channel source to 7.1 channels.

To receive a Dolby Digital or PCM (raw digital data) signal on your surround receiver, you must have the optical jack, or TOS Link, output connected to your receiver. The TiVo does not have coaxial digital output, so you must connect with the optical cable, which is virtually identical for sound quality, but it is quite a bit more fragile.

Satellite Dishes and Antennas

If you are a DirecTV subscriber (whether new or not), it is highly recommended that you have a professional DirecTV installer hook up the required larger oval dish you will need, as well as a recommended over-the-air (OTA) antenna to receive local digital channels. DirecTV now requires a triple low-noise blocker (LNB) dish, with most of the HDTV material now being received on the third or middle satellite. Recently DirecTV activated a brand new satellite that will allow for the addition of more HD and local channels. The triple LNB dish usually has a built-in four-way multiplexer, allowing the hookup of four DirecTV receivers, all getting independent control.

The OTA antenna is a necessity until all local channels broadcast in HD on DirecTV. The HD versions of Fox and CBS are currently available only on DirecTV in markets that are considered "owned and operated," meaning that the parent network also owns the local affiliate. This is only a small part of the United States, so the majority of consumers are currently forced to use a standard VHF/UHF antenna to pick these channels up. AntennaWeb (http://antennaweb.org) can be a great resource for setting up your antenna, choosing the right equipment, and finding comments and reviews of the equipment available on the market today.

Most consumers think of VHF/UHF antennas in their antiquated form: huge monstrosities that tower over your neighborhood, wreaking havoc on your homeowner's association. In fact, these monstrosities still give some of the very best performance for picking up local digital channels, but they no longer need to sit high atop your roof—they can often be easily mounted in your attic. Of course, there are now many antennas marketed specifically for HD; some of them are amplified, and some of them are simply old-school metal pterodactyls with new HD labels.

The amplified antennas really do nothing for performance, nor do the antennas designed to attach directly to your HD DirecTV dish. Newer HD antennas, though, such as the Winegard SquareShooter, not only increase performance, but look pretty cool to boot. The SquareShooter (http://www.winegard.com/) has out-performed nearly every antenna tested in most locations, and it is a 12 inch square that sits atop a normal DirecTV Dish mounting arm, or it can be mounted on the side of your home or a chimney.

Connecting the Equipment

When you unpack your HD-TiVo, you will notice that the HR10-250 ships with component video cables, which can be used if your television does not support HDCP, as well as an inexpensive HDMI cable and a backwards-compatible HDMI to DVI cable. The HDMI to DVI cable has the smaller USB-sized HDMI connection on one end and a standard DVI male connector on the other. This allows you to connect a monitor or display that has a DVI jack using only that cable and still get the video signal from the HR10-250.

NOTE *Videophiles will want to upgrade the cables, but the necessity of doing so is unclear. Most experts feel that the quality of digital video does not really depend on the quality of cable used; you either receive a digital video signal or you don't.*

After choosing the proper video connection between your HR10-250 and your TV, you will need to hook up a third-party optical cable between your TiVo and your surround receiver, as well as plug in your two satellite RG6 cables and your single antenna cable. The TiVo splits the OTA signal digitally and uses two tuners so that you can record two programs at once, but it does require two separate satellite cables. You can still use the unit with just one cable, but you can only record one show at a time, and if you are recording a channel, you can't watch a different one.

Once your TiVo is properly installed, move on to the Guided Setup, which varies very little from the normal DirecTV DVR setup. Follow these steps:

Your HR10-250 will ask you what output resolution you prefer, and give you the choice of 480P, 720P, or 1080I. You will also input the shape of your television, either 16x9 Widescreen or 4:3 Standard. Once you have set your resolution, the unit will go through its setup and perform multiple phone tests and downloads. The HR10-250

How to ... Connect More Than Four Receivers to Your Dish

More than 70 percent of DirecTV households in the United States already use all four of their multiplexer's lines to power four independent receivers. That leaves you one short for the HR10-250 (which requires two lines).

To solve this problem, you need to install of a secondary piece of equipment—a larger multiplexer. Terk, probably the most famous installation accessory manufacturer, makes a 5-in, 8-out multiswitcher that does the job in most situations. You use the 4 lines from your current built-in multiplexer in your dish and the antenna line as input for the multiswitcher, and then each of the eight outputs carry both the antenna and satellite signal. You simply use a special splitter to separate the two signals out at the end of each line.

will do a series of tests on satellite inputs, as well as test and scan for channels with the OTA antenna input.

1. Choose your connection type, as shown in Figure 12-1.

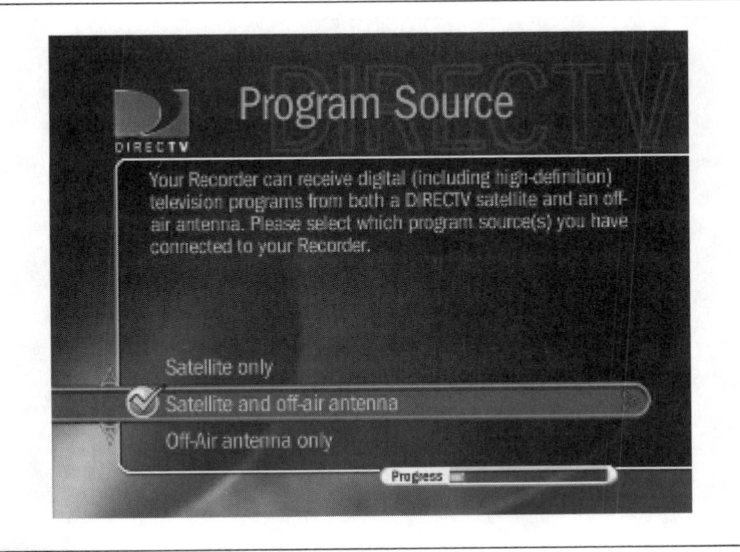

FIGURE 12-1 Choosing your connection type

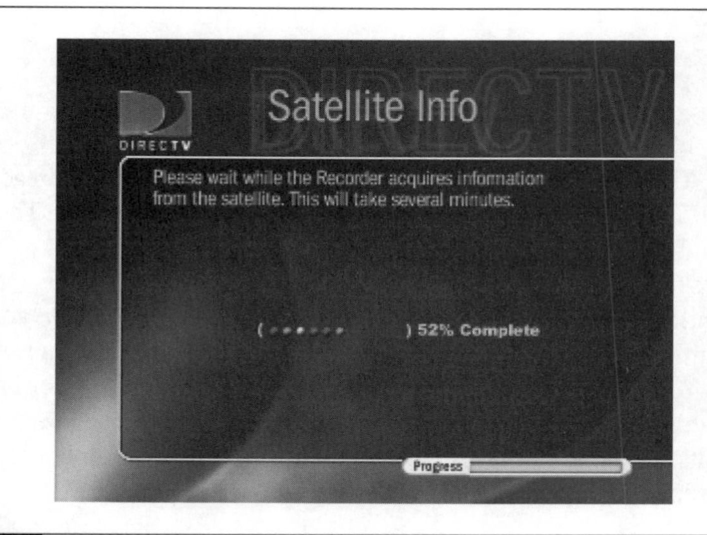

FIGURE 12-2 The DVR acquires setup information.

2. The DVR will retrieve any relevant setup information, as shown in Figure 12-2.
3. Select your dish type, as shown in Figure 12-3. To get all HDTV programming on DirecTV, choose the 3 LNB dish.

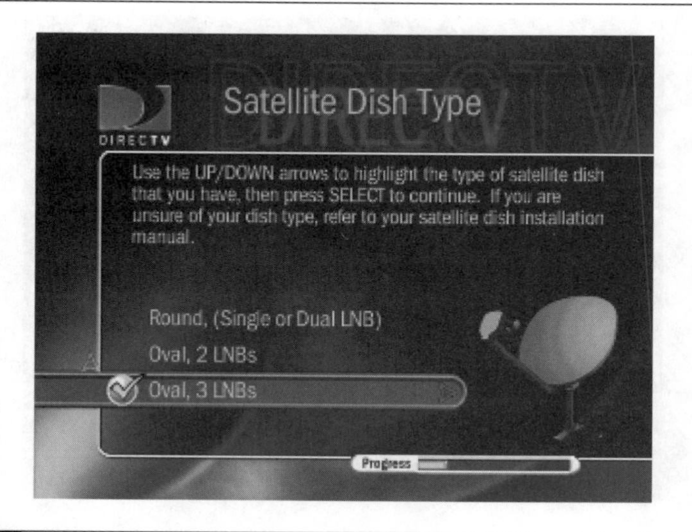

FIGURE 12-3 Selecting your dish type

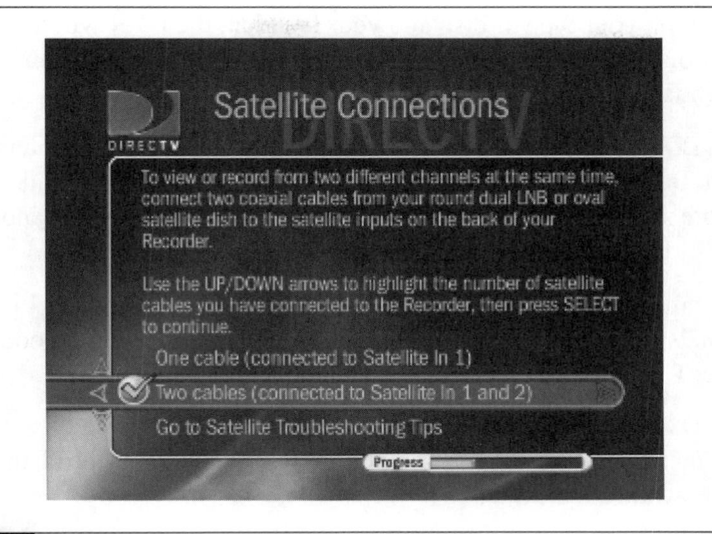

FIGURE 12-4 Specifying the number of satellite cables attached to your receiver

4. Specify the number of satellite cables attached to your receiver, as shown in Figure 12-4. If you have two satellite cables, you will be able to record two different channels at once.
5. Enter your Zip code, as shown in Figure 12-5. This tells the DVR what local programming to give you, and it helps accurately set your time zone.

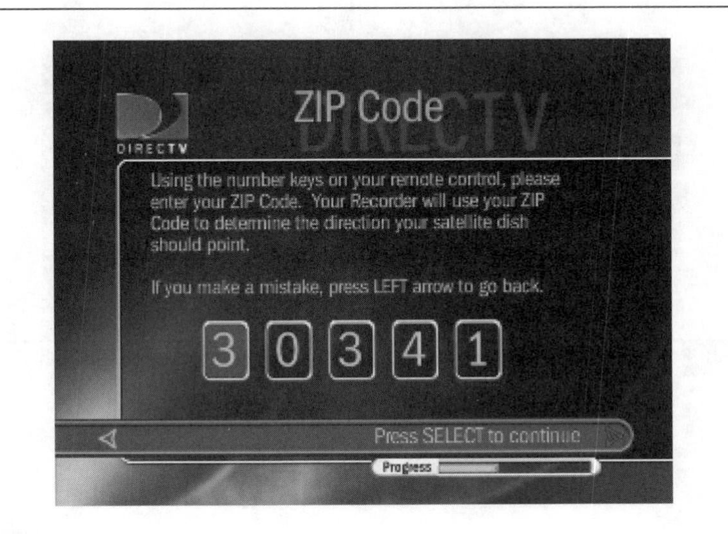

FIGURE 12-5 Entering your Zip code

6. Based on your type of dish and your location, the DVR will tell you where your dish should be pointing, as shown in Figure 12-6. This information is essential to properly installing your HD dish.
7. The DVR then provides more information about properly mounting the HD dish. Be sure you have all your required setup information and tools handy before starting. One person should be at your television while another person moves the dish.
8. As someone aims the dish, check the test signal, as shown in Figure 12-7. Signals from all triple LNBs must be over 75 to maintain a good quality DirecTV signal.
9. The DVR now verifies the settings you entered for the dish type, multiswitch, and number of cables. When that is done, it downloads the channel lineups, as shown in Figure 12-8.
10. Once you have completed the Guided Setup press the DirecTV button and go to the menu. Select the Messages & Setup option, as shown in Figure 12-9.

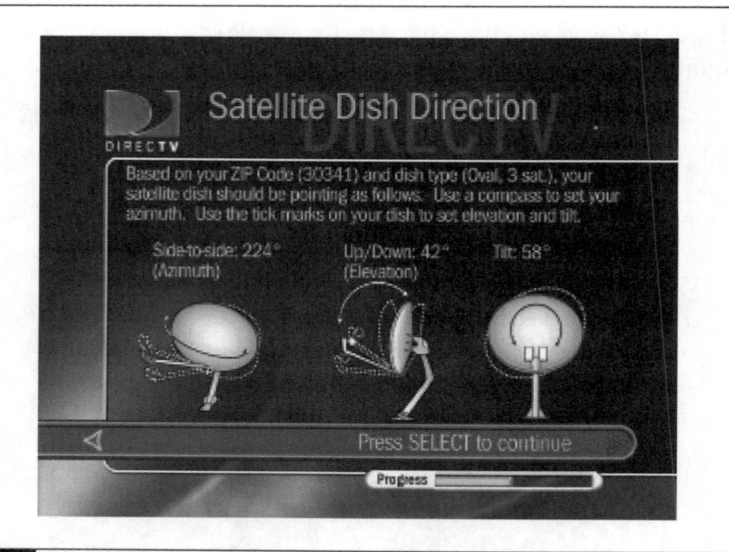

FIGURE 12-6 The DVR displays information about where your HD dish should be pointing.

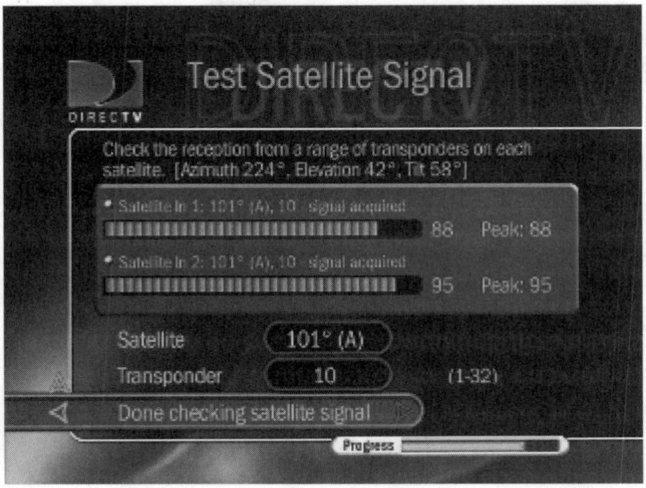

FIGURE 12-7 Checking the signal strength

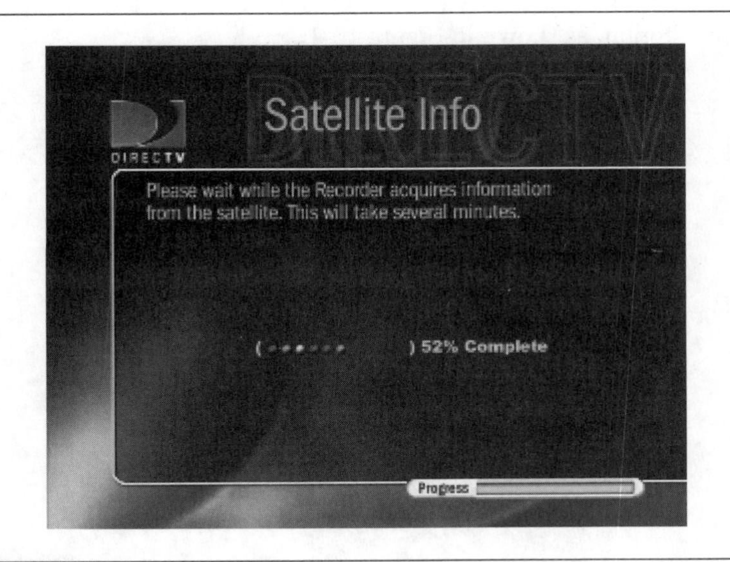

FIGURE 12-8 The DVR downloads the channel lineups.

12

FIGURE 12-9 Selecting the Messages & Setup option

11. Select Settings and then Audio to get to the Dolby Digital menu. Select Dolby Digital, as shown in Figure 12-10.

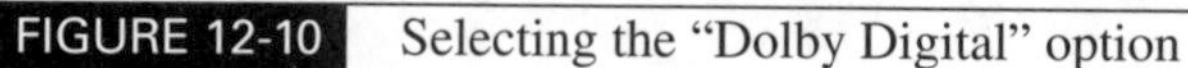

FIGURE 12-10 Selecting the "Dolby Digital" option

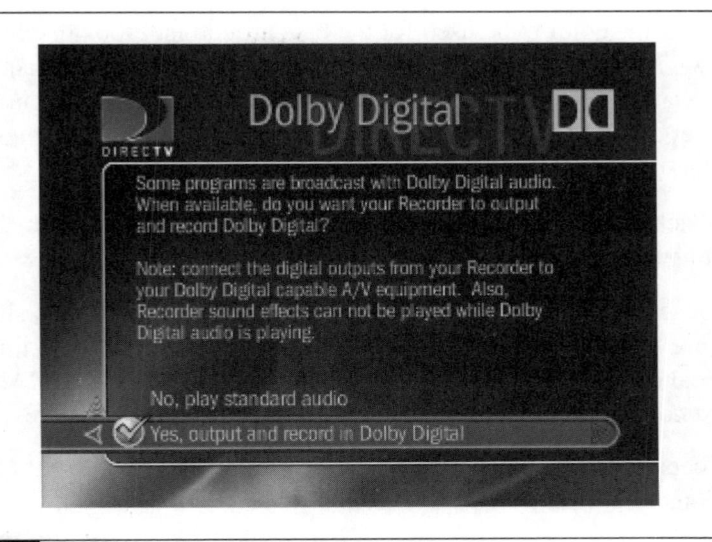

FIGURE 12-11 Selecting Dolby Digital audio

12. Select the Yes, Output and Record in Dolby Digital option only if you are using a Dolby Digital receiver or surround-sound processor, as shown in Figure 12-11.
13. Go back to the Settings menu and select the Video option. Then select the Letterbox Color option, as shown in Figure 12-12.

FIGURE 12-12 Selecting the Letterbox Color option

14. Specify the color to be used for the bars around letterbox displays, as shown in Figure 12-13. If you have a plasma or CRT-based displays, you will want to choose gray bars rather than black, as they will burn in much slower. DLP, LCD, and D-ILA owners can choose whichever color they prefer, as neither will burn in.
15. Go back to the Video Settings menu and select the Output Format—Current Setting option.
16. Choose the video output mode appropriate for your TV, as shown in Figure 12-14. The vast majority of HDTVs are now 720P, so this is the logical choice for plasma, DLP, or LCD owners. Analog HDTV owners must consult their owner's manual to determine their native resolution.
17. Go back to the Video Settings menu and select the TV Aspect Correction option.
18. If you want to keep the 4:3 image (1.33:1) in its proper ratio of height to width on a 16:9 display, choose the Panel option, which will give you bars on the side of the image (see Figure 12-15).

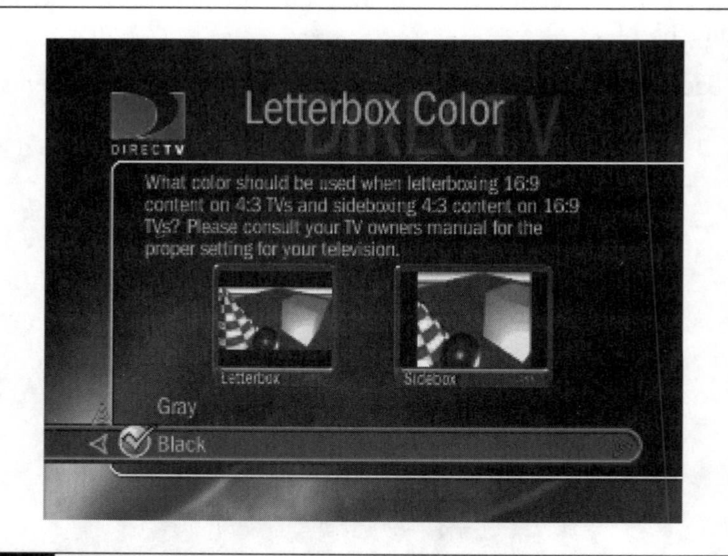

FIGURE 12-13 Choosing a color for the bars framing letterbox displays

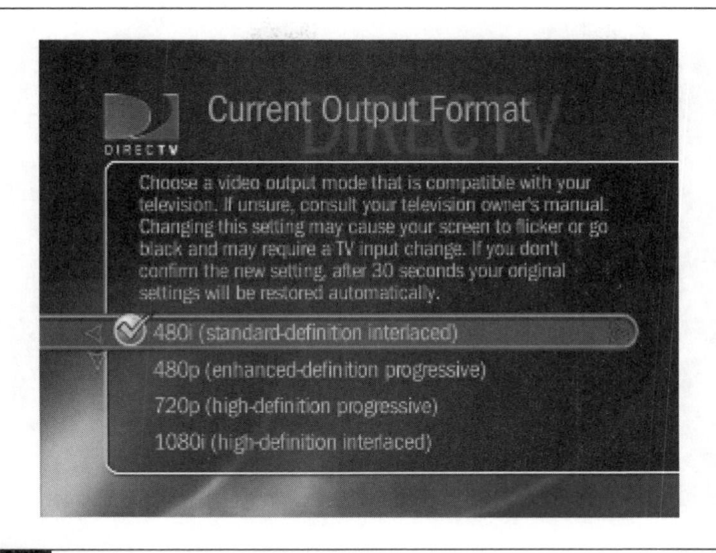

FIGURE 12-14 Choosing the video output mode

19. Go back to the Video Settings menu and select the TV Aspect Ratio option.
20. Select the aspect ratio of your TV, as shown in Figure 12-16. Very few HDTVs left on the market today are in the 4:3 (1.33:1) aspect ratio that has been used for the past 60 years; the majority of new sets are 16:9 widescreen TVs.

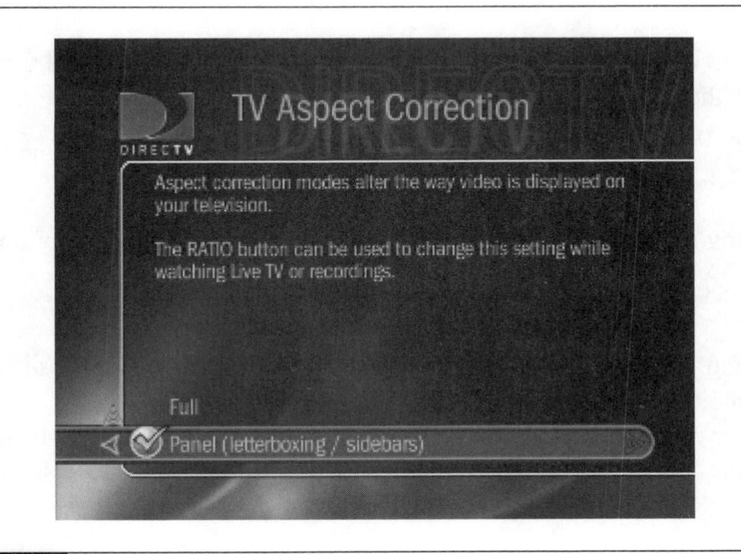

FIGURE 12-15 Setting the aspect correction for 4:3 images

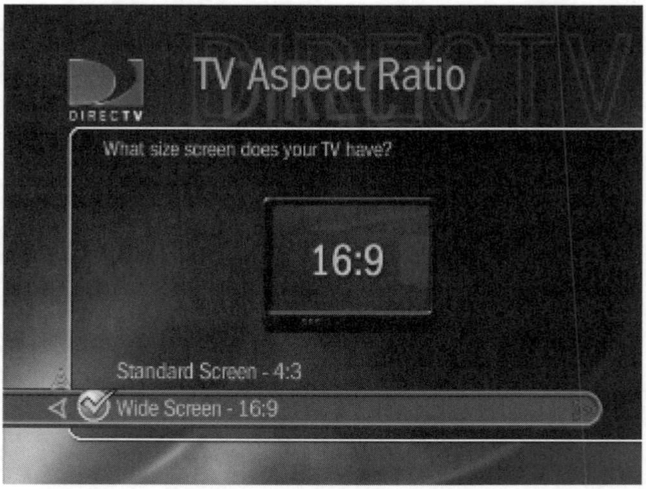

FIGURE 12-16 Selecting the aspect ratio of your TV

Finding High-Definition and Dolby Digital 5.1 Shows

Everything is set up and you are finally ready to roll! Navigating the DirecTV DVR guides is a breeze. The first thing you can do to save yourself a lot of time in the long run is to eliminate the channels you will never want to watch. Follow these steps:

1. Go to the TiVo or DirecTV Central, and select Setup, Channel Changing, and then Channels You Receive.
2. Select each and every channel you receive, and delete the channels that you receive but don't watch—this is a tedious task.
3. Bring up your guide and press ENTER to change your guide options.
4. Select the Channels You Receive guide.

Once you've done this, you won't have to repeatedly surf past those annoying "how to order" screens, and it will save you significant time.

Most consumers who have an HDTV will want to watch a lot of HD material. To make this easier, you can add the HD channels to your favorites list and then breeze through just the HD channels with a push of the Favorites button. Currently, DirecTV offers the following HD channels:

- Channel 73 ESPN HD
- Channel 76 Discovery HD
- Channel 78 HDNet Movies
- Channel 79 HDNet
- Channel 80 CBS-HD
- Channel 81 CBS HD
- Channel 85 PPV HD/Spice/Playboy HD
- Channel 88 HD HBO
- Channel 91 Showtime HD

Currently HDNet, HDNet Movies, and Discovery HD offer the only 24-hour-a-day HD lineup, with ESPN offering only a handful of HD shows a day, and HBO and Showtime offering most of their movies and original programming in high definition. Beware, though, because HBO and Showtime occasionally crop or matte films that were shot in 2.35:1, the widest common aspect ratio used in shooting film. This is done because studios are still concerned over sending perfect HD copies of their films out to consumers, and most consumers, even with an HDTV, still prefer to have their screen filled with the image rather than have black bars on the top and bottom. Even with a widescreen display, this aspect ratio requires letterboxing to show the entire image.

HDNet offers a little something for everyone, from the top-rated *Bikini Destinations* to movie trailers and soccer in HD. HDNet also has select professional and college sports on occasion, as well as original programming and reruns of mostly cancelled network shows in HD, such as *Bette* (starring Bette Midler) and *Wolf Lake* (starring Tim Matheson and Lou Diamond Philips). HDNet also offers older shows, such as *Hogan's Heroes* and *Charlie's Angels,* in HDTV, both looking simply breathtaking in comparison to their old SD counterparts. HDNet Movies offers mostly B-run movies, with very few major motion pictures except those that are fairly old. Unlike HBO or Showtime, HDNet Movies shows every film in its original aspect ratio.

Discovery HD is by far the most popular HD channel available today and is well worth the $9.95 price for the HD package. It is loaded with original programming including such hits as *Trading Spaces,* the *Jeff Corwin Experience,* and a whole

How to ... Directly Access Your Local Digital OTA Channels

To access your local digital channels directly by their channel numbers (for instance 10-2), you press the individual numbers and use the "-"button for the hyphen. (The "-" button on the HR10-250 shares space with the Advance button.) So to enter 10-2, you would press the 1 button, then the 0 button, then the Advance button, and then 2.

host of other Discovery goodies. This is the one HD channel you can count on 24 hours a day for excellent quality, jaw-dropping visuals, and quality programming to boot. Most brick-and-mortar retail stores use Discovery HD to demo their HD sets, and for good reason.

Local HD channels vary from market to market, but the FCC now requires each station to broadcast at least 480P or better material for at least a portion of every day. Most stations broadcast a signal 24 hours a day, but HD material is usually only broadcast during prime time or for major sporting events. A few daytime shows, such as soap operas, broadcast in HD, but that still makes up the smallest percentage of the market. Local digital channels are broadcast mostly on UHF, with only a few digital channels appearing on VHF.

TiVo is the coolest product on the market today for one reason: it is the most user-friendly consumer application that has ever been brought to market. With the HD TiVo, you can create all the same WishLists and Season Passes that you can on every TiVo, and you have the added benefit of being able to specify HDTV and Dolby Digital as subcategories. To create a new WishList, follow these steps:

1. From your DirecTV Central menu, select the Pick Programs to Record option.
2. Select the Search By WishList option.
3. Select the Create a New WishList option.
4. Select the Category Only option, as this will allow you to highlight just the HDTV programming.
5. Select the Audio & Video option.
6. Scroll down to the bottom of the list and select HDTV or Dolby Digital, depending on your programming choice.

You can choose to auto-record every HD show, or to pick and choose the ones you want based on your newly created WishList. The WishList only searches a few days ahead at a time, instead of the full two weeks the guide displays, because auto-recording every HD channel for two weeks would quickly fill your hard drive to capacity.

Getting the Most Out of Your Hard Drive

The new HR10-250 comes standard with a 250GB hard drive, which is currently the granddaddy of large hard drive configurations in a TiVo unit. It's enough to hold a whopping 30 hours of high bit-rate HDTV, or nearly 200 hours of standard definition programming. That sounds like a lot, but keep in mind that most folks are going to record a ton of HD material, so you need to keep the meter running.

The DirecTV DVR does not tell you how much space is free on your hard drive, nor does it allow for adjusting the recording quality, as most of the stand-alone units do. It records everything via a "time shift," which is a perfect copy every time. This means that you will need to do some simple math to get an idea of how much recording space remains.

Using an HDTV WishList is the easiest way to record lots of HDTV programming, but you must use caution because multiple channels show 24 hours of HD material a day, and you can fill up the hard drive with the TiVo's dual tuners in 15 hours. If this happens, material will be deleted, or your TiVo will simply stop recording until you delete the excess.

It is simpler to not have your WishList auto-record everything. Instead, choose the View Upcoming feature to pick and choose what you do and do not want to record. When you are making your choices, look for the Dolby Surround description, as this means the movie will not be Dolby Digital 5.1, but two-channel surround—this can be available on movies that have a DVD counterpart containing a newer Dolby Digital 5.1 soundtrack. One marathon of HD HBO's *Deadwood* and *Sopranos* back to back in HDTV could fill your hard drive.

The future of hard drive expansion via the USB ports is unclear, so many consumers will turn to services such as WeaKnees (http://www.weaknees.com) or PTVupgrade (http://www.ptvupgrade.com) to add bigger hard drives to their TiVos. Warranty issues aside, this is relatively straightforward and easy to accomplish, but the buyer must beware: the HD TiVo runs far hotter than any other TiVo unit to date, and you must have a second fan for the second hard drive or you will melt your new pride and joy relatively quickly. Companies such as WeaKnees will sell you a kit that contains everything you need, including a second 300GB hard drive, a harness to hold it, and a fan. This can expand your storage space from 35 HD hours (or 200 hours of standard programming) to 70 HD hours (or 400 hours of standard programming), which is enough space to record days of HD material at a time.

The future of the HD TiVo will hopefully include factory-authorized hard drive additions via the USB 2.0 port, allowing for at least two additional drives, and the possibility of chaining over a dozen drives together. Companies are doing this now for DVD media servers, such as the Molino media server, which ships with a built-in terabyte hard drive and also USB and FireWire (IEEE 1394) ports for adding an endless amount of external storage. Unfortunately, without hacking into your TiVo, it is unlikely that consumers will see the ability to move HDTV files from their TiVo to an optical disc, such as the upcoming Blu-ray Disc or HD-DVD-R, any time soon. Currently, in Japan, DirecTV's Japanese counterpart licenses Sony's Blu-ray recorder with a built-in satellite tuner and the ability to record your favorite HD shows directly to the hard drive, and to move them later to your Blu-ray Disc for archiving and future playback. This unit will not work in the United States, and it will most likely not have a U.S. cousin compatible with DirecTV anytime in the near future.

Voices from the Community

HD TiVo Is the Most Highly Anticipated Consumer Product in Years

I first heard about TiVo from a friend who had a state-of-the-art home theater and a whole house full of high-end audio equipment. He told me that the only piece he would never sell was his TiVo. That got my attention. And after buying one, I realized that his praise was justified.

When I first saw HD TV at a store, I was totally mesmerized by the 3-D nature of the images. When I purchased one, my only lament was that not all broadcasts were in HD, and that my TiVo could not record even the ones that were.

HD TiVo is like having your cake and eating it too. The opportunity to record in HD and at the same time watch another channel is a dream come true. I am counting the days until I can replace my standard TiVo with the HD version.

—Richard Kraus,
Health Store Owner,
Atlanta, Georgia

Chapter 13 The Future of DVRs

How to...

- Look Critically at the Numbers Behind the TiVo Phenomenon
- Understand the Changes in TiVo's Place on the Home Network
- Keep up with the Upcoming Changes in DVD and DVR Technology
- Understand The Advertising Wars Over TiVo
- Determine What's in Store for the Longer Term

As happens with any technology that is on the cutting edge of the consumer marketplace, there has been a lot of speculation about how TiVo will continue to evolve, what kind of services it will provide, and even whether it will still be around in the coming years. However, like Apple Computers, TiVo has been declared dead before—and come back stronger than ever. For example, Peter Ausnit, a digital and interactive television markets analyst at Deutsche Banc Alex. Brown, based in San Francisco, said in 2001, "I continue to believe that TiVo will vanish from the Earth within 12 months from now."

If you've read this far in the book, you're on your way to becoming one of the "TiVo faithful" and you're likely hooked on this new way of experiencing television. So it might come as a shock to you that this amazing new service might be going the way of the Edsel or the electric typewriter—a fascinating, useful item, but one whose purpose was either ahead of its time, badly marketed, or simply left behind in the rough-and-tumble world of the consumer electronics marketplace. So, the best way to start off a chapter on TiVo's future is to ask if it has much of one: *Is TiVo going to fade away or endure over the next decade?*

The Numbers Behind the TiVo Phenomenon

In retrospect, Peter Ausnit's 2001 opinion has seemed overly pessimistic. After all, TiVo's financial numbers have continued to be fairly strong. In 2003, TiVo's revenue was $141 million, compared with $96 million in 2002. TiVo also announced a bold plan in late 2003 to effectively double the size of its subscription base, and is willing to spend upwards of $50 million to do it.

Information is scarce as to how the $50 million will be spent. Not surprisingly, the powers at TiVo want to keep their upcoming strategies private until they are implemented. However, one of TiVo's remaining hurdles is to create a mass market awareness of its product, so those beyond the tech-savvy audience will know what it's all about.

A guess is that it will likely involve a major marketing push to increase consumer awareness of the product, combined with a grassroots or "viral" marketing strategy to ensure that consumers who are happy using TiVo's service are given an incentive to recruit other users. As of this writing, the first phase could be well under way, given TiVo's deals with Sony and DirecTV, providing much of the fuel needed to power the media message that will saturate the market.

Some Speculation on TiVo's Strategies

With its stated goal of reaching sustained profitability by the end of fiscal year 2006, TiVo has set itself some high hurdles, but it also appears energized enough to meet the goals. Mike Ramsay, the chief executive of TiVo, has stated that TiVo's main goal has been to focus on subscription revenue and sustained profitability. Furthermore, he states that given the "acceleration in the market and the strongest cash balance we've had in three years, we believe the time is right to take advantage of the exciting growth opportunities that lie ahead of us."

So far, so good. One part of Ramsay's strategy of pursuing growth opportunities via corporate alliances has succeeded brilliantly. TiVo added a third of a million subscribers in the end quarter of 2003—a record that easily smashed the 2002 record of 100,000 subscribers. Two thirds of these subscribers—about 200,000—came from the partnership with satellite provider DirecTV in creating combination DirecTV-TiVo boxes, as shown in Figure 13-1. This brings the company's total number of subscribers to its highest level yet, 1.3 million.

All's Well (Mostly) in TiVo-Land

As a company, TiVo has been seeing growth that, when plotted on a graph, looks close to what start-up analysts dream about when evaluating a company—the "hockey stick" pattern, showing strong increases. Yet like any start-up—and TiVo in many ways is still a start-up at heart—there are darker spots in the picture. For example, the net loss for 2003 was $32 million, or 48 cents per share.

This came about in spite of the explosive growth that the company experienced that year. On the one hand, this could be the investment needed to gain critical market share before competitors enter the same market. On the other hand, one always has to ensure that a company isn't paying too dearly to gather ground that could be lost again, or that won't produce enough profit to make the investment worthwhile. One of the lessons of the dot-com crash in 2000–2001 was simply that profit and cash flows do indeed matter, and can't be replaced by deficit-financed growth to capture market share.

FIGURE 13-1 A combination DirecTV-TiVo box

Patent Issues

TiVo also faces problems that bedevil many companies that bring a new product to market. For example, TiVo has locked horns in nasty patent disputes in the past, draining the company of cash that could have been used to develop new technology or reach more consumers. Right now, there are still ripples being felt from its dispute with SonicBlue, the maker of ReplayTV. ReplayTV is a device that allows users to trade shows over the Internet, and is itself under heavy scrutiny from the entertainment industry.

Threats to the TiVo

As the idea of the TiVo service penetrates the mainstream consumer electronics market, TiVo can't help but attract competitors who can make DVRs for customers who are prepared to buy and don't particularly care about brand names. Unfortunately for TiVo, the actual production of a DVR isn't financially or technologically prohibitive. Once a TiVo-like form of operating system is developed, all a potential rival has to do is add a hard drive and some graphics chips, and a reasonable copy of the TiVo technology can be achieved, though without TiVo's well lauded user friendly interface.

Wired magazine has noted that it took Linux programmer Isaac Richards and a small team of like-minded coders less than one year to create MythTV. MythTV is a full-featured, open source knockoff of the TiVo system. To make matters worse, Richards and his friends created this in their spare time, demonstrating that TiVo's technology was not only duplicable, but could be reproduced easily, which marked TiVo as a target for unscrupulous companies that would ignore patent law and legal injunctions to get their product to a hungry market.

UltimateTV DVR

And where there's an open software market, you can expect Bill Gates to be somewhere in the picture. Sure enough, Microsoft has been implementing what it calls its UltimateTV DVR technology in its Windows-based PC Media Center platform. Available on DirecTV receivers, the UltimateTV service from Microsoft digitally records satellite television in a manner very similar to TiVo. Like TiVo's box, the UltimateTV unit uses an internal hard drive to automatically store whatever TV show you're watching. What distinguishes UltimateTV service from TiVo is the following:

- A 30-second skip, good for short jumps and skipping commercials.
- Automatic time "padding" when setting up recording schedules helps cover for sloppy network scheduling.
- The viewer can control more than 225 channels of DirecTV satellite programming.

However, UltimateTV still lacks a few of TiVo's more important features:

- A straightforward interface, requiring fewer screens and button presses to get to your choices.
- Scheduling priorities that the user can customize.

- Better detection of program reruns.
- User control over the "no-repeat" rule.

Comcast Scientific Atlanta Explorer 8000

The cable set-top box makers themselves are also getting into the act now. Considered one of the greatest potential threats, Scientific-Atlanta and Motorola are manufacturing items such as the Explorer 8000 HD, shown in Figure 13-2.

Comcast cable service has begun to provide this basic DVR in select markets around the country. The main appeal of this unit to its audience has been no up-front cost and a low $9.95 monthly fee for the DVR service. Additional DVRs cost an additional $29.95 per month, though these fees will vary widely depending on the geographic area and consumer specials offered.

FIGURE 13-2 A potential threat to TiVo—the Scientific-Atlanta Explorer 8000HD

Basic initial setup consists of hooking the DVR up in place of a cable box. The box gets program lineup information from the cable. Since the Comcast DVR, like the DirecTV DVR, is both a DVR and a decoder box, no separate cable box is required as it is with stand-alone TiVo and ReplayTV DVRs.

Interestingly enough, Scientific Atlanta's entry into this market seems to have attracted only the bargain-hunting or entry-level consumers looking for any kind of DVR that could have lower up-front costs than TiVo. While it has lower costs, however, this model lags behind in two of TiVo's strongest areas: the user interface and the quality of the picture.

As a whole, the Scientific Atlanta Explorer's user interface is crude, with low resolution resulting in blocky text and an unfinished feel. Worse yet, the low resolution limits the amount of information that can be displayed on the screen, requiring the user to make more choices to get the same result as on a TiVo.

Because the Scientific Atlanta Explorer 8000 is designed with the cable companies in mind, naturally the picture quality of the recorded shows will depend entirely on the quality of the signal from the cable company. Like most cable companies, Comcast provides a mix of analog and digital channels, and the recording quality varies with the signal type. On analog channels, the low resolution can make hazy details even hazier, though digital channels naturally look much crisper.

Sony DVR Technology

Sony was one of the original manufacturers of the TiVo box. However, this one-time TiVo builder is now bundling its own DVR software into PCs and DVD players.

Sony has come up with what some are calling the first "MegaTiVo" of the future by unveiling a new DVR called the Type X. This monster TiVo will come with an impressive 1 *terabyte* of storage coupled to *seven* TV tuners. This means that the Type X will have the capability to record up to seven different shows at the same time—though given the state of cable television, it might be difficult to find seven TV shows worth recording simultaneously, and later watching!

The seven-tuner feature might be more useful (instead of being simply a technological extravagance) as a result of HDTV's evolution. One idea that has been floated now and again in HDTV circles is to allow broadcasters to tape and send multiple camera angles of a single event to your television, thereby allowing you to choose the angle from which you see the final scoring Super Bowl touchdown. If a single event is being broadcast with seven angles to choose from, having seven tuners recording the multiple streams of data seems downright practical.

The Commoditization of the TiVo

The promise of standard-issue DVR cable boxes could lead to a real commoditization of DVRs in TiVo's near future, which would mean that TiVo would lose its uniqueness as a product and a company. At present, one of TiVo's strengths has been its sense of brand awareness, what some would call *community*.

As a company grows, it can retain that consumer mystique and dedication with some effort—witness the success of Apple, or the eBay auction site, for example. However, TiVo will likely suffer if it gets into a position where it has to compete for new subscribers who can purchase a similar DVR machine for less. TiVo's problem is that neither the software nor the TiVo box itself are difficult for a competitor to produce. Ultimately, if the product market changes into a version of the consumer PC market's commodity war (and while unlikely, this could happen), then that price could lower than what TiVo needs to make to survive.

The TiVo-less TiVo

Finally, what development could impact the future of TiVo more than getting the TiVo-like features without the TiVo box? Software that offers TiVo-like recording for your computer is a small but growing phenomenon. Of course, tuner cards that promise to turn your computer into your personal television have been around for years. For example, for about $140, the SnapStream Personal Video Station 3 allows you to record your favorite shows and duplicate many of TiVo's basic features, but without the monthly subscription charges. You do need a fairly new PC with plenty of processor speed—a chip with a clock speed of 733 MHz or faster is recommended to run the software imaging properly.

This option is best-suited for those who are comfortable installing components and software on their computer, as compared with the plug and play simplicity of the TiVo. Typically, a user will have to install a tuner card, the software it requires, and a Windows-compatible driver so that the card will work with the video recording software.

The software product SnapStream provides a TV-like, full-screen presentation of live and recorded video as well as a program guide. Although it is not as smooth as TiVo's excellent graphical user interface, the guide is operable by remote control. SnapStream also includes features that until now have only been associated with a TiVo or TiVo-like box:

- A 30-second skip function, which can be handy for skipping commercials
- A built-in streaming video server that allows you to watch live or recorded TV from any computer connected to your home network

- The ability to stream recorded shows to you across a broadband connection when you're away from home
- Archiving playback on many gadgets, including DVD players and PocketPCs

NOTE *While SnapStream software emulates some of TiVo's features, one of TiVo's signature features is missing: the recommendations. Whereas TiVo's sophisticated system can suggest additional programming based on your prior choices, SnapStream can only record what you tell it to.*

TiVo's Place on the Home Network

The latest round of developments has focused on wireless home networking. As a result, many companies have committed to designing new products that allow consumers with multiple computers in their homes to access network resources, such as central home servers, color laser printers, and the most popular complement to wireless networking, broadband access.

TiVo originally positioned its service to be the hub for home entertainment, based on these home networking trends, but TiVo now wants the DVRs to simply be a part of the network. According to Brodie Keast, one of TiVo's senior vice presidents, market research determined that consumers found it awkward to have to manage digital content from more than one place on the system. TiVo was also surprised by the popularity of 802.11b wireless home networks. With the computer as the clear digital content center of the system, it was decided that the best way for TiVo subscribers to access and share digital content was to simply piggyback on the rapidly expanding market of wireless networking.

Given TiVo's ability to work easily with wireless adapters, such as the line from Linksys, it's fair to say that this positioning of the TiVo box on the system will allow TiVo to continue to develop as a valuable network component. People may not yet be ready to do their computing on their TiVo, but as home networking and home theatres continue to merge, people will certainly rely more on the TiVo to deliver their programs without a wired system.

Upcoming Changes in DVD Technology

One new feature for TiVos in 2004 was the combining of the TiVo service with the ability to play and burn DVDs. As a result, developments in the DVD field will directly affect the use of the TiVo in the years to come.

Did you know?

The Market for TiVo Upgrades

One of the signs that TiVo has come of age and has begun to attract more attention as a mature technology is the growing aftermarket in TiVo upgrades. WeaKnees.com, a TiVo upgrade and parts company, takes TiVo Series2 systems and adds additional hard drive space, bumping up the storage capacity from 80 hours to as many as 320 hours of standard television, or the ability to store programs broadcast in the extremely data-rich format of high-definition television.

One combo sold by WeaKnees.com is the TwinBreeze and Advanced Cooling Pak. It's a sort of "TiVo on Steroids" that comes with two 120GB drives, an extra fan, and an extra fan bracket. The setup is designed to keep the unit cooler than standard models, while at the same time keeping the fan units quieter than those in a standard TiVo.

Another combo is the PowerTrip Supply Saver, which staggers the start-up time of a dual drive system so that the factory-installed TiVo power supply can still support both drives at the same time.

And perhaps the best argument for allowing a firm like WeaKnees.com to do your upgrade is that if you open the TiVo case yourself, you void the company warranty. By comparison, WeaKnees.com takes over the TiVo warranty and extends it by another three months, eliminating the uncertainty and risk when you want to upgrade your TiVo's capacity or decrease its noise level.

The latest development was the release of DVD-RW discs that record at up to four times the normal playback speed. This format, which is called the "plus" format, was rolled out by Philips Electronics and Hewlett-Packard. The catch is that these discs cannot be recorded on drives that have a top speed of 1X or 2X.

Unsafe at Any (Okay, Just 4X) Speed

According to Pioneer Electronics, the disc incompatibility stems from the need to make rewritable media that is more sensitive to laser light for higher-speed recording. As of this writing, rewritable discs supporting 4X speed are starting to enter the market. However, drives that are able to support the 4X rewritable technology have been on sale for a while, and this will mask some of the impact felt by consumers.

Media makers will likely put warning labels on higher-speed discs to alert consumers to the incompatibility problem. Trying to record on the new 4X discs with older writers may result in the disc being ejected by the hardware, or damage to the data on the disc. If you own an older DVD player or burner, keep an eye out for this label when you go shopping for DVD-RWs.

Double-Layer DVDs

Although 4X speed is one major development in DVD technology, the real race has been to come out with the first double-layer DVD. A double-layer DVD would allow you to cram more data onto the disc. Traditionally, DVDs have held 4.7GB, but this changed when Sony introduced double-layer DVD burners that use the plus format. This new format can store up to 8.5GB on one disc.

Optical Discs

As if all that weren't confusing enough for consumers, the big contest may be in determining what medium will eventually replace today's DVDs. Next generation optical disc technology may be the answer. To date, there have been impressive gains in speed, reliability, and data storage on this medium. It has also proven to be very useful when storing data-intensive HDTV programs. Should HDTV break into the mainstream market in a big way (and indications are that it already is starting to happen), this new format may be the way we store and watch movies in 2010.

High-Definition Television and TiVo

As High-Definition TV (HDTV) expands into the mainstream marketplace, DVRs will respond by increasing the amount of hard drive space available on their boxes. This development is absolutely necessary because signals for HDTV carry far more information per frame than the standard television signal.

NOTE *One HDTV standard in the United States (1080i) calls for images that are 1,920 pixels by 1,080 pixels, refreshed 60 times a second. By comparison, standard analog TV in the United States uses a much wimpier 500 dots by 525 dots. End result? A picture that will be so sharp, you'll feel closer and more "in the action" on the screen than before.*

DVRs configured for HDTV will carry 250GB hard drives, which should provide enough room to store 30 hours of high-definition programming. To put the information load in perspective, this huge hard drive would be able to store over 200 hours of programming sent over a standard broadcast.

The Good and Bad of the Coming HDTV Tide

The good news about super-sized TiVo hard drives designed to store HDTV is not only that they'll be able to store more programming than ever, but also that the price of larger and larger hard drives has fallen more quickly, relatively speaking, than any other computer component in the last ten years.

On the flip side, the introduction of a new recording medium has predictably brought pressure from Hollywood, which senses both a major revenue opportunity and a real chance to stem the tide of copyright violation, which has become epidemic in recent years. Accordingly, manufacturers are planning to include new copy-protection mechanisms that will prevent users from recording programs and sharing them via other networked devices and over broadband Internet.

Content protection will have two layers. The first focuses on the interface that transfers the digital signal from the DVR to the display. This is Digital Visual Interface (DVI), which is designed to work with High-Bandwidth Digital Content Protection (HDCP). HDCP is software that will encrypt the digital signal and allow only an authorized device to play the protected content.

The second part of the content protection will be a preinstalled chip set in the DVR, which will check to see if the HDTV display is authorized to play the program. If the authorization is okay, the DVR will play properly. However, if the display doesn't sense the appropriate code, the DVR will either not play the content back, or (to the even greater frustration of the copyright pirate), it may play at a much lower, grainier resolution, eliminating any gains from the copyright theft.

The Up-and-Comers: Emerging HDTV Devices

Which new devices are likely to become the must-have, HDTV stocking stuffers? Well, HDTV users are already the type of buyers who like cutting-edge equipment, and they typically demand the latest and greatest without respect to price. With that in mind, here are some of the offerings as of early 2004:

- **EchoStar Dish Network** In mid-December 2003, this innovative company began selling the first high-definition DVR, with the ability to record programming from its own satellite TV network. DishPlayer's product was also the first to use the soon-to-be-standard upgraded 250GB hard drive, and it comes with three tuners. This will allow viewers to watch one show while recording two others at the same time. It will also be able to record standard-definition programming. Expected retail price will be around $1,000.

- **DirecTV** TiVo formally entered an agreement to develop and sell an HDTV DVR, again with a 250GB hard drive. Going one better than the EchoStar machine, the new DirecTV HD DVR has four tuners. Two of the tuners can receive DirecTV programming, and two receive over-the-air programming. The device can record two programs at once.
- **Scientific-Atlanta** As of December 2003, the Explorer 8000HD-DVR became available to Time Warner digital cable subscribers in the test market of Green Bay, Wisconsin. Depending on its reception by the consumers in this market, the company plans to roll out new boxes in late 2004 or early 2005.
- **LG** Like EchoStar's unit, LG's HDTV DVR will have three tuners. Unlike the other HDTV DVRs, LG breaks with the new 250GB convention by providing a comparatively small 120GB hard drive to record high-definition cable broadcasts. This gives the LG unit the ability to record 12 hours of high-definition, or 120 hours of analog television.

The Advertising Wars

TiVo's impact on the advertising industry has been nothing short of titanic, in psychological if not raw financial terms. Television advertising's effectiveness, already diminished by consumers' ability to easily change channels during commercials with their remote controls, stands to lose further ground with TiVo. (And if you're one of the TiVo faithful, that was probably one of the top two reasons you went ahead and bought a TiVo—to avoid the commercials!)

Did you know?

Jay Leno in High Definition

Subscribers to satellite television service will be able to record high-definition programs like *The Tonight Show* and *CSI,* as well as premium programming, such as HBO and ESPN (which now have separate channels broadcast in high definition).

Did you know?

And the best part about having a TiVo is…

According to Forrest Research, an advertising-agency analyst firm, more than 50 percent of DVR users surveyed said that skipping commercials is their favorite feature!

The changes are already beginning to show. In the coming years, expect more—a lot more—product placement on your favorite shows. Another phenomenon you'll see more of is one that you may have already noticed at the supermarket—advertisements on the handle of your shopping cart, on the floor tiles, or on the grocery separator bar at the checkout. These are ad placements in what marketers call "captive transit spaces." Places you have to pass by or where you are literally "trapped" for a short period of time. Expect to see more ads on the gas pump as you fill up, on the windows of the local bus, and on the back of the driver's seat in taxi cabs.

However, stop and consider for a moment: although TiVo's service helps you skip annoying ads, is TiVo itself anti-commercial? It's hard to argue with the company's own literature on the matter. For example, TiVo's own web site (http://www.tivo.com/5.5.asp) has this to say to advertisers:

> TiVo works with a growing number of agencies and clients from many industries to deliver branded entertainment and special offers to TiVo viewers. This new type of advertising and entertainment is delivered via the unique and powerful promotional platform of TiVo...
>
> As an advertiser you'll:
>
> - Reach and create a quality connection with a desirable demographic
> - Combine the creative impact of long-form TV advertising with the quantitative measurement of on-line advertising
> - Leverage unique lead generation capabilities
> - Enhance existing or new marketing programs

In short, TiVo hasn't exactly played hard to get when it comes to demonstrating that it can serve up users in prepackaged, captive form to corporate marketers. For example, in 2003, TiVo added a special service—placing full-length film trailers onto users' hard drives. Not that this was an unwelcome or intrusive addition; by TiVo's own estimates, seven out of 10 viewers spent the three-and-a-half minutes voluntarily watching previews for the third *Austin Powers* movie. The trailers were a success, and TiVo proved its point to Madison Avenue: advertising through TiVo works, and works quite well.

The Marketing Empire Strikes Back! Video-to-Video

With the first test of placing trailers on people's TiVos a smashing success, TiVo plans to step up efforts to woo advertisers who want to get their messages across to people who are savvier than the general public about avoiding ads. The "next big thing" TiVo brings to this area is known as Video-to-Video.V2V, as it is sometimes called, allows marketing clips to be promoted via small icons that appear on the screen as viewers fast-forward past regular ads. This service allows viewers to press a button on their remote control and watch a three-minute video describing products and services.

If successful, this new feature could turn out to be a cash cow for TiVo. Rather than going head to head with advertising firms who foresee DVRs negatively impacting television advertising when a projected 79 million consumers acquire them, TiVo could turn into one of the industry's more valuable allies. Little wonder that TiVo plans to license the technology behind Video-to-Video to advertisers, and tap into this potentially lucrative and industry-changing technology.

What About Privacy Concerns?

Imagine that after a hard day's work, you switch on your TiVo and see that your V2V service has a slew of small icons on the screen as you fast-forward past regular ads. At the press of a button, you can choose to watch three-minute videos about products that will improve your golf game, "How To" series on home improvement, and the latest consumer research on the new environmentally-friendly SUV. Amazingly enough, two of these ads are exactly in line with your current hobbies, and sure enough, you've been thinking about getting one of those SUVs ever since it made its debut on *Seinfield 2050: The Sequel*. How does TiVo know your tastes to a "T"?

Whether you call it market sampling or invasion of privacy usually depends on whether you know exactly what data is being collected, and whether you can voluntarily and easily opt out. As of this writing, TiVo has made a pledge not to sell information that identifies specific viewer households. However, the company does plan to make the data available on a per-neighborhood or residential zone system, broken down into various demographic and general marketing categories. At present, TiVo may not know that you like golf, home improvement, or the car that Seinfeld drives in 2050. But as it stands, if you happen to live in a community that has interests in golf courses and do-it-yourself stores like Home Depot, and where the numbers of SUVs per person are in a ratio of 1:1, then you can be sure that the marketers buying TiVo's information will ensure that you see ads relating to these interests and to sales at these stores.

According to TiVo, the tracking system is set up to preserve an individual's privacy. The system's software allows the machine to send two files back to TiVo headquarters. The first logs a serial number to identify the machine and neighborhood location. However, no viewing data or remote control selections are logged. The second file contains viewing info and your remote selections. This log is time-stamped, but it has no serial number to identify the machine.

Would it be possible to decode and group these two pieces of data? Given today's sophisticated and intelligent coders, certainly. One can argue that marketers really don't care what you, as an individual, really watch in any case—they need to make sales calls based on hundreds or thousands of households, not just yours. On the other hand, if viewing habits become public record, they could easily cause privacy-related problems. Who would want the world to know if they ever ordered adult entertainment, or that they liked violent action films.

The bottom line is that if you're concerned about your privacy, it's best to watch what services like TiVo are doing to actively protect the balance between privacy and marketing. Beyond that, it's up to each consumer to decide what they will allow others to know, and what services they can do without if someone is watching them.

What's in Store for the Longer Term

Beyond the next few years, it's open speculation as to where TiVo will go. Some predictions can be made with a fair amount of confidence, if one just assumes that certain demographic or market trends will continue. For example, whether or not TiVo as a brand continues to 2010, 2020, or 2050, the DVR is likely to stay with us, and will continue to force television and its advertisers to cope with increased shifts in the way marketing is done.

It's also safe to say that as the price of computer hardware continues to fall, the capabilities and quality of the available devices will rise. By the end of the decade, it's easy to think that we'll be able to purchase DVRs with multi-terabyte hard drives. Certainly we'll see a continued change in the recording media. Just as DVDs and MP3 players are smaller than VCR tapes, perhaps we'll see optical discs change to

Voices from the Community

Forward to the Future—TiVo, DVDs, and Pizza?

One thing I can't wait for TiVo to add is a service where I can order the DVD of a movie that I'm watching, directly from the remote. When the WishList finds me great, obscure movies that I enjoy on my TiVo, I don't want to hog my hard drive space with them, but it's easy to forget to order them the next time I run up to use my computer. This way, if TiVo offered me a simple option to order without searching online, typing in credit card numbers, and such, I think I'd take advantage. Plus, if I have friends over who love the movie we saw, it could make my gift-giving a lot easier!

Another feature that I'm dying for TiVo to add is the ability to order pizza while I'm watching TV. When I get home from work and get engrossed in watching a few episodes of *24* back-to-back, I can forget about fixing dinner, and by the time I'm done, it's too late to really eat anything substantial right before I go to bed. If TiVo made it simple to order a pizza and handle the billing, I could continue to watch my favorite show, pause it when the pizza guy comes, and not even have to run upstairs to get my wallet. Keep those features coming, TiVo!

—Joel Elad, Marketing and Sales Director for Top Cow Productions, a comic book and entertainment company, and hopeless TiVo addict

fingernail-size crystal chips, or even to pure electronic form that can be accesses via a retinal or thumbprint scan.

And just as Hollywood will continue to try to prevent mass illegal copying with various controls, expect equally motivated and intelligent people to try to circumvent these controls and make anonymous distribution and copying easier. The question isn't whether there will be a Napster-like company popping up in the future, but when and where the company will be.

Of course, TiVo's future is by no means certain. Like the Apple Macintosh, it's a machine that appeared to fill just the right need at just the right time. TiVo may also become more of the central system that its developers once envisioned. Should television become the main channel for video email and home-to-home video conferencing—especially if telecommuting continues to grow in popularity—then TiVo may continue to find itself offering yet more items to service this slice of the demographic pie.

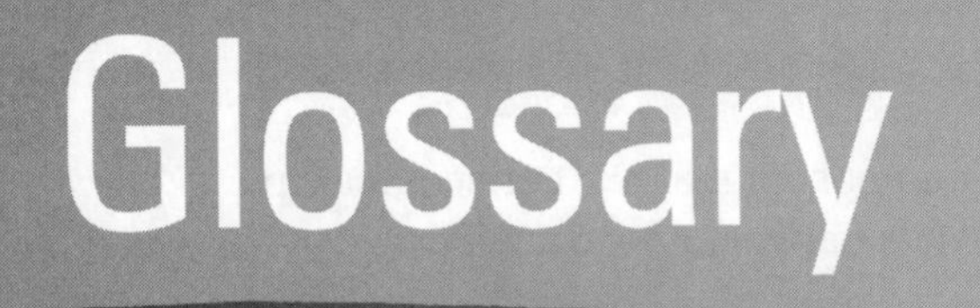

Glossary

Now Showing
NFL Today
Will and Grace
Guiding Light
Law and Order
World of Nature
Monday 10/15
Thursday 10/18
Thursday 11/16
Tuesday 11/20
Saturday 12/5
Friday 12/10
Wednesday 12/14
Tuesday 12/30
press ENTER to change)
8:46
7:30
10:00

A/V receiver (audio/visual receiver) A stereo component that can connect multiple A/V devices. The A/V receiver can switch signals between those devices.

Coaxial cable *See* RF coaxial A/V cable.

Composite A/V cable A color-coded, three-pronged A/V cable. The yellow wire carries video and the red and white wires carry right and left audio.

Daily call The TiVo's regularly scheduled phone call to retrieve program guide data and keep the TiVo service up to date.

Dolby Digital (AC-3) sound A six-channel surround sound system from Dolby. The six channels consist of five full-range channels and one narrowband bass channel. Dolby Digital requires the use of an optical audio cable.

Dolby Pro Logic Surround Sound An enhanced version of Dolby Surround Sound that adds an additional center channel.

Dolby Surround Sound The standard surround-sound system using four-speakers: left and right main channels and left and right surround channels in the rear of the home theatre system.

DVR (digital video recorder) A TiVo is a DVR, as are digital recorders distributed by satellite and cable TV providers.

Guided Setup A series of steps you follow when first setting up your TiVo. During this process, you choose a local phone number for calls to the TiVo service, choose a program provider and program listings, and download program guide information.

Green screen Often referred to as the *Green Screen of Death,* this is a screen image with a message not to disconnect the TiVo while it attempts to repair a software problem. This is frequently an indication of hard drive failure.

Hack A software or hardware modification you make to your TiVo to provide new or expanded capabilities.

HDCP (High Definition Copy Protection) The method of encryption used for transmitting and protecting HDTV streams. Your display must be HDCP-compatible to decode the signal.

HDMI (High Definition Multimedia Interface) A digital one-cable solution for passing HDTV to a single input on an HDTV display.

HDTV (high-definition TV) A broadcasting standard that provides television viewers with higher resolution programming than has been available on traditional TV. You must have an HDTV set to benefit from the increased resolution. DirecTV recently released a TiVo DVR that lets you record HDTV programming.

Home Media Features An add-on option for the Series 2 TiVo units that gives you extra features, such as Remote Scheduling, Multi-Room Viewing, and the ability to view and play photos and music files on your TiVo.

IDE cable A flat ribbon cable used to connect a hard drive to a circuit board.

Instant reply The ability to quickly play back a portion of recently viewed programming by using the TiVo remote control.

IR (infrared) A line-of-sight communication medium used by devices to transmit commands to other devices. Most remote controls use IR to send requests to an A/V device.

IR code An infrared code for controlling a particular cable or satellite box.

IR control cable A cable that connects to the back of your DVR and transmits an IR signal to change channels on your cable or satellite box.

Jumper A small plastic and metal fitting designed to make an electrical connection between two contact points, as on a SCSI hard drive.

Live TV Regular programming that you can watch as it is broadcast. Live TV can be accessed either by pressing the Live TV button on the remote or by selecting it from the TiVo Central menu.

Multi-Room Viewing A Home Media Features option that lets you exchange recorded programs among TiVos located in the same house. You can watch a program on one TiVo that was originally recorded on another TiVo, probably located in another room.

Now Playing list A list of recorded programs currently available for viewing on your TiVo.

Parental controls Settings that allow you to lock out certain channels and restrict programs containing unsuitable content unless a password is entered. The parental control settings can be accessed from the My Preferences menu.

PowerTrip A device used to delay power getting to some equipment, such as a second hard drive, in order to delay the start-up of that equipment.

Program guide data Information about programs airing on channels from a particular cable or satellite provider. Program guide data typically includes the date and time a program will air, as well as a short description and other information, such as actors' and directors' names.

PVR (personal video recorder) *See* DVR.

RCA cable *See* Composite A/V cable.

Remote Scheduling A Home Media Features option that lets you schedule programs for recording from TiVo's Web site. Your TiVo must be connected to a broadband Internet connection for this to work.

RF coaxial A/V cable A single A/V cable that transmits a combined audio and video signal.

RF splitter A device that allows a cable signal to be split before being connected to a television or DVR.

SAP (Second audio program) An alternate audio track available on some programs.

Serial control cable A cable that connects to your DVR and your satellite receiver and transmits channel-change commands.

S-Video cable An A/V cable that transmits video. The red and white lines of a composite A/V cable must also be connected between the devices in order to transmit the audio.

Season Pass A standing request to record a particular show whenever an episode is shown.

Season Pass Manager The screen where you can change and delete a Season Pass you've previously requested.

Series1 The original TiVo model, which is no longer sold by TiVo. Refurbished and used models are available through places like online auction sites. The Series1 TiVo is easier to hack.

Series2 The second-generation TiVo model that TiVo and its manufacturers are currently selling. It's more difficult to hack, but it contains new features, such as the Home Media Features. DirecTV is also currently offering a Series2 unit, but the Home Media Features is not available for DirecTV customers.

Suggestions A feature of the TiVo service that locates and records programs you might like to view. It only records suggestions when extra hard-drive space is available and no other programs are scheduled for recording at that time. Suggestions are based on the programs you record and on your ratings of programs, indicated by using the thumbs-up and thumbs-down buttons on your remote control.

TiVo Central (or DirecTV Central) The top-level menu, which allows you to access all of TiVo's functionality.

Thumbs Up The green button on TiVo's remote control that you push to give a program a positive rating. You can give a program up to three thumbs up. TiVo uses the ratings to suggest programs you might like, based on your past viewing habits.

Thumbs Down The red button on TiVo's remote control that you push to give a program a negative rating. You can give a program up to three thumbs down. TiVo uses the ratings to suggest programs you might like, based on your past viewing habits.

Torx A variety of screw head with a characteristic star shape, commonly used in TiVo assembly.

Wireless modem A device that allows a phone signal to be transmitted over radio waves.

WishList A TiVo feature that finds and records programs that include your favorite actor, director, or other customized criteria.

Index

Numbers

3Com, USB-to-Ethernet adapter offered by, 98
4-pin connectors in S-Video, lining up, 10
4X discs, availability of, 334–335
9-pin data connectors, using with TiVo-DVD Recorders, 198–199, 201
15-pin data connectors, using with TiVo-DVD Recorders, 198–199, 201
18X recording speed of DVR-810H, meaning of, 208
30-second skip feature, changing Jump to End button to, 232
33 variables, meaning of, 284–285, 287, 289–290
222 variables, meaning of, 284–285, 287, 289–290
250GB hard drives, significance of, 324
480P resolution, explanation of, 186, 300–301, 308
640x480 versus 1920x1080 resolution, 302
720P resolution, explanation of, 301
1080i HDTV standard significance of, 335
1111 variables, meaning of, 284–285, 287, 289–290
8000HD-DVR HDTV device, availability of, 337

Symbols

* (asterisk), using with Keyword WishLists, 59
/ (slashes), omitting, 60
___ variables, meaning of, 284–285, 287, 289–290
' (apostrophes), omitting, 60
" (quotes), using when searching phrases, 60

A

AboCom, USB-to-Ethernet adapters offered by, 98
accessories
- resource for, 108
- setting up for TiVo-DVD Recorders, 188–191

Accton, USB-to-Ethernet adapters offered by, 98
Actor WishLists, creating, 59
ADMtek, USB-to-Ethernet adapters offered by, 98
advanced search feature, using with TV listings, 144–146
advertising industry, impact of TiVos on, 337–339
Allied Telesyn, USB-to-Ethernet adapter offered by, 98
alphanumeric passwords, entering, 107
antennas, using with satellite dishes, 310–311
AOL TiVo feature, using for first time, 159
AOL TV Listings screen, selecting programs from, 159
AOL's Remote Scheduling, using, 157–164
apostrophes ('), omitting, 60
Architecture 1-6 hard-drive upgrades, overview of, 258–259
Arrange by Type option, effect of, 129
aspect ratio
- determining for TVs, 308
- relationship to HDTV, 301
- selecting for HD TiVos, 319, 321

asterisk (*), using with Keyword WishLists, 59
ATSC (American Television Standards Committee), relationship to HDTV, 300
audio cables, setting up for TiVo-DVD Recorders, 188–191
audio information, displaying on TiVo-DVD Recorders, 216
Audio Settings screen, displaying for HD TiVos, 317
audio signals, optimizing, 202, 204
AutoTest mode, enabling, 236
A/V cables, setting up for TiVo-DVD Recorders, 188–191
A/V receivers
- connecting TiVo-DVD Recorders to, 202–204
- solving cabling problems with, 7–8

B

backdoor codes for Series1 TiVos, list of, 231
backdoor mode
- purpose of, 229
- unlocking, 230–231

backups
- performing, 283–284
- restoring to one or two drives, 284–286

Basic (Analog) TV lineup, description of, 24
Basic recording quality, using, 46, 48
basic search feature, using with TV listings, 143–144
Belkin, USB-to-Ethernet adapter offered by, 98
Best recording quality, using, 48
Billionton, USB-to-Ethernet adapters offered by, 98
bit rates, relationship to TiVo-DVD Recorders, 210
black screen, appearance of, 247
blue screen, troubleshooting, 247
boot disks, creating for DiskUtil, 291–292
bridges, linking phone cords with, 193
broadband versus phone lines, 140–141

Browse by Channel menu, navigating, 53–54, 149
Browse by Time menu, navigating, 52–53
browsing log files, 235
browsing music files, 129–130
"burn-in" effect, relationship to plasma screens and CRTs, 301
buttons on remote control, descriptions of, 32–33

C

cable boxes
 entering brand of, 25
 hooking up TiVos to, 10–12
Cable Lineup screen, displaying, 24
cables
 connecting to, 202–204
 setting up for TiVo-DVD Recorders, 188–191
 using with external modems, 240
calls, unsuccessful completion of, 247
"captive transit spaces," explanation of, 338
Cat5 crossover Ethernet cables
 benefits of, 181
 using to stack TiVos, 179
CDs (compact discs)
 burning or copying, 207
 playing on TiVo-DVD Recorders, 212–213
Chan button on remote control, effect of, 33
Change Dialing options, selecting, 195
channels
 browsing TV listings by, 146–149
 changing in live TV, 37
 eliminating, 321
 receiving, 147
 selecting, 25
 watching and recording simultaneously, 14–15
channels 3 and 4, choosing between, 10
Choose Keywords screen, displaying, 60
Clear and Delete Everything option, advisory about, 83
Clear button on remote control, effect of, 33
Clear-Enter-Clear codes, overview of, 234–236
Clips on Disk section, accessing, 234
clock, enabling, 233
coaxial cable
 chaining items with, 7
 daisy-chaining, 12–13
 description of, 6
codes. *See also* software hacks
 Clear-Enter-Clear codes, 234–236
 SPS (Select-Play-Select) codes, 231–233
 triple-thumb codes, 233–234
comma, substituting when setting up external modems, 241
Compaq, USB-to-Ethernet adapter offered by, 98
component video cables
 benefits of, 202, 204
 using with TiVo-DVD Recorders, 189–190
composite video cables
 description of, 6
 using, 16
 using with TiVo-DVD Recorders, 189
"Computer Input Only" DVI ports, relationship to HD TiVos, 308
configurations, testing and using, 16
Confirmation screen, displaying, 219
Conflicts screen
 displaying, 47
 displaying for Season Passes, 56–57
Connection to TiVo DVR screen, displaying, 25
connections. *See* HD TiVos; network connections; wired connections; wireless connections
content protection, relationship to HDTV, 336
copy protection
 and HDTV, 304–305
 and TiVo-DVD Recorders, 211
Copying to DVD screen, displaying, 220
cords, reaching phone jacks with, 193
crossover Cat5 Ethernet cables
 benefits of, 181
 using to stack TiVos, 179
crunching progress, checking on, 27
Current Output Format screen, displaying for HD TiVos, 320

D

daily calls, unsuccessful completion of, 247
daisy-chaining coaxial cable, 12–13
data mining, explanation of, 85
data-collection system, opting out of, 88–90
DCDi (Directional Correlation Deinterlacing) technology, enhancing pictures with, 207
Delete Recording? screen, displaying, 68
Delete This Recording? screen, displaying, 67
desktop software. *See* TiVo Desktop software
DHCP client ID networks, creating network connections on, 108–109
DHCP networks, creating network connections on, 99–101
dial tones
 troubleshooting access to, 194–195
 troubleshooting static on, 196
dialing options, configuring, 22–23
Dialtone Detection option, troubleshooting dial-tone problems with, 194–195
Digital Basic lineup, description of, 24
Digital Extended Basic lineup, description of, 24
dip switches, setting for US Robotics modems, 240
directional pad button on remote control
 effect of, 32
 saving time with, 36
DirecTiVo, toll-free number for, 18
Director WishLists, creating, 59
Director's Clapboard icon, explanation of, 35
DirecTV
 dishes and antennas used with, 310–311
 HD channels offered by, 322
 lifetime service agreement offered by, 250

DIRECTV Central screen, displaying for HD TiVos, 317
DirecTV DVR guides, navigating, 321
DirecTV HDTV device, availability of, 337
DirecTV HR10-250 HD TiVos, availability of, 306–307
DirecTV-TiVo boxes, success of, 327–328
disc drives, removing condensation from, 211
disc tray, protecting, 212
Discovery HD, channels offered by, 322–323
discs, cleaning, 211
DiskUtil, creating boot disk of, 291–292
D-Link, USB-to-Ethernet adapters offered by, 98
DLP (Digital Light Projection), explanation of, 301
Dolby Digital
configuring for HD TiVos, 317–318
types of, 309–310
Dolby Digital 5.1 shows, finding, 321–324
Done Creating WishList screen, displaying, 61
Don't Do Anything option, effect of, 169
double-layer DVDs, availability of, 335. *See also* DVDs (digital video discs)
drive brackets, using in hard-drive upgrades, 275–277
drive screws, using in hard-drive upgrades, 276
D-Theater copy protection, development of, 304
DVD banners, using with TiVo-DVD Recorders, 214–215
DVD burners, hooking up TiVos to, 12–13
DVD Menu Screen
accessing from remote, 212–213
navigating, 214
DVD settings, changing for TiVo-DVD Recorders, 215
DVD technology, upcoming changes in, 333–335
DVD-R (DVD-recordable)
archiving with, 186
description of, 185
versus DVD-RW, 209
DVD-RAM format, description of, 185
DVD-RW (DVD-rewritable)
description of, 185
"plus" type of, 334
DVDs (digital video discs). *See also* double-layer DVDs
burning or copying, 207
copying programs from TiVo-DVD Recorders to, 217–220
copy-protection on, 304
and file sizes, 220
moving between frames on, 215–216
naming, 219
and parental controls, 220
playing on TiVo-DVD Recorders, 212–213
recording from TiVos, 218
recording from video cameras, 223–224
recording from videotapes, 221–223
terminology related to, 185
DVI (Digital Visual Interface), explanation of, 336
DVI ports, Computer Input Only type of, 308
DVR switch on remote control, effect of, 33

E

EchoStar Dish Network HDTV device, availability of, 336
EDTV (extended definition TV), explanation of, 300–301
Elsa, USB-to-Ethernet adapter offered by, 98
e-mail confirmations, receiving for scheduling requests, 153–155, 157, 162, 164
encryption
setting for peer-to-peer networks, 178
WEP (Wired Equivalent Privacy), 107
wireless encryption levels, 108
Enter button on remote control, effect of, 33
episodes, recording, 150
error messages
Failed. Unknown error, 179
"no dialtone", 197
The Series2 80MB DVR could not be reached error, 180
software hack for, 234
Ethernet
advantages and disadvantages of, 97
connecting TiVo to, 99
explanation of, 97
Explorer 8000 versus TiVos, 330–331
Extended Basic (Analog) lineup, description of, 24
external modems
benefits of, 197–198
configuring, 240
connecting to TiVos, 241
materials required for, 239–240

F

factory drives, adding drives to, 286–288
Failed. Unknown error. message, displaying when using crossover connections, 179
fan noise, testing, 246
Fast Ethernet networks, speed of, 174
Fast Forward button on remote control, effect of, 33
fast forward speeds, adjusting, 238
FF/RW, turning off f correction in, 235
film versus HDTV (High-Definition TV), 302
firewall ports, opening, 115
"Firmware Update Required" message, meaning of, 104
folders, organizing music files in, 120–121
formatting settings
losing when formatting hard drives, 288–291
saving when formatting hard drives, 288–289
FPDs (Fixed Pixel Displays), examples of, 301

frames, moving between in DVDs, 215–216
FTC (Federal Trade Commission), decision made by, 88–89

G

game consoles, connecting TiVo-DVD Recorders to, 202–204
gateway addresses, entering for wireless static IP networks, 110
Gefen Technologies, development of HDMI by, 305
Getting Program screen, displaying on remote TiVos, 170
granularity of information, significance of, 84
Grid Guide, changing to, 39–40
grounding, importance during upgrades, 256–257
GSOD (Green Screen of Death), occurrence of, 245–246
Guide button on remote control, effect of, 32
Guide Options screen, displaying, 40
Guided Setup, configuring TiVos with, 19–27

H

hard drive space versus recording quality, 46–48
hard drives
- adding versus replacing of, 250–251
- deciding on space requirements for, 248–250
- denoting sectors on, 284–285, 287, 289–290
- determining presence of, 250
- diagnosing, 244–245
- formatting and installation concerns, 251–252
- future expansion of, 324
- handling, 257–258
- models of, 249
- optimizing use of, 324–325
- previewing installation instructions for, 259
- researching, 252–254
- symptoms of failure of, 245–247
- unlocking, 291–292
- upgrading, 243–244

hard-drive formatting options
- adding drives to TiVo factory drives, 286–288
- copying from TiVo drive to single or dual drives, 288–291
- overview of, 282–283
- restoring a backup to one or two drives, 284–286

hard-drive upgrade kits
- features of, 252–253
- parts provided in, 254–255
- searching for vendors of, 253–255
- warranties provided with, 255

hard-drive upgrades. *See also* upgrades
- hands-off options for, 255–256
- parts summary for, 275–280
- reasons for, 243–248
- taking safety measures for, 256–257
- Web resources for, 293

Hawking, USB-to-Ethernet adapter offered by, 98
HD channels, offerings by Direc TV, 322
HD DVRs, overview of, 303–305
HD (high-definition) sets, types of, 301
HD shows, finding, 321–324
HD TiVos. *See also* Hughes HR10-250 TiVos
- choosing connection type for, 312
- choosing letterbox color bars for, 319
- choosing video output mode for, 320
- connections for, 312
- creating WishLists for, 323
- finding best deals on, 307
- future of, 325
- preparing for setup of, 307–310
- purchasing, 306–307
- selecting aspect ratio for, 319
- selecting Messages & Setup option for, 317
- selecting video settings for, 318
- setting aspect correction for, 320
- specifying satellite cables for, 314

HDCP (High-Bandwidth Digital Copy Protection), development of, 304–305
HDMI (High Definition Multimedia Interface) development of, 305
HDMI to DVI cables, using with HD TiVos, 311
HDNet, channels offered by, 322
HDTV broadcasts, recording two at once, 306
HDTV devices
- 8000HD-DVR, 337
- DirecTV, 337
- EchoStar Dish Network, 336
- LG's HDTV DVR, 337

HDTV (High-Definition TV)
- evolution of, 303
- versus film, 302
- growing popularity of, 303
- native rate of, 308
- overview of, 300–302
- relationship to DVR technology, 331, 335–337

HDTV programming, recording, 324
hexadecimal passwords, entering, 107–108
hidden Showcases, making visible, 234. *See also* TiVo Showcases
High recording quality, using, 48
high-definition programming, recording of, 299
HMF (Home Media Features), Multi-Room Viewing feature in, 166
"hockey stick" growth pattern, significance of, 327
Home Control connectors, using with TiVo-DVD Recorders, 199
home entertainment centers, wiring basics for, 6–8
Home Media option, features of, 95
home networks
- components of, 96–97
- connecting TiVo to, 95–96
- TiVo's place in, 333

home theaters, hooking up TiVos to, 15
Hughes GXCEBOT hard-drive upgrades
- disassembly steps for, 262–263
- drive formatting note for, 263
- parts required for, 262
- reassembly steps for, 264–265

Hughes hard drives, features of, 249
Hughes HDVR2/3 and SD-DVR-40/120 hard-drive upgrades
- disassembly steps for, 268–269
- drive formatting note for, 269–270
- parts required for, 268
- reassembly steps for, 270

Hughes HR10-250 hard-drive upgrades. *See also* HD TiVos
- disassembly steps for, 271–272
- drive formatting note for, 272
- parts required for, 271
- reassembly steps for, 272–273

Hughes HR10-250 TiVos
- configuring, 311–321
- equipment included with, 311–312
- hard drives in, 324

I

I in HD resolution, explanation of, 300
icons
- identifying, 66
- overview of, 35
- for recording, 48

IDE cables, using in hard-drive upgrades, 276–277
"In" label on devices, meaning of, 188
In Progress screen, displaying, 49
Include Subfolders setting, changing in Music Play Options screen, 128
indexing progress, checking on, 27
info banners, using with live TV, 37
Info button on remote control, effect of, 32
information sites, Web addresses for, 292–293
information-collection system, opting out of, 88–90
Instant Replay feature, using, 33, 43
interlaced images, scanning of, 300
Internet
- performing remote scheduling by means of, 138–141
- recording over, 96
- surfing throughout the house, 96

Internet streams, listening to, 131
IP addresses
- assigning for use with wired DHCP networks, 99
- assigning for use with wired static IP networks, 103
- assigning for wireless static IP networks, 109–110

Ipreview feature, using, 38, 78–79
IR control cable, connecting, 11–12
IR (infrared) control cables, connecting to TiVo-DVD Recorders, 199–200, 202, 203
IR Test Instructions screen, displaying, 26
italic fonts, applying, 234

J

jacks on composite A/V cables, using with TiVo-DVD Recorders, 189
JavaHMO versus TiVo Desktop software, 116–118
Jump to End button on remote control
- changing to 30-second skip button, 232
- effect of, 33

jumpers, using in hard-drive upgrades, 278–279
JVC, development of D-Theater by, 304

K

Keep Until option, setting, 67
Keyword WishLists, creating, 59
kit and part dealers, Web resources for, 293

L

LCD (Liquid Crystal Display), explanation of, 301
L-COS (Liquid Crystal on Silicon), explanation of, 301
Letterbox Color screen, displaying for HD TiVos, 319
LG's HDTV DVR, availability of, 337
lifetime TiVo service
- cost of, 18
- obtaining from DirecTV, 250

lights, absence of, 247
Linksys
- network adapters offered by, 104
- USB-to-Ethernet adapters offered by, 98

live TV
- accessing, 36–37
- capacity of buffer for, 41
- changing channels and volume in, 37
- manipulating, 41–42
- pausing, 42
- recording from, 49–50
- rewinding, 42
- using info banners with, 37–38
- using Ipreview button with, 38

live TV buffer
- jumping to end or beginning of, 43
- purpose of, 30–31

Live TV button on remote control, effect of, 32
LNB (low-noise blocker) satellite dishes, using, 310
log files, browsing, 235
"Low Speed Data" connectors, using with TiVo-DVD Recorders, 199

M

Macintosh users, system requirements for TiVo Desktop software, 111
Macrovision, purpose of, 304
Mail icon, explanation of, 35
manual recordings, setting up, 55–56
Maxtor hard drives, using, 253
"May not record" message, appearance of, 149
Medium recording quality, using, 46, 48
"MegaTiVo," Type X DVR as, 331
Menu Screen
- accessing from remote, 212–213
- navigating, 214

menus
- Music & Photos, 35
- navigating on TiVos, 20
- Now Playing, 35
- Showcases & TV Guide, 35
- TiVo Central, 34–35

Messages & Setup feature, description of, 36
MFS Tools software
acquiring, 282
backing up hard drives with, 283–284
features of, 280–282
mini-discs, playing on TiVo-DVD Recorders, 212
modems. *See* external modems
monthly TiVo service, cost of, 18
MoodLogic software, using, 130, 132–135
MP3 format
converting music files to, 123
relationship to TiVo-DVD Recorders, 210
Multi-Room Viewing
features of, 166
setting up, 166–167
troubleshooting problems with, 180–181
multiswitcher by Terk, features of, 312
Music & Photos menu, features of, 35, 123–124
music files, 129–130
converting, 123
faster loading of, 121
listening to, 123–129
listing, 129
mixing with MoodLogic, 132–135
organizing, 120–121
publishing, 121–123
storage of, 120
troubleshooting, 129
music folders
creating, 120–121
selecting, 125
Music Play Options screen
selecting from, 127–128
using with TiVo-DVD Recorders, 216–217
music servers, selecting, 124
music settings, changing, 127–128
Mute button on remote control, effect of, 32

N

Name Recording screen, displaying, 222
network adapters, purchasing, 104
network connections
creating on wired DHCP networks, 99–101
creating on wired static IP networks, 101–104
creating on wireless DHCP client ID networks, 108–109
creating on wireless static IP networks, 109–110
and speeds, 174–175
network masks
assigning for use with wired static IP networks, 103
entering for wireless static IP networks, 109
network node, explanation of, 97
network settings, configuring for wireless home networks, 105–108
"no dialtone" message, occurrence of, 197
Node Navigator
using to enable advanced WishLists, 237
viewing, 236
noise, presence inside TiVos, 245
Now Playing icons, descriptions of, 66
Now Playing menu, features of, 35
Now Playing on TiVo screen
displaying, 65, 168
jumping to, 238
sorting in OS 3.0, 237
Now Playing on TiVo screen, displaying, 174
Now Playing Options screen, displaying, 68
NTSC (National Television Standards Committee), relationship to HDTV, 300
number buttons on remote control, effect of, 33

O

operating system, shutting down, 236
optical digital cables and jacks
benefits of, 202, 204
using with TiVo-DVD Recorders, 190–191
optical discs, future of, 335
OTA channels, accessing, 323
OTA (over-the-air) antennas, using with satellite dishes, 310–311
"Out" label on devices, meaning of, 188
overshoot correction, turning off, 235

P

P in HD resolution, explanation of, 300
parental controls
advisory about setting of, 214
restrictions on setting of, 220
part and kit dealers, Web resources for, 293
passwords
entering for peer-to-peer networks, 178
hexadecimal and alphanumeric types of, 107
patent issues, significance of, 328
Pause button on remote control, effect of, 33
PC users, system requirements for TiVo Desktop software, 110
PC-created DVDs, playing on TiVo-DVD Recorders, 209–210
peer-to-peer networks, creating, 177–178. *See also* wireless connections
permissions, setting up for Multi-Room Viewing, 167
personal information, removing from TiVo's database, 87
Personal Video Station 3, significance of, 332
personalization technologies, significance of, 75
Philips DSR7x hard-drive upgrades
disassembly steps for, 268–269
drive formatting note for, 269–270
parts required for, 268
reassembly steps for, 270
Philips DSR6000 hard-drive upgrades
disassembly steps for, 262–263
drive formatting note for, 263
parts required for, 262
reassembly steps for, 264–265
Philips hard drives, features of, 249
Philips PTV and HDR hard-drive upgrades
disassembly steps for, 260–261
drive formatting note for, 261

parts required for, 260
reassembly steps for, 261–262
Phone & Network Setup screen, displaying, 179
Phone Available option, troubleshooting dial-tone problems with, 194
phone calls, making while TiVo-DVD Recorders are using phone lines, 192
phone connections, selecting, 194–195
phone cords, reaching phone jacks with, 193
phone jacks
plugging into, 8
reaching with cords, 193
phone lines
versus broadband, 140–141
connecting TiVo-DVD Recorders to, 191–192
reclaiming, 96
splitting when using TiVo-DVD Recorders, 192–193
troubleshooting problems with, 193–197
phone splitters, using, 8
phone-dialing options, configuring, 22–23
Pick Programs to Record screen, displaying, 51
Pick Shows to Record feature, description of, 36
picture, enhancing with higher quality connections, 15–16
Pioneer TiVo-DVD recorders
connecting to cables or satellite boxes, 198–18
connecting to phone lines, 191–16
considering use of, 184–5
example setups for, 200-22
setting up, 187–9
Pioneer 810H and 57H upgrades
disassembly steps for, 273–274
drive formatting note for, 274
parts required for, 273
reassembly steps for, 274–275
Pioneer DVR-57H, recording speed of, 208
Pioneer DVR-810H, recording speed of, 208
piracy concerns, relationship to HDTV, 303–304
pixellation, explanation of, 245
plasma screens, disadvantages of, 301
Play button on remote control, effect of, 33
playlists
converting, 123
selecting, 126
"plus" DVD-RW format, significance of, 334
Plus service. *See* TiVo Plus service
power splitters, using in hard-drive upgrades, 278–279
power surges, responding to, 197
PowerTrips, features of, 278, 280, 334
"Press THUMBS UP" message, appearance of, 38
Priority recording option, choosing, 152–153
privacy concerns, considering with V2V (Video-to-Video), 339–340
Privacy Foundation, findings of, 88
privacy of information, improving level of, 85–86
privacy policies
overview of, 86–88
reviewing, 18
reviewing online, 21
Program Call screen, displaying, 27
Program Info page, displaying, 150
program information, viewing, 43–44. *See also* transferred programs; TV programs
Program screen, displaying during transfers, 173
"program source," explanation of, 21
Program Source screen, selecting connection type from, 312
programming settings
losing when formatting hard drives, 288–291
saving when formatting hard drives, 288–289
programs. *See* TV programs
progressive scanning, explanation of, 186, 300
PTVupgrade, Web address for, 324
PureCinema 74 progressive scan, explanation of, 186

Q

Quality recording option, choosing, 153
quotes ("), using when searching phrases, 60

R

rating system
buttons for, 32
changing, 83
resetting, 83
using, 81–84
RCA cables.
description of, 6
using, 16
using with TiVo-DVD Recorders, 189
RCA DVR 39/40 hard-drive upgrades
disassembly steps for, 268–269
drive formatting note for, 269–270
parts required for, 268
reassembly steps for, 270
RCA hard drives, features of, 249
RCA RC930 Caller-ID-Compatible Wireless Modem Jack, features of, 105
rebooting, frequent occurrence of, 245
Record button on remote control, effect of, 33
Record by Time or Channel option, choosing, 52–53
Record from Camera/VCR? screen, displaying, 221
recorded programs
displaying on remote TiVos, 169
storing with Multi-Room Viewing feature, 166
recording history, viewing, 69–70
recording icons, explanations of, 48
recording options
confirming, 151–153
setting, 44–46
Recording Options screen
displaying, 222
displaying for Season Passes, 54
recording quality level
changing, 80
versus hard drive space, 46–48

recording screen, bringing up for Ipreview, 79
recording TV programs, 149
Recording Type option in AOL, effect of, 162
recording type, picking, 56
recordings. *See also* scheduled recordings; WishLists
  deleting, 67
  determining saving options for, 163
  making by time and channel, 52–53
  making for upcoming episodes, 150
  making manually, 55–56
  making while watching TV, 49–50
  managing, 67
  scheduling, 36
  scheduling for features programs, 76
  scheduling with AOL's remote scheduling, 157–164
  setting up from TiVo Central, 50–51
  viewing upcoming recordings, 69–70
"Region Codes" on DVDs, explanation of, 208–209
remote control
  accessing DVD screen with, 212–213
  navigating with buttons on, 32–33, 128
remote control codes
  Clear-Enter-Clear codes, 234–236
  SPS (Select-Play-Select) codes, 231–233
  triple-thumb codes, 233–234
remote scheduling
  with AOL, 157–164
  limitations of, 139–140
  performing over Internet, 138–141
remote shortcuts, using, 238–239
Repeat setting, changing in Music Play Options screen, 128
ReplayTV versus TiVos, 232
resolutions for HD broadcasts, examples of, 300–302
Rewind button on remote control, effect of, 33
rewind speeds, adjusting, 238
RF cable, description of, 6
RF ports, plugging coaxial cables into, 9–10
RF program sources and satellite boxes, connecting to, 201–202
RIAA (Recording Industry Association of America), lawsuits launched by, 303–304
router addresses, entering for wireless static IP networks, 110

S

Samsung SIR-S4120 and SIR-S4040
  hard-drive upgrades
    disassembly steps for, 268–269
    drive formatting note for, 269–270
    parts required for, 268
    reassembly steps for, 270
Sat 1 or 2, lack of input on, 247
satellite boxes, connecting TiVo-DVD Recorders to, 199–204
Satellite Connections screen, displaying when configuring HD TiVos, 314
Satellite Dish Direction screen, displaying for HD TiVos, 315
satellite dishes
  configuring for HD TiVos, 313
  connecting receivers to, 312
  hooking up TiVos to, 10–12
  using antennas with, 310–311
Satellite Info screen, displaying for HD TiVos, 313, 316
satellite TV, benefits of subscribing to, 337
scan rates, relationship to HDTV, 301–302, 308
Schedule It! page, displaying, 151
scheduled recordings, managing, 69–71. *See also* recordings; WishLists
Scheduled Suggestions. *See also* TiVo Suggestions
  displaying in To Do list, 234, 235
  turning off in To Do list, 235
Scheduling Complete page, displaying, 152
scheduling conflicts, occurrence of, 140
screens, navigating, 20
Search by Title screen, displaying, 51–52
search feature, using with TV listings, 143–146
Search Using a WishList screen, displaying, 58
searching lineup data with WishLists, 62
Season Pass Manager, using, 70–71
Season Passes
  creating, 53–54, 56–57
  creating in AOL, 162
  editing, 71
  handling conflicts with, 71
  obtaining remotely, 154–157
sectors on hard drives, denoting number of, 284–285, 287, 289–290
Select button on remote control, effect of, 32
serial cables, connecting, 11
serial control cables, connecting TiVo-DVD Recorders to, 198–199, 201, 203
Series1 TiVos, examples of, 229–230
The Series2 80MB DVR could not be reached error, occurrence of, 180
Series2 units
  selecting from pull-down menus, 142
  setting up Multi-Room View on, 166–167
service number, locating, 18
Setup Call screen, displaying, 23
Showcases & TV Guide menu, features of, 35
Showcases feature. *See also* hidden Showcases
shows. *See* TV programs
Shuffle setting, changing in Music Play Options screen, 128
Siemens, USB-to-Ethernet adapter offered by, 98
signal strength, increasing, 180
Skip Fwd and Skip Back buttons, effects of, 216
slashes (/), omitting, 60

Slow button on remote control, effect of, 33
smartBridges, USB-to-Ethernet adapter offered by, 98
SMC, USB-to-Ethernet adapters offered by, 98
SnapStream Personal Video Station 3, significance of, 332
software. *See* TiVo Desktop software; TiVo software
software hacks. *See also* codes
 adjusting fast forward and rewind speeds, 238
 AutoTest mode, 236
 browsing log files, 235
 Clear-Enter-Clear codes, 234–236
 Clips on Disk, 234
 enabling advanced WishLists, 237–238
 error messages, 234
 explanation of, 229
 hidden Showcases, 234
 italic fonts, 234
 quicker timeline bar, 233
 rebooting and resetting TiVos, 236
 rebuilding suggestions, 235
 remote shortcuts, 238–239
 scheduled suggestions and hidden recordings, 234
 shutting down operating system, 236
 sorting Now Playing in OS 3.0, 237
 SPS (Select-Play-Select) codes, 231–233
 TiVo Clock, 233
 TiVo Status, 233
 triple-thumb codes, 233–234
 turning off f correction in FF/RW, 235
 turning on and off Scheduled Suggestions in To Do list, 235
 viewing Node Navigator, 236
SOHOware, USB-to-Ethernet adapter offered by, 98
songs
 categorizing in MoodLogic, 133
 selecting, 126
Sony DVR technology, overview of, 331
Sony SAT T-60 hard-drive upgrades
 disassembly steps for, 262–263
 drive formatting note for, 263
 parts required for, 262
 reassembly steps for, 264–265
Sony SVR2000 hard-drive upgrades
 disassembly steps for, 260–261
 drive formatting note for, 261
 parts required for, 260
 reassembly steps for, 261–262
Sony SVR3000 hard-drive upgrades
 disassembly steps for, 266
 drive formatting note for, 266
 parts required for, 265
 reassembly steps for, 267–268
sorting Now Playing in OS 3.0, 237
sound and picture synchronization, problems with, 247
space, upgrading, 243
speed of network connections, significance of, 174–175
SPS (Select-Play-Select) codes
 entering and disabling, 231
 overview of, 7–8
SquareShooter antenna, Web address for, 311
Star icon, explanation of, 35
static guards, using as safety measure for upgrades, 256–257
static IP networks, creating network connections on, 101–104
status mode, enabling, 233
Stop Transfer screen, displaying, 174
storage capacity, increasing by stacking TiVos, 178–179, 181
streaming audio, listening to, 131
streaming programs, resuming playing of, 173
"strike energy," relationship to hard drives, 278
suggestions. *See* TiVo Suggestions
suggestions engine, dynamics of, 74–75
support site, Web address for, 129
surge protectors, using, 197
surround-sound receivers, relationship to Dolby Digital 5.1 for HD TiVos, 309–310
S-Video cables
 benefits of, 202, 204
 description of, 6
 hooking up TiVos to, 10
 price range of, 16
 receiving better pictures with, 15–16
 using with TiVo-DVD Recorders, 189
system requirements, determining for TiVo Desktop software, 110–111

T

TCP port 2190, checking open status of, 115
TCP ports 8080 through 8089, checking open status of, 115
TCP/IP settings, choosing for wired DHCP networks, 99–100
televisions. *See* live TV; TVs
Terk multiswitcher, features of, 312
Test Satellite Signal screen, displaying for HD TiVos, 316
third-party TiVo forum, Web address for, 254
Thumbs-Up and Thumbs-Down ratings
 buttons for, 32
 changing, 83
 resetting, 83
 using, 81–84
thunderstorms, potential impact on TiVo connections, 197
timeline bar, removing, 233
Title WishLists, creating, 59
TiVo boxes, troubleshooting dial-tone problems with, 194–195
TiVo button, using shortcuts with, 32, 239
TiVo Central
 accessing for Thumbs-Up and Thumbs-Down ratings, 83
 accessing for TiVolution Magazine, 78
 accessing for TiVolution Showcases, 77
 accessing to copy videotapes to TiVo hard drives, 221
 accessing to create peer-to-peer networks, 177
 choices in, 34–35
 setting up recordings from, 50–51

TiVo Central Online
accessing for Multi-Room View setup, 167
Web address for, 141
TiVo Character icon, explanation of, 35
TiVo Clock, enabling, 233
TiVo configurations, testing and using, 16
TiVo-DVD recorders
connecting to cables or satellite boxes, 198–18
connecting to phone lines, 191–16
considering, 184–5
example setups for, 200-22
setting up, 187–9
TiVo Desktop software
downloading, 112–115
versus JavaHMO, 116–118
system requirements for, 110–111
troubleshooting, 129
TiVo Guide
changing to Grid Guide from, 39–40
recording from, 50
using, 39–41
TiVo Plus service
contacting, 88–89
features of, 80
upgrading to, 201
TiVo Publisher software
features of, 114
using with music, 122
TiVo Series2 units
selecting from pull-down menus, 142
setting up Multi-Room View on, 166–167
TiVo Server software, features of, 114
TiVo service number, locating, 18
TiVo services
Ipreview, 78–79
picking, 18–19
Tivo Showcases, 76–77
TiVolution Magazine, 77–78
TiVo setup configurations, examples of, 200
TiVo Showcases. *See also* hidden Showcases
accessing, 63
description of, 62
features of, 76
success of, 78
viewing, 77
TiVo software, backing up prior to formatting hard drive, 283–284
TiVo status mode, enabling, 233
TiVo Suggestions. *See also* Scheduled Suggestions
features of, 62
overview of, 80
quirks of, 81
rebuilding, 235
recording of, 30–31, 80
TiVo TCD23x, TCD24x, and TCD54x hard-drive upgrades
disassembly steps for, 268–269
drive formatting note for, 269–270
notes for, 270–271
parts required for, 268
reassembly steps for, 270
TiVo TCD130040 and TXD140060 hard-drive upgrades
disassembly steps for, 262–263
drive formatting note for, 263
parts required for, 262
reassembly steps for, 264–265
TiVo upgrades, market for, 334
TiVo version 3.0 and earlier, unlocking backdoor mode on, 230–231
TiVo-DVD Recorder setup
of cables and accessories, 188
overview of, 187
TiVo-DVD Recorders
audio format for, 208–209
changing DVD settings for, 215
connecting to cable or satellite boxes, 198–200
connecting to cables, satellite boxes, A/V receivers, and game consoles, 202–204
connecting to phone lines, 191–192
considering use of, 184–187
copying programs from hard drive to DVDs, 217–220
copy-protection issues related to, 211
disc compatibility for playback on, 208–209
disc compatibility for recording on, 208
disc compatibility with other DVD players, 210
displaying audio information on, 216
displaying Music Play options for, 216–217
inserting and playing discs on, 212–213
making phone calls during use of phone lines, 192
MP3 format issues related to, 210
MP3/audio information and playback settings for, 215–220
navigating DVD Menu screen with, 214
noise-reduction circuit on, 207
playing PC-created DVDs on, 208–209
recording in Video Mode on, 210
recording speed of DVR-810H, 208
troubleshooting connection problems with, 195–196
troubleshooting physical problems occurring to, 211
troubleshooting problems connecting to TiVo service, 197
using DVD banners with, 214
using navigation buttons with, 215–216
video format for, 208–209
Tivo-less TiVo, explanation of, 332–333
TiVolution Magazine
features of, 64, 77
viewing, 78–79
TiVomatic. *See* Ipreview feature
TiVos. *See also* HD TiVos
activating, 17–19
benefits of, 323
commoditization of, 332

configuring for external modems, 241
configuring with Guided Setup, 19–27
connecting to home network, 95–96
connecting via wireless connections, 176–178
equipment sold with, 5
financial numbers related to, 326–327
finding accessories for, 108
functionality of, 30–31
future of, 340–341
hooking up to cable boxes or satellites, 10–12
hooking up to home theaters, 13, 15
hooking up to multiple TV sources, 12
hooking up to power, 8
hooking up to TVs without cable boxes, 8–10
hooking up to VCRs or DVD burners to archive shows, 12–13
hooking up to watch one channel while recording another, 14–15
identifying model number of, 229–230
identifying running versions of, 230
inside of, 259
leaving unplugged while hooking up, 5
navigation buttons on, 215–216
online store for, 5
patent issues related to, 328
rebooting and resetting, 236
recording DVDs from, 218
registering, 17
versus ReplayTV, 232
restarting and resetting, 83–84
selecting for recording programs in AOL, 162
sending and receiving programs from simultaneously, 171
setting to record over Internet, 96
setting up for Multi-Room Viewing, 166–167
stacking to increase storage capacity, 178–179, 181
thumbing through, 82
toll-free number for, 87
transferring TV programs between, 168–175
unpacking, 4–5
using remote control buttons on, 128

TiVo's competitors
- Atlanta Explorer 8000, 330–331
- overview of, 329
- Sony DVR technology, 331
- UltimateTV DVR, 329–330

TiVos strategies, speculation on, 327
TiVo's Webmaster, contacting, 87
To Do List screen
- displaying, 69
- displaying Scheduled Suggestions in, 234, 235

toll-free numbers
- for DirecTiVo, 18
- for removing information TiVo's database, 87

Torx screws, using in hard-drive upgrades, 280–281
tracking system, purpose of, 340
transfer speeds, comparing, 175
transferred programs, playing, 175–176. *See also* program information; TV programs
triple-thumb codes, using, 233–234
troubleshooting
- physical problems occurring to TiVo-DVD Recorders, 211
- problems with Multi-Room viewing, 180–181
- problems with phone lines, 193–197
- TiVo Desktop software, 129
- Web resource for, 293

TV Aspect Correction screen, displaying for HD TiVos, 320
TV Input button on remote control, effect of, 32
TV lineup, picking, 24
TV listings
- accessing with TiVolution Magazine, 77–78
- browsing by channel, 146–148
- manipulating in AOL, 159
- searching on TiVo Central Online, 141–146
- using advanced search feature with, 144–146
- using basic search feature with, 143–144

TV Power button on remote control, effect of, 32
TV programs. *See also* program information; transferred programs
- determining Keep Until period for, 67
- dynamics of recording of, 30
- estimating transfer time for, 172–174
- finding in TiVo Guide, 39–41
- pausing during remote transfer of, 172
- playing and resuming, 65–67
- rating with Thumbs-Up and Thumbs-Down, 81–84
- recording and viewing with TiVo Showcases, 76
- recording manually, 55–56
- recording remotely over Internet, 138–141
- recording routinely with Season Passes, 154–157
- recording with Ipreview, 78–79
- resuming viewing after pausing, 176–177
- scheduling for recording, 149–151
- searching by title, 51–52
- stopping during transfer, 173
- transferring between two TiVos, 168–175
- transferring with Multi-Room Viewing feature, 166
- troubleshooting problems with transfers of, 180–181
- using rating system with, 80
- viewing promotions for, 38
- viewing settings for, 70

TV sources
- determining, 21
- hooking up TiVos to, 12

TVs. *See also* live TV
determining aspect ratio of, 308
hooking up TiVos to, 8–10
setting aspect ratio for use with HD TiVos, 321
TwinBreeze and Advanced Cooling Pak, features of, 334
Type X DVR, future of, 331

U

UDP port 2190, checking open status of, 115
UltimateTV DVR versus TiVos, 329–330
underscore variables (____), meaning of, 284–285, 287, 289–290
Upcoming Program screen, displaying, 50
upcoming recordings, viewing, 69–70
Upcoming Showings screen, displaying, 47
upgrades, market for, 334. *See also* hard-drive upgrades
UPS (uninterruptible power supply), using, 197
US Robotics modems, setting up, 240
USB-to-Ethernet adapters
using, 174–175
using to stack TiVos, 179
vendors of, 98

V

V2V (Video-to-Video), overview of, 339–340
VCR tapes, transferring content from, 185
VCRs, hooking up TiVos to, 12–13
VHF/UHF antennas, using, 310
video cables, setting up for TiVo-DVD Recorders, 188–191
video cameras, recording DVDs from, 223–224
Video Mode, using with TiVo-DVD Recorders, 210
video quality, comparing, 7
Video Recording Quality screen, displaying, 244
Video Settings screen, displaying for HD TiVos, 318
video signals, optimizing, 202, 204
videotapes, recording DVDs from, 221–223
Voices from the Community, 63
A Fan of the Desktop Photo Feature Speaks Up, 111
Forward to the Future—TiVo, DVDs, and Pizza?, 341
HD TiVo Is the Most Highly Anticipated Consumer Product in Years, 325
No-Fuss Freedom of Copying from TiVo, 206
Pioneer—the TiVo That Does It All, 187
Thumbs Up for TiVo, 82
The TiVo Revolution Begins, 9
Touchdowns for TiVo!, 232
Using TiVo for a Practical Joke, 163
A Yellow Submarine It's Not, but TiVo Helps Nonetheless, 136
You Can't TiVo Your Cat Yet, 181
Vol button on remote control, effect of, 33
volume, changing in live TV, 37

W

warranties, advisory about, 257
Watch from the beginning option, selecting, 176–177
WeaKnees.com
combos sold by, 334
obtaining hard drives from, 324
viewing hard-drive installation instructions at, 259
Web sites
accessories for TiVos, 108
Antenna Web, 310
boot disks, 291
conversion programs for music files, 123
DirecTV, 307
hard drives, upgrading, 250
hard-drive addition versus replacement, 251
information sites, 292–293
JavaHMO, 116
kit dealers, 293
Linksys, 104
MFS Tools software, 282
MoodLogic software for TiVos, 130–131
part dealers, 293
Privacy Policy, 86–87
privacy policy for TiVos, 21
PTVupgrade, 324
registering TiVos, 17
setting up, 307
SquareShooter antenna, 311
third-party TiVo forum, 254
TiVo, 112
TiVo Central Online, 141
TiVo models, 230
TiVo Plus service, 201
TiVo's support site, 129
troubleshooting guide, 293
WeaKnees.com, 293, 324
WeaKnees.com for hard-drive installation instructions, 259
Webmaster for TiVo, contacting, 87
"Welcome. Powering Up..." screen, getting stuck on, 245
Welcome! screen, navigating from, 20
WEP (Wired Equivalent Privacy) encryption, configuring for wireless home networks, 107
Will Get Later screen, displaying, 171
Window button on remote control, effect of, 32
wired connections
setting up, 98–104
versus wireless connections, 97
wired DHCP networks, creating network connections on, 99–101
wired static IP networks, creating network connections on, 101–104
wireless adapters, advisory about placement of, 104
Wireless Checklist, significance of, 105

wireless connections. *See also* peer-to-peer networks
benefits of, 96
connecting TiVos by means of, 176–178
setting up, 104
wireless DHCP client ID networks, creating network connections on, 108–109
wireless encryption, selecting level of, 108
wireless home networks, configuring network settings for, 105–108
wireless static IP networks, creating network connections on, 109–110
wiring
of home entertainment centers, 6–8
testing, 196
testing after configuring TiVo configurations, 16
WishLists. *See also* recordings; scheduled recordings
creating, 57–61
creating for HD TiVos, 323–324
editing and deleting, 62
enabling, 237–238
searching lineup data with, 62
types of, 58

X

XXX variables, meaning of, 284–285, 287, 289–290

Y

YYY variables, meaning of, 284–285, 287, 289–290

Z

ZIP code, entering for HD TiVos, 314

INTERNATIONAL CONTACT INFORMATION

AUSTRALIA
McGraw-Hill Book Company
Australia Pty. Ltd.
TEL +61-2-9900-1800
FAX +61-2-9878-8881
http://www.mcgraw-hill.com.au
books-it_sydney@mcgraw-hill.com

CANADA
McGraw-Hill Ryerson Ltd.
TEL +905-430-5000
FAX +905-430-5020
http://www.mcgraw-hill.ca

GREECE, MIDDLE EAST, & AFRICA
(Excluding South Africa)
McGraw-Hill Hellas
TEL +30-210-6560-990
TEL +30-210-6560-993
TEL +30-210-6560-994
FAX +30-210-6545-525

MEXICO (Also serving Latin America)
McGraw-Hill Interamericana Editores
S.A. de C.V.
TEL +525-1500-5108
FAX +525-117-1589
http://www.mcgraw-hill.com.mx
carlos_ruiz@mcgraw-hill.com

SINGAPORE (Serving Asia)
McGraw-Hill Book Company
TEL +65-6863-1580
FAX +65-6862-3354
http://www.mcgraw-hill.com.sg
mghasia@mcgraw-hill.com

SOUTH AFRICA
McGraw-Hill South Africa
TEL +27-11-622-7512
FAX +27-11-622-9045
robyn_swanepoel@mcgraw-hill.com

SPAIN
McGraw-Hill/
Interamericana de España, S.A.U.
TEL +34-91-180-3000
FAX +34-91-372-8513
http://www.mcgraw-hill.es
professional@mcgraw-hill.es

UNITED KINGDOM, NORTHERN,
EASTERN, & CENTRAL EUROPE
McGraw-Hill Education Europe
TEL +44-1-628-502500
FAX +44-1-628-770224
http://www.mcgraw-hill.co.uk
emea_queries@mcgraw-hill.com

ALL OTHER INQUIRIES Contact:
McGraw-Hill/Osborne
TEL +1-510-420-7700
FAX +1-510-420-7703
http://www.osborne.com
omg_international@mcgraw-hill.com